THE WORDSMITHS

Oscar Hammerstein 2nd
and Alan Jay Lerner

STEPHEN CITRON

New York
OXFORD UNIVERSITY PRESS
1995

Oxford University Press

Oxford   New York   Toronto
Delhi   Bombay   Calcutta   Madras   Karachi
Kuala Lumpur   Singapore   Hong Kong   Tokyo
Nairobi   Dar es Salaam   Cape Town
Melbourne   Auckland   Madrid

and associated companies in
Berlin   Ibadan

Published by Oxford University Press, Inc.,
198 Madison Avenue, New York, New York 10016

Oxford is a registered trademark of Oxford University Press

Library of Congress Cataloging-in-Publication Data
Citron, Stephen
The wordsmiths : Oscar Hammerstein 2nd and Alan Jay Lerner / Stephen Citron.
p. cm.—(The great songwriters series)
Includes bibliographical references and index.
ISBN   0-19-508386-5
1. Hammerstein, Oscar, 1895–1960.   2. Lerner, Alan Jay, 1918–86.
3. Librettists—United States—Biography.   I. Title.   II. Series.
Citron, Stephen. Great songwriters series.
ML403.C56   1995
782.1'4'0922—dc20
[B]   94–20279

1 3 5 7 9 8 6 4 2

Printed in the United States of America
on acid-free paper

For
MITCH DOUGLAS
whose friendship I have
long cherished
and for whose help
in the preparation of this book
I am deeply beholden

Acknowledgments

W RITING A REVISIONIST biography is always fraught with the challenge of digging into the past and stirring up memories that may or may not be accurate. So much that is written about theatrical subjects, especially when written by imaginative theater people themselves, is error laden. Of course, it is the biographer's duty to check and recheck facts and reportage—whether it comes from a newspaper, an earlier biography, or someone's memory.

Fortunately, I have had much help concerning the output of my subjects from the past, for both Hammerstein and Lerner played out their lives on an open stage. Practically everything they did was newsworthy—from the announcements of forthcoming shows to critiques—sometimes raves, sometimes post mortems. Beyond their works for the musical theater, many of their exploits were newsworthy as well—from Lerner's seven sensational divorces to the denial of a passport to Hammerstein during the McCarthy era.

In my quest to separate the true reportage from the fancifully imaginative, I have been fortunate in obtaining the help of both families. Oscar Hammerstein's three children, William, Alice, and James, have opened their homes, their files, and their memories to help me get a true picture of life with and away from father. Alan Lerner's daughters, too, Liza Bibb and Jennifer Fraser, gave me a long interview during which we discussed their father from their earliest memories until his death. Cousins Will Lerner and Barbara Lerner Barrett were helpful in separating the real from the fabrication in Lerner's childhood. Thanks also to Sara Ziff, James Stearns, and James Bettman, who knew Lerner at Columbia Grammar School; and to LeMoyne Sylvester, the archivist at Choate-Rosemary Hall School, who gave insight into Lerner's formative years while a student there and who put me in touch with roommates John O. Jones, F. Sigel Workman, and Carleton Tobey. David Sykes and Gail Pottle did the same for the Bedales School in England and initiated a correspondence between myself and Justin Brooke, one of Lerner's classmates. At Harvard, my thanks to Robin McElhenny and Jeanne T. Newlin for chronicling Lerner's undergraduate years.

Of Lerner's wives, some of whom spoke to me preferring not be quoted, Marion Bell was the most forthcoming, and for her frank corroboration of many of the rumors that have surrounded Alan Jay Lerner and his eight marriages, I thank her. Perhaps the ghosts can now be laid to rest.

Liza Lerner Bibb told me Judy Insel, her father's secretary, knew him better than anyone. Unfortunately, Ms. Insel's information was unavailable to me, but it was propitious that Stephanie Sheahan, who for many years was Lerner's "other secretary" and perhaps more perceptive to the inner man, gave me a long interview.

Since this book is almost more concerned with each man's creative efforts rather that his private life, excerpts from lyrics and librettos were necessary to make their points. For this generous use I am grateful to the Rodgers & Hammerstein Organization—Bert Fink, Director of Special Projects, as well as Nicole Gilette, Maxine Lang, Stephen Cole, and the managing director, Ted Chapin. David Grossberg is in charge of the Alan Jay Lerner estate, and he and his staff were most helpful to me. Mr. Grossberg saw to my being able to examine every lyric, libretto, and notation of Lerner's considerable oeuvre. Many thanks also to Steve Clark, Caroline Underwood, Kevin White, and Jim Henney at Warner-Chappel Music and to Michael Kerker at ASCAP.

Otto Harbach partnered Oscar Hammerstein on eleven works (more than any other collaborator, including Richard Rodgers, with whom he wrote eight musicals). Fortunately Otto's son William has kept recordings and tapes—a complete oral history—and manuscripts that outline the collaboration. Bill Harbach spent many hours with me, going over all the Harbach-Hammerstein musicals. William Hammerstein, too, has been forthcoming with files, clippings, and letters concerning his father's work after Harbach, with Kern and Richard Rodgers. Thanks, too, to Jonathan Schwartz for information on his father, Arthur's, collaborations with Hammerstein. Gerard Kenney collaborated with Alan Lerner on his last, uncompleted show, *My Man Godfrey*. For several candid conversations, my thanks.

Richard Lewine, who produced the R & H *Cinderella* in 1957, gave me much information about Hammerstein's working process. By an uncanny coincidence, he was also pianist-arranger of Alan Lerner's first musicals during his undergraduate days at Harvard. Since both Hammerstein and Lerner were presidents of the Dramatists Guild and since Lewine served during the latter's term (and beyond) as vice president of that organization, he had much information to impart. Speaking of the Dramatists Guild, its Executive Director, David LeVine, was most forthcoming with information relating to both Hammerstein's and Lerner's tenures.

Thanks to Jane Powell, who has total recall of the time she starred in Lerner's *Royal Wedding*, and Kitty Carlisle Hart for her fund of information on her and her husband Moss Hart's input into both *My Fair Lady* and *Camelot*. Agnes de Mille talked to me at length about her work on Hammerstein's *Oklahoma!*, *Carousel*, and *Allegro* as well as Lerner's *Brigadoon*. The late great choreographer was candid and insightful.

Others who gave me long interviews were Miles Kreuger, who has written the definitive book on *Show Boat* and was a close friend of Alan Lerner as well, and Arnold Michaelis, who shared with me a wonderful hour he recorded with Hammerstein. Thanks to Dr. Maurice Huberman for his help in translating Bizet's *Carmen* into Hammerstein's *Carmen Jones*, and to Thomas Z. Shepard, who has produced and/or recorded many of the works of both Hammerstein and Lerner, for his keen insight into their musicals that he so generously shared with me.

Perhaps the one man who knew Alan Lerner better than any other is Stone "Bud" Widney, who played an integral part in every show from 1949 until his death in 1986. Bud discussed each show with me at length. Lerner called Widney his production manager, but added that "it is like calling a bottle of Dom Perignon a cold drink."

Many others gave me help in particular areas of these great lyric-librettists' lives and works. They are listed below in alphabetical order. My deep thanks to them all: Elizabeth Alexander, Elizabeth Auman, Mel Arbite, Billy Barnes, Don Black, Ed Cramer, Gemze De Lapp, Carl De Milia, Dr. Leonard Diamond, Hugh Fordin, Madeline Gilford, Jane Howard Hammerstein, Luther Henderson, John and Jane Hess, George Hetherington, Mark Horowitz, Robert Jorgen, Irwin Karp, Ruth Leon, Jack Lawrence, Jerome Lawrence, John Lisko, Roddy McDowell, Philip McKinley, Terry McNealy, Frank Military, Sheridan Morley, John Pattnosh, Sandra Payne, Dory Previn, Rex Reed, Kate Rivers, Jenny Roscoe, Helena Scott, Charles Sens, Rose Tobias Shaw, Paul and Vivian Shiers, Herman Silverman, Fred Mustard Stewart, Morton Weyler, Barry and Fran Weissler, Ray White, Claudia Widgery, and Yuriko.

I feel twice blessed in having the editorial assistance of two great editors, both of whom are specialists in their fields, and amateurs—in the original sense of the word—of musical theater: in London, Christopher Sinclair-Stevenson, who believed in the The Great Songwriters series from the start, and Sheldon Meyer at Oxford University Press in New York, who seems to have edited every worthwhile musical theater book in print. For her fine copyediting—although that sounds like too prosaic a term for her considerable contribution to this book—I'd like to thank Joellyn Ausanka.

My son, Alexander Citron, a fine musician and teacher gave, me much assistance in assembling and editing my manuscript and always made helpful comments on the text. To him my deep thanks.

I have reserved these final lines to thank my wife, Anne Edwards, without whose help this book could never have been written. Herself a gifted biographer, she took time from her creative work to read through the manuscript and make copious notes, always in a nonintrusive, questioning, and constructive manner. I have expressed my love and gratitude to her in every one of my books. For none is she more deserving of a sweeping bow.

New Milford, Connecticut   S. C.
March 1995

Contents

THE WORDSMITHS

Prelude

THE STORY HAS OFTEN BEEN TOLD. Sometimes the protagonist is Dorothy Hammerstein outwrestling the Kern aficionados as to the reasons behind the success of "Ol' Man River"; at others it is Nancy Olson Lerner announcing flatly that her husband's words gave "I Could Have Danced All Night" imperishable classic status. But whoever is responsible matters not a jot. Nor does it matter that the tale is probably apocryphal:

Mrs. Hammerstein—using her as example because of seniority—having overheard her dinner table companion congratulate Mrs. Kern on her husband's having written "Ol' Man River," interrupted with "Jerry Kern didn't write 'Ol' Man River,' *my* husband did! What Kern wrote," she added, singing, "was dum–dum–*dee*–dum."

This old tale only serves to accent how often the composer walks away with the honors for a memorable song while the lyricist is overlooked. Imagine then the unenviable position of the lyric-librettist, who is responsible for two-thirds of what the public has paid to see: a believable book and literate lyrics (the other third being memorable music). Certainly he is credited if the show is a hit, but more often it is he who comes in for the most blame when and if the musical flops.

It has always been so. One went to see the latest Victor Herbert, Friml, or Romberg musical. Only theater trivia experts recall the names of the other collaborators of megahits *No, No, Nanette* or *Hit the Deck!* besides Vincent Youmans. Even today the marquees over Broadway and Shaftsbury Avenue scream out the name of Andrew Lloyd Webber while various associates—Don Black, Richard Stilgoe, Trevor Nunn, Charles Hart, yes, even T. S. Eliot— are only listed *inside* the covers of the program book sold in the theater lobby.

Until Oscar Hammerstein created a string of glittering hits with his last associate Richard Rodgers, he was, incredibly, in somewhat the same position. Although he had contributed lyrics and libretto to many successes and was long and highly esteemed by the critical faculty, the public at large applauded *Rose-Marie* as being an operetta by Rudolph Friml. Hammerstein conceived the books, adapted the stories, and wrote the lyrics for every one of the following, yet it was Kern's *Sunny, Show Boat,* and *Music in the Air,* Gershwin's *Song of the Flame,* and Romberg's *Desert Song* and *New Moon.* Such is the power of music that unless the names of both composer *and* lyricist are teamed on a program, the public only remembers the one who wrote the tune.

Hammerstein's name only became a household word when he replaced Lorenz Hart as Richard Rodgers' collaborator and business partner. Soon the Rodgers and Hammerstein connection, familiarly known as R & H, would become the most famous collaboration in theater—musical and non—worldwide.[1]

Alan Jay Lerner was much luckier. He began his career as lyricist with a collaborator, and his name in the musical theater was never overlooked. Although the first show they wrote, *The Life of the Party,* did not even make it to Broadway, the team of Frederick Loewe and Alan Jay Lerner soon did, and their alliterative pairing—sometimes stormy, sometimes smooth, but always imaginative—as Lerner and Loewe (often, Alan and "Fritz" and sometimes even L & L) was to last off and on for eighteen years and achieve a fame only rivaled by that of R & H. It is curious that they referred to their collaboration in an order contrary to usual precedence of composer to lyricist. Kander and Ebb, Bock and Harnick, Schwartz and Dietz, George and Ira Gershwin, even Rodgers and Hart—or Hammerstein—all mentioned the musician's name first.

No small reason for Lerner's prominence in the team was the fact that up until the time of *My Fair Lady* all the libretti of his musicals had been based on his original ideas, prompting the critics to mention his name in conjunction with the book and to refer to Lerner again when they discussed the lyrics. And with his extravagant personality, eight marriages, alimony, custody, and court trials, Lerner was frequently headline news. Thus his name soon became better known than that of his retiring collaborator.

Although he created many musical perennials and his *My Fair Lady* is generally considered to stand at the apex of what was known as the "Golden Age of the Musical," Lerner was nowhere near as prolific as his colleague Oscar Hammerstein. He did, however, write book and lyrics for thirteen musicals (seven with Loewe), screenplays for four musicals (two of which won the Academy Award), a splendid biographic-memoir, and perhaps the best history extant of American musical theater.

---

[1] The Rodgers & Hammerstein Organization has continued to grow in the generation after the death of the two original partners. Besides their individual and collective works it now controls and administrates the works of Irving Berlin, Stephen Sondheim, and others.

Hammerstein, an early bloomer, a quicker worker, and a much less neurotic artist, was responsible for the lyrics to thirty-five musicals, only half of which were successful on Broadway. Except for two of these he wrote the libretto as well. One must add to that the two musicals he wrote for London's West End, the twelve films and television musicals to which he was a major contributor, and the three serious plays he co-wrote. Because he was often sparked by another's idea, be it Otto Harbach's, Jerome Kern's, Josh Logan's, or even Richard Rodgers's, his name was often listed as co-librettist.

Sometimes he even supervised the whole project. Holding onto the directorial reins to two shows, he was able to come up with one hit[2] and one miss[3]— not a bad average for the mercurial musical theater. To round out Hammerstein's prodigious total of fifty-five works, one must add his first and last solo effort, a drama called *The Light* which succumbed in New Haven on its way to Broadway after four performances.

Thus in the span of some forty-one years, in a career that was launched in 1919 and cut short by cancer in 1960, Oscar Hammerstein retains the record for sheer productivity. The only challengers who came close were his own collaborators, Jerome Kern and Richard Rodgers.

Oscar Clendenning Greeley Hammerstein,[4] having been born in 1895, was twenty-three years old when Alan Jay Lerner was born into a wealthy mercantile New York family. By the time Lerner began to write his fledgling songs for the Harvard Hasty Pudding shows, Hammerstein was in his early forties and had a string of successes to his credit. Because he produced such varied works and was competent in drama, farce, and musical comedy, he was the idol of the small but serious group of budding libretto-lyricists trying to break out of the mold of post-depression girlie musicals whose inane plots were calculated only to amuse the tired businessman. It follows that Lerner, used to the best, should look up to him and try to emulate him.

Alan learned that the long-legged Hammerstein strode around the room mut-

---

[2] *Good Boy*, which Hammerstein co-wrote with Otto Harbach and Henry Meyers, had music by Herbert Stothart and Harry Ruby. The lyrics were by Bert Kalmar. It featured "boop-boop-a-doop" girl Helen Kane in the song that launched her career with the show's only hit, "I Wanna Be Loved By You," which captured Broadway's fancy for eight months before the show closed.

[3] In *The Gang's All Here* Oscar had the assistance of Russel Crouse and Morrie Ryskind. The music was by Lewis Gensler, with lyrics by Owen Murphy and Robert A. Simon. In 1931, with not a single memorable song, the show eked out twenty-three Broadway performances. Doyen critic John Mason Brown shot the most devastating salvo: ". . . the production proved as painful as it was long."

[4] Clendenning was the name of the minister who married his parents; Greeley was the maiden name of the minister's wife, whose father, Horace, was supposed to have coined the phrase, "Go West, young man."

tering the lyric he was creating in the meter it was to be performed. Oscar fitted these words to his own melody—dreadful, according to his wife, Dorothy, to whom he would occasionally sing a work-in-progress. Then he would jot his ideas down while standing at a high bookkeeper's desk and begin pacing again. It was less hectic than sitting his six foot two frame down at an ordinary desk and then popping up again to resume his gait. Lerner assumed that if the system worked for Hammerstein it might do as well for himself, and he bought a similar antique. He used it only occasionally. Five foot six inch Alan preferred to pace with clipboard in hand. More often he would sit motionless in a chair, cogitating while biting ballpoint pens or his fingers, jotting lyrics, strewing the papers on the floor awaiting a secretary to collect and transcribe them.

Another area where the younger man tried to copy his idol was in disseminating the considerable experience of their craft. Both men were natural-born teachers who loved to pass on what they knew of the musical theater. Late in his life Lerner was to develop an advisory relationship with Maury Yeston (*Nine, Grand Hotel*) similar to the one Stephen Sondheim had with his mentor, Hammerstein.

"When I was young," Sondheim reported, "Oscar Hammerstein told me to come to his place from time to time to show him my work and for him to give me some friendly advice."

Lerner learned much about stagecraft from the older man. Whereas formerly the second act was frequently written while the first was already in rehearsal, Hammerstein taught that the construction of the drama must be airtight before the score is readied or actors are invited in. From *Carousel,* which premiered before Lerner's first moderately successful musical, he learned that soliloquy, unacceptable in straight drama since Shakespeare's time, can be the turning point of a musical play. Hammerstein's eight-minute scene was a daring departure from the usual three-minute song that could be fitted onto a recording. As all good students learn from a groundbreaker, in *My Fair Lady* Lerner went even farther than his mentor. Here he gives us four of them: "Why Can't the English," I'm an Ordinary Man," "A Hymn to Him," and the final turning point of the entire evening, "I've Grown Accustomed to Her Face." Lerner even gives Liza her own amusing soliloquy, "Just You Wait, 'enry 'iggins." Actually, the entire play depends on Higgins's or Liza's opinions shared with the audience.

Lerner also absorbed the power of underscoring—music under dialogue—from Hammerstein, using a technique long accepted in cinema whereby the music (always in reprise) heightens the emotion. Oscar first broke ground in *The King and I* by eliminating the sung finale. It was a bold idea to have plot, dialogue, and underscoring take over. Lerner took the idea and ran with it, combining it with soliloquy as King Arthur's magnificent ending to the first act of *Camelot* or the moving finale of *Lolita.*

Both Hammerstein and Lerner believed in polishing a lyric until it gleamed. And although Hammerstein's lyrics were never plodding or businesslike, his creative manner fortunately was. However, the solid two weeks of work that went into the eight typewritten pages that comprise the "Soliloquy" are nothing compared with the weeks that André Previn says would often elapse before the composer was given Lerner's completed lyric. "It took him longer to finish a lyric than it took an elephant to have a baby," Previn says. "He would finish a song with the exception of two lines and then not hand it in until he had finished those lines," which, Previn adds, "could be months." Lerner could be maddeningly slow in his search for perfection, and because of insecurity he refused to deliver a lyric to his collaborator until it satisfied him—no matter how it delayed the production.

From the above it is easy to see that although Hammerstein and Lerner had great enthusiasm for each other's talent, their personalities were not at all alike. In truth, Hammerstein's unruffleable personality meshed well with the pragmatic one of Richard Rodgers, and he probably would have had even smoother sailing had he been working with Frederick Loewe, who was closer to his own age. But Lerner's neuroticism rubbed Rodgers's businesslike mind and sense of dispatch like sandpaper.[5]

Yet in spite of the differences in age, temperament, background, and life-styles, these two giants had two things that draw them together, making them unique as the tandem subjects of a book on musicals.

They revolutionized them.

By creating a libretto—these days called the book—and then musicalizing extended sections of their words—actually robbing from themselves—they were able to produce seamless works that moved from song to play to song to play. Their other unique quality was that they created characters one can accept. They themselves believed deeply in what they were creating. They both tried to let their characters speak in language that was authentic to themselves, the plot, and its timeframe. No tired businessman musical was produced by either man.

Although Lerner was more heavily drawn to whimsy, and Hammerstein to the poetic ideal, they both understood that in modern theater the play and characters need to be believable for the songs to have a chance to touch one's heart.

Hammerstein learned this from Kern and by the late 1920s was the only

---

[5] In 1961, some eight months after the death of Hammerstein, Rodgers and Lerner did form a partnership to work on I Picked a Daisy (later On a Clear Day You Can See Forever). Some two years later it was dissolved amid scathing recriminations on both sides. Burton Lane took over the composition of the score, and the work finally reached Broadway in 1965.

practitioner of the "book show" whose works seem valid today. Then he passed the flame—not the entire torch—to Lerner.

Rodgers and Hammerstein's musicals habitually opened two years apart; *Oklahoma!* premiered in 1943, *Carousel* in 1945. It seems that Lerner and Loewe tried to follow this orderly procedure by presenting their first show, the imaginative *The Day Before Spring,* in 1945, and their first real hit, *Brigadoon,* in 1947. Throughout Hammerstein's seventeen-year relationship with Rodgers, he was able to maintain the pattern of a biannual production, but Lerner with Loewe or (as frequently happened) without him was too mercurial to stick to that kind of production schedule. Perhaps that's the key to the difference between the two men.

Hammerstein I

1847–1895

OSCAR HAMMERSTEIN was obsessed with opera. His namesake and grandson had an equivalent passion, but his was for musicals. The younger Oscar was later to develop that form into the believable musical play and revolutionize the musical theater to such an extent that, since he always called himself Oscar Hammerstein 2nd, his grandfather had to be referred to by the dynastic title of Oscar I. At the mention of the Hammerstein name today's generation immediately thinks of the grandson, usually mentally hyphenated with the name of his last collaborator, Richard Rodgers.

Young Oscar's career was an abundant one, and if it was not aimed at opera, it certainly was pointed early toward operetta. After producing early successes like *Tickle Me, Wildflower,* and *Rose-Marie* while he was still in his twenties, he was to collaborate in the next decade on operettas with the most distinguished composers of his time—Kern, Gershwin, Friml, Romberg, and Kalmán—producing such milestones as *Sunny, The Desert Song, The New Moon,* and *Show Boat.* But by 1934, when for five years (except for *Music in the Air*) he had written nothing but a series of flops, he began to look inward away from his usual operetta-like plots embroiling fictional princesses with commoners or romances set in exotic locales and to think of setting down the true history of his illustrious namesake.

It was a wise decision, but even that honest story was to be stymied when the screenplay he wrote about his celebrated grandfather—remembered today as a builder of theaters and impresario of grand opera—was shelved by MGM. But while he was in Hollywood he attended a concert performance of Bizet's *Carmen* at the Hollywood Bowl. Musical comedy, not opera, was his chief joy, but perhaps he went to that massive Hollywood arena because his mind was so deeply searching for the key to Oscar I's enigmatic passion, a self-destructive obsession with tenors and especially divas that had forced him into several bankruptcies.

Watching this masterpiece presented without staging, costumes, or settings, sung in a language foreign to its audience, and heard in a cavernous stadium with noisy planes intermittently drowning out the music, Oscar was amazed that it still seemed to hold its audience spellbound. He fell in love with *Carmen* and became as entranced by opera in English and for "everybody" as his grandfather. Immediately he wrote to Sam Katz, music supervisor at MGM, that he thought a cinematic setting of this opera and others like *Butterfly*, *Pagliacci*, *Faust*, or *Boheme* could not fail at the box office.

After a story-lunch during which he gained Sam Katz's approval of "sugar-coated translated opera," Oscar's first idea was to adapt the Carmen story for the screen. To this end he created a Jezebel-like scenario which he called *Lulu Belle*, and invited Duke Ellington to write an original score. They had hardly begun working together when Oscar noted "he [Ellington] was late for an appointment one day, and since he had a night-club life and I like to work at nine o'clock in the morning, I didn't think it would be a very happy collaboration."

But during the next eight years the thought of adapting *Carmen* to a modern setting never left him, and from time to time he dabbled at a lyric here, a bit of dialogue there, much as one might pick up and put down a crossword puzzle. Since he had long ago decided to use Bizet's music exactly as it was written, his collaborator was always at hand yet never demanding when the obligation of earning a living on Broadway or when a contracted Hollywood deadline took priority. *Carmen*, or *Carmen Jones* as he called her, became his private project in a career that encompassed forgettable musical films like *The Night Is Young*, *Give Us This Night*, *Swing High*, *Swing Low*, and *The Lady Objects*, with perhaps *High, Wide and Handsome* the only imaginative scenario of all his films. He picked up the project in earnest again, as he so candidly stated, "merely because I was out of a job. Hollywood had offered me nothing, and nobody was asking me to write a play here in New York. It was the lowest part of my slump, and so for the first time in my life I wrote something on spec."

Having been born in 1895, he was now fast approaching middle age and felt written out. None of the original stories he had created appealed to the public. Indeed, throughout his life he was to do far better with adaptations than with original plots. Now, updating opera seemed a lifeline for Oscar, and he was to declare to anyone who would listen that "adapting *Carmen* gave me more pleasure than anything I had written heretofore. I loved working on it. I took my time. There was no deadline."

As for his recent dismal record, it had not been established through lack of industry, for in the last half decade he had created six musicals for the stage and two scenarios for the screen. He just seemed out of step with a country emerging from the depression. Although he was to turn out competent scenarios and libretti which went on to have splendid productions and much ballyhoo, great success was to elude him, and he was to become even more of a misfit throughout the late 1930s and early '40s.

He did not consider the commercial possibilities of *Carmen Jones* until the early '40s, after war had broken out in Europe. This, coupled with the failure of his much touted last collaboration with Jerome Kern, *Very Warm for May*, galvanized the opera in his mind, and he immediately saw the possibilities of transferring the locale from a cigarette factory in Seville to a parachute manufacturing plant in the South. Now Don José could remain what he was in Bizet's opera—a sensitive man smitten with love for a wilful woman. His being an ordinary corporal brought relevance to the story, for now the toreador, Escamilio, would become his adversary, a warlike, prizefighting champion, his name transformed into Husky Miller. There was sufficient fervor and zest in these characters to put them on a stage filled with serious drama and more than enough resemblance in the self-sufficient Carmen Jones for the millions of women now working in defense industries to identify with. *Carmen Jones*—call it opera or musical—held sufficient private inner turmoil to stand up to the grand passions playing outside the theaters—the fall of France, the invasion of Poland, Pearl Harbor.

Carmen Jones was completed in 1941, but because of the projected difficulty of finding the more than one hundred black singing-actors, it received many rejections. Even canny lyricist Howard Dietz termed it "the most turned-down show in town," implying that Hammerstein had lost his wits, for "no one in their right mind would sit down and write Bizet's *Carmen* as a modern American musical with an all-black cast."

At last it was optioned by producer Max Gordon who, it seems, had neither the money nor the desire to produce it. It lay upon his desk for many months until November 1942, six months before the opening of *Oklahoma!*, when Oscar mentioned the project and his annoyance at Gordon's delay to Billy Rose, Broadway's diminutive dynamo.[1] Rose asked to see the libretto, and Oscar sent it over the next day with an apologetic note saying that his adaptation sounded much better with the music. Whether Rose, who was, by the way, an excellent composer of light music, was able to hear the words being sung to Bizet's tunes in his mind's ear or not, he telephoned Oscar excitedly within a few hours and promised a production at once.

---

[1] Although best known for the lavish *Aquacades* starring his wife Eleanor Holm, Rose, who always had a taste for popular "culture," had produced Rodgers and Hart's circus extravaganza, *Jumbo*, in 1935. After the success of *Carmen Jones* he was to produce *Seven Lively Arts* the next year. For the latter arty revue Rose went so far as to commission a ballet by Igor Stravinsky.

Of course the tumultuous reception accorded *Oklahoma!*, assuring that Hammerstein's name and fame would regain their former cachet, absent since the days of *Rose-Marie* and *Show Boat*, gave Rose further insurance for the success of his "opera," and he hastened to get *Carmen Jones* on stage. Now Hammerstein was to become again the preeminent librettist of the musical theater until the end of his life.[2]

On December 2, 1943, *Carmen Jones*, which had been Oscar's artistic sustenance during his least successful period, moved into the Broadway Theatre for a long and successful run. Although the other operatic adaptations he had dreamed about were to remain stillborn,[3] through this labor of love, this modernization of the libretto of Meilhac and Halevy, Oscar 2nd was certainly completing his grandfather's dream.

Born in Stettin, a city in the part of Germany known as Pomerania, near the Polish border, on May 8, 1847, Oscar Hammerstein was an iconoclast, a dreamer, and a misfit with little sense of reality from his childhood. The family fortunes increased, and by the time Oscar was eight, his ambitious father, Abraham, a strict disciplinarian who began as a building contractor, had become a trader on the stock exchange and had moved his family to Berlin. Young Oscar had piano instruction from his mother at an early age, and now the knowledge-hungry youngster requested and was given private violin lessons. His father had little time for his burgeoning brood of five children and especially the demands of his eldest, the daydreaming Oscar, who wanted to enter the Berlin Conservatory.

At last through the intervention of his mother, by the time he was twelve the boy was allowed to audition and enroll there. However, although he practiced diligently, it was soon obvious to all, especially to his dictatorial father, that Oscar's talent was not exceptional.

But in the course of his studies, Oscar attended operas with the rest of his class and each time came away feverishly exhilarated. His father, his siblings, and friends had no patience for his babbling as to how *he* would direct a performance

---

[2] Remembering his past failures while thumbing his nose at the show business fraternity, he bought a page in the annual *Variety* issue and placed the following bold announcement: HOLIDAY GREETINGS FROM OSCAR HAMMERSTEIN 2ND AUTHOR OF *SUNNY RIVER* (6 WEEKS AT THE ST. JAMES) *VERY WARM FOR MAY* (7 WEEKS AT THE ALVIN) *THREE SISTERS* (7 WEEKS AT THE DRURY LANE) *BALL AT THE SAVOY* (5 WEEKS AT THE DRURY LANE) *FREE FOR ALL* (3 WEEKS AT THE MANHATTAN). "I'VE DONE IT BEFORE AND I CAN DO IT AGAIN."

[3] In spite of its acclaim, *Carmen Jones* seems to have spawned no revolution in musical theater. Its only clear-cut issue was *My Darlin' Aida*, staged a decade later, which set the Egyptian opera in the South of the Confederacy. Pharaoh became General Farrow; Radames was called Raymond. It was all ponderous, a bit silly, and ultimately unsuccessful.

of *Fidelio* or *Die Zauberflöt*, but his sensitive mother, Bertha, applauded these passionate recitals of his proposed staging of the great operatic masterpieces.

Unfortunately with her death two years later, when Oscar was fourteen, and the remarriage of his father to his counterpart with regard to regimen and chastisement, life in this Prussian-Jewish household became unbearable. At fifteen Oscar seemed trapped in an alien atmosphere that only ridiculed his grandiose operatic schemes. There seemed nothing to look forward to but nightly political discussions at the dinner table and next year's service in the German army. Since Germany was embarking on an aggressive warlike policy, young Oscar, small of frame and Jewish to boot, knew he was in for a lifetime of derision from his strongly patriotic and militaristic countrymen. This was not to mention his thwarted operatic dreams.

In his maturity, Oscar often told the highly unlikely story of how he broke away from the family yoke after having been given an especially violent beating with the strap of his ice skate. It seems he had skipped his violin practice for a turn on the frozen lake nearby, and on his return his father, furious at the breach of discipline, had whipped him with the skate's leather thong and sent him directly to bed. As he lay there he formulated a plan which he carried out the next morning.

Rising early, he walked to the business section of Berlin and pawned his violin, using the money to buy a third-class ticket to Hamburg and thence to Hull, England. There he was taken aboard the *Isaac Webb* and with other immigrants headed for New York. Oscar's own accounts of the journey vary. In some, he worked his way across the Atlantic; in others he nearly died when a storm threw the ship off course and the crossing took eighty-nine days during which Oscar reported he was confined with all the other steerage passengers to the crowded airless quarters below decks. In any case, although it seems unlikely that in the mid-nineteenth century a fifteen-year-old without papers could make his way from Germany through Britain and thence to America, the proof is that a penniless, German-speaking Oscar Hammerstein arrived in New York in the winter of 1863, the middle of the Civil War. Because of the conflict there was a dearth of laborers and the young immigrant immediately found work in a cigar factory.

Industrious, frugal, studious, and imaginative, he thrived from the beginning. Within a year he invented a mold that would insure the uniformity of cigars. Soon he designed a machine that would make a dozen cigars at a time, and shortly after that he developed a device that stripped tobacco by air suction—the forerunner of the modern vacuum cleaner. He sold this invention outright for the then enormous sum of $6000, not being aware that there were many millions to be made if he patented and leased this tool.

In 1868, his English now only reflecting a trace of German accent, he fell in love with Rose Blau, the sister of one of his coworkers at the factory, and when she reached her seventeenth birthday they were married. Rose bore him a son who died in infancy and then successively four more boys: Harry, Arthur, William (who was to father Oscar 2nd), and Abraham Lincoln Hammerstein, named in the late President's honor a decade after his assassination. Shortly after the birth of this son, Rose died.

By that time Oscar had founded *The United States Tobacco Journal,* which soon became an indispensable publication for cigar trade information.[4] In later years he would develop his characteristic appearance by which he was caricatured in papers all over the world. The pointed beard, the smoking cigar, the Prince Albert coat worn with striped trousers, and the ever present top hat which he never removed and which was designed to augment his rotund five foot four inch frame all contributed to make him a recognizable original.

Realizing that Harlem was acres of open land just waiting to be developed, he invested his savings in an an apartment building. Soon his deferred dream of an opera house was to take flight and in 1888, on the pretext that the tenants of his apartment house needed cultural entertainment in the area, he sold his interests in the thriving tobacco journal and designed and built the Harlem Opera House.[5] Now he was to become a full-fledged impresario, hoping to lure the stars of the Metropolitan Opera to appear in this northern Manhattan wilderness. For money, often ruinous sums, he was able to inveigle singers of the caliber of Lili Lehmann and conductors like Walter Damrosch to appear, but he never could muster the glittering social throngs he sought to fill the plush velvet loges he had designed.

Yet with a confidence that only the greatest land barons or fools seem to possess, he decided that Harlem needed not one showplace but two. He put up his investments, which had increased to twenty-four apartment houses and thirty private residences, and against all logic, he erected the Columbus Theatre on 125th Street near Lexington Avenue. Now the throngs who attended the Columbus were able to make up for his heavy losses in opera. So he was to establish a lifelong pattern.

Two years later he launched the Manhattan Opera House (on land where Macy's Department Store now stands). But his most visionary edifice was the Olympia, a giant complex of four theaters on the east side of Broadway between 44th and 45th Streets.

Using his sons as aides, Oscar oversaw the minutest details of construction

---

[4] The journal is still published. Oscar Hammerstein I is listed on its masthead as founder.

[5] The Harlem Opera House and Harlem Music Hall (1893) have become part of the Apollo Theater, still standing today and famous as the home of "soul music."

of all his buildings, inventing architectural details to insure acoustical perfection and comfortable seating. The Olympia presented opera comique, burlesque, and vaudeville in three of its auditoriums, and in the fourth introduced cabaret entertainment to the New Yorkers in a roof-garden restaurant and promenade.

But Oscar was always in and out of bankruptcy. One theater was sold to pay for another. Three years after the Olympia opened its doors, with the New York Life Insurance Company foreclosing on the mortgage, the president of the company wrote Oscar a note stating that henceforth they would be running the theaters. Oscar snapped a letter back to the director, saying, "I am in receipt of your letter which is now before me and in a few minutes will be behind me."

Within a year he rebounded, and, mortgaging everything he had, he built the Victoria on a corner of Times Square. The Victoria was almost completed when he ran out of money, and the contractor refused to finish the roof until Oscar came up with the lacking $2000. According to an article in the *New York World,* he met a woman on the streetcar who had appeared in one of his operettas. "He looked depressed," the report continues, "and a sympathetic note in her voice encouraged him to tell her his troubles. At 23rd Street she asked him to get off the car with her. On that corner was the Garfield Bank, where she had her savings—$2000 in cash which she gave him, refusing a receipt and telling him to repay her whenever he could. That loan made possible the completion of what turned out to be one of the world's most successful and memorable theaters."

At the opening performances Oscar explained why he named his theater to the press: "I have named my theater Hammerstein's Victoria because I have been victorious over mine enemies—those dirty bloodsuckers at New York Life."

The Victoria opened originally as an operetta house but quickly became the famous for headline vaudeville. Oscar hired all sorts of freak acts to satiate the public's morbid curiosity. Soon anyone who had figured prominently in a scandal was eligible for at least a split week there. Hammerstein even made his home there, living in two small dusty rooms with a cot and washbasin.

Oscar I built twelve glorious theaters in all, five of which he designated as opera houses. Many of his auditoriums are still in use today. Nor did he confine his construction to New York. Gone may be the Philadelphia Opera House at Broad and Poplar and the London Opera House in Kingsway which became the Stoll Theatre, but the New York landmark of the Manhattan Opera House attests to the durability of his vision.

Two of Oscar's theaters became very important to the life and career of his namesake: the Victoria, the theater Oscar 2nd's father, Willie, managed so adeptly, and the glorious Harlem Opera House, which was the setting for young Oscar's first foray into the theatre.

Hammerstein 2nd

1896–1919

IF THE VICTORIA, Broadway's latest showplace, was to become New York's most outrageous venue for circus-like acts, the Olympia, built almost a decade before was to remain its most elegant. The latter opened its doors in 1895, the year of Oscar Hammerstein 2nd's[1] birth. Its opening that year had little commemorative value for Oscar I's namesake—his firstborn male grandchild—it was mere coincidence. Building the Olympia was not an act of celebration, but an egotistical venture for the grandfather.

Shortly after the Olympia opened its doors he mounted his own operetta, *Marguerite,* which played havoc with the Faust legend. Not only did he direct and produce the elaborate production, but he created the libretto, lyrics, *and* music for it. It was just the kind of musical potboiler masquerading as art that his grandson would spend his lifetime trying to eradicate from the stages of the world.

If Oscar I was a supreme egotist, it must be admitted that once they grew to manhood, his sons Arthur and especially Willie (he was never called by his given name of William) recognized this and throughout their lives showed him benevolence mixed with fury; an overwhelming love-hate relationship with "the old man."

Willie's early marriage out of the Jewish faith to Scottish-Presbyterian Alice Nimmo was his first attempt to be "noticed" by his father. Naming his firstborn Oscar was another ambivalent move, for in the Ashkenazi or German-Jewish tradition it is considered not an honor, but near heresy to name after a living relative—no matter how famous. Additionally, Willie and Alice gave the child two unwieldy, unnecessary, and rather pretentious Gentile names: Clendenning

---

[1] Throughout his life Oscar Hammerstein preferred to sign "2nd" after his name (rather than the pretentious "II"), even when his fame eclipsed that of his celebrated grandfather.

and Greeley. If he had noticed, this would have irked the grandfather even more.

But if Willie wanted to impress his father with his emancipation and iconoclasm while kowtowing to his gray eminence, young Oscar did not. When as a child he was presented to the rotund man with the top hat who smelled heavily of dead cigars, the boy was terrified. He wondered why "the old man," as the members of his family called him, was so much more revered than his beloved Grandfather Nimmo. "He seemed a much nicer man," Oscar recalled, "with snow-white wavy hair and kind blue eyes. He bought me hard candy and took me for walks in the park."

Even though Oscar 2nd and his namesake were to have little intercourse and eventually the younger was to be embarrassed by the antics of the diminutive giant, one can see their similar qualities. Each was an unabashed romantic, and both were idealists. Although they had different ways of showing it, both men were passionate about musicals, and, most of all, each devoted his life to injecting art as they viewed it into the lyric theater.

As a young boy Oscar was drawn to his mother's side of the family, especially the Nimmos, with whom he lived from the time he was five. It was then that young Oscar, who only saw his father Willie as a figure who left early and returned home late from his work as manager of the one of his grandfather's theaters, moved with the family to two apartments on West 112th Street.

The decision to move was not solely Willie's. His wife, Alice, was addicted to moving, and by the time Oscar reached his fifteenth birthday, the family had moved eight more times. It should be reported that it was a time of burgeoning apartment building in New York, when landlords would offer two, sometimes three months free rental to new tenants. Frugal New Yorkers, and Alice Hammerstein, true to her Scottish ancestry, was certainly that, since "the old man" paid his sons a pittance, frequently pulled up roots and settled into new, freshly painted quarters.

Incomprehensibly, at the time of this move Willie and Alice lived in the upstairs flat along with Alice's sister, Aunt Mousie, and two-year old baby brother Reggie, while Oscar—nicknamed "Ockie" all his life because little Reggie was unable to pronounce "Oscar"—shared his grandmother's bed. Oscar saw little of the family upstairs but spent the greater part of each day with Grandfather Nimmo, who was devoted to watercolor and oil painting and slept in the room next door.

James and Janet Nimmo had had a falling out some years before over another woman in James's life, and although they had not gone so far to divorce, Janet's bed was "off limits" to James. Oscar's nightly presence there was perhaps necessary insurance against any infraction of the rules.

The lack of communication between his grandmother and her spouse must have appealed to the sensitive young boy who often saw his mother and aunt in the upstairs apartment "in physical conflict trying to beat each other's brains out." Fistfighting was not confined to the women in the family, for Oscar remembered how his father was "very jealous of my mother. . . . He kept her at home and took her out very seldom and when they did go out there was always some fight afterward because some man had been looking at her, and he thought that perhaps the man wouldn't have been looking at her if she hadn't looked back."

"I have also seen my father and mother in physical conflict," Oscar confessed to his own son William, adding that he found a way to stop fighting. "I would start to cry. . . . I learned this when I was very young. . . . It wasn't hard to do because it did shake me up a lot and I felt like crying anyway."

Life was much more ordered in the flat below. Here the imaginative young boy thrived on the stories his grandfather told of life in the old country and enjoyed the equality with which James Nimmo treated him. They would share a whiskey and egg punch every morning, then frequently stroll to a nearby park for an afternoon of sketching, and invariably would top the evening off by splitting a bottle of Guinness. The youngster enjoyed sitting in on the conversations of his elders and was a natural performer.

It was a time when precocious youngsters like Oscar were encouraged to "recite," and Oscar needed little urging to perform his specialty, "Little Black Boy," a maudlin and racist piece (by today's standards) about a black boy whose soul turns white when he dies. His mother and grandmother applauded Oscar's histrionics. Not so his father.

Willie was adamant about not wanting either of his sons to have even a passing acquaintance with theatrics. He even found attending theater boring; to him it was merely a business he had been born into. He had little respect for the performers he booked onto his stages and felt theater folk were fast and unwholesome. He went so far as to forbid his wife from associating with his stepsister, Stella, who was beginning her career as an actress. But he saved his real fury for opera, for with something like the right of eminent domain, "the old man" would unexpectedly swoop down on the box office of whatever theater Willie might be managing and take the receipts to invest in whatever diva or opera production might take his current fancy.

Oscar was bright and artistic. Reggie, nineteen months younger, chose not to compete, so he became the family clown. As Oscar excelled in his studies and was skipped ahead a grade or two, so Reggie failed to pass into the next grade, and the gulf between the brothers widened. But it must be noted that all through his life adult life in the theater Oscar was able to find a place as casting director or stage manager for his younger sibling.

Although Willie was firmly against it, in 1903, when young Oscar was eight, he and Reggie were taken by Alice to attend their first play. The show chosen was *The Fisher Maiden,* a comic opera full of mistaken identities and predictabilities. This fiasco, which originally played briefly at the Victoria, had been transferred to the Harlem venue, and since the Hammersteins now had moved to East 116th Street, the Harlem Opera House was their local theater. A mature Oscar was to describe it as their "subway circuit—except there was no subway as yet." But even though *The Fisher Maiden* met with little success in sophisticated Manhattan, it was a roaring success in Harlem and the boys loved it.

The mature Oscar was to deplore his parents' and Reggie's lack of education. Part of his resentment for his impresario grandfather arose in him because his own father—actually, all of Oscar I's brood—were never given the opportunity to finish high school. They were thrown into the business end of theater management, becoming box office attendants or even bricklayers who would know how to *build* theaters. Oscar 2nd missed being born into an "atmosphere of culture. The only cultural achievement in the family was the dubious painting of my Grandpa Nimmo, and the brilliant operatic career of my Grandfather Hammerstein. Everybody hated this latter accomplishment because it kept the family from becoming rich."

But achieving wealth was not Oscar 2nd's goal. Knowledge was. Even as a child young Oscar became an avid reader. "My mother would go down to Barnetts and buy four or five books. I would read two in one day," he was to recall, and, pointing out his Scottish mother's frugality, he added, "and then she would take them back saying I had read them before and exchange them for two others." It was about this time that Oscar was given piano lessons. Romantic show-off that he was, and gifted with a good ear, he enjoyed playing for all his relatives and copied the teacher's gestures and interpretation. Miraculously he became a good sight reader rather than an ear player. The training was to serve him well throughout his career, for his sensitivity to musical line was to make him equally adept at setting words *to* music or creating a lyric that would sing from the page.

As for their religious beliefs, it was a curious family. Grandmother Nimmo was Presbyterian, but she rarely went to church. Yet at Easter time she would insist the whole family attend the Catholic cathedral because, she said, "the floral displays are put on better than in any of the other churches." Alice attended the Episcopalian Church of All Angels regularly, and, although far from religious, served her family no meat on Friday. She also insisted her boys fast completely on Good Friday. Willie, although he considered himself Jewish, only stepped into a synagogue in the case of a show-business colleague's funeral.

This lack of religious zeal makes the family decision to have Oscar, nine, and Reggie, eight, circumcised in 1904 all the more incredible. The boys were

Ten-year-old Oscar Hammerstein 2nd in his first dress suit. Photo courtesy of The Rodgers and Hammerstein Organization (R & H).

told it was for "sanitary reasons," and it must be noted that Alice, up to date with the latest medical procedures, seems to have suggested the surgery. Oscar always believed their family doctor, being a little low in funds, had forced the operation on his gullible mother and rather disinterested father.

But Willie cannot be held too accountable for his preoccupation. Naturally cold, reserved, and cynical, he was rarely able to unbend. Only his ambition, which had surfaced when he was in his thirties when "the old man" put him in complete charge of the Victoria, created any fire in him. Here he seemed to have the magic touch to combine scandal, comedy, and headliners into one program, and soon he had become so successful that competitors Keith, Proctor, and Pastor ceded him all of mid-Manhattan in order to keep him from combining with any other chain.

Besides acts of the caliber of the Dolly Sisters, Fanny Brice, Weber and Fields, Al Jolson, Will Rogers, and Houdini, at "the Corner" under Willie one could see celebrities like Evelyn Thaw, who had been the love object in a headline murder case, or the talentless Lady Frances Hope, whose act climaxed in holding up the renowned Hope Diamond. Each program spotlighted such "freaks" as Sober Sue—a $1000 prize was offered to anyone who could make her smile. It was later discovered that her facial muscles were atrophied. Word on the street was that Willie was not above creating his own hoopla, that he himself persuaded two of his chorines to shoot an out-of-work actor, thereafter billing the girls as "The Shooting Stars."

No wonder he forbade young Oscar to think of a career in the theater. To him it was all cheap humbug. A blowup happened in the summer of 1909 when neither Willie nor Alice took kindly to Oscar and Reggie's involvement with "*Buffalo Bill Cody's Combined Wild West Show.*" The boys, who were mesmerized by the show which climaxed with an Indian raid on a stagecoach, made contact with the stage manager, asking to see the show again and again. Soon they were invited to be involved in every performance by acting as the two terrified young hostages in the stagecoach.

Yet far more devastating than his parents' imprecation against performing on stage was Alice's death when Oscar was fifteen. An early champion of birth control, she died of peritonitis due to the primitive abortive methods in use those days. Thirty-five at the time, Alice was well past the age when women who had teenagers in the house considered it seemly to have a newborn.

Oscar's method of dealing with the death of his mother was classically stoic. "I went for long walks and thought it over," he was to write years later, "and I began to adjust myself. I never felt like going to anybody for help." He felt his mother's death crystallized his own attitude toward death. "I never feel shaken by death," he was to say, "as I would have been if this had not happened to me when I was fifteen. I received the shock and took it, and sort of resisted as an enemy, the grief that comes after death, rather than giving way to it. I get stubborn and say this is not going to lick me because it didn't then." Certainly this attitude is patently shown in Billy Bigelow's triumph over death in *Carousel* and in the lyric of "You'll Never Walk Alone." It shines forth even more triumphantly in the moving final scene of *The King and I.*

Less than a year later Willie was to marry Alice's sister, Mousie, ostensibly to bring up his two sons. This was a common ritual of the times when childbirth was frequently deadly and most families had a maiden aunt ready to step into the breach. Still, it must be added that Aunt Mousie, although cruder and more brusque than his mother, was as loving a substitute as could be found, and Oscar the man continued to think of Mousie with love and care for her for the rest of her life.

Soon he was to register at Columbia College, and although his heart would have yearned to enter some form of theater, he obeyed his father's wishes and enrolled in a pre-law course. It was about this time that the family drew closer together. Uncle Arthur had become a successful producer, having brought out *The Firefly* and *High Jinx*, and he and Willie could rail against the operatic insanities of "the old man." At last Willie, in poor health, the coffers of his theater frequently raided by his father, felt he must resign from the Victoria.

Young Oscar, displaying his newly acquired legal prowess, was to join his father and uncle in drafting a statement that began: "I left the Victoria Theatre because I hoped by so doing I could save my father from himself."

The year 1914 was not only the fateful year war broke out in Europe but was also a devastatingly calamitous time for the Hammerstein clan. Willie succumbed to Bright's Disease on June 10th, only three months after Abraham Lincoln Hammerstein died. Harry Hammerstein, Oscar I's firstborn, was to pass away less than two months later.

In his essay *Some Kind of a Grandfather,* Oscar was to write of their confrontation and near communication when he walked into a bedroom of his apartment after the death of his father and found the bereft old man crumpled in a chair. "He looked small and beaten. This made me feel strong. Surprising myself with an unnatural valor, I walked right over to him and said: 'How do you do, Grandfather?' He shook my hand limply. I sat down near him. Nothing more was said by either of us. The continued silence gave me the strange feeling that I had not come into the room at all."

Willie's funeral in the Harlem synagogue Temple Israel was attended by more than a thousand dignitaries from the theater. It was in the Orthodox Jewish tradition, with the casket rolled down the aisle. So beloved was he by his showfolk friends that the Mayor Jimmy Walker decreed the lights be dimmed on Broadway and taps be sounded. Taps would be heard again and the "Great White Way" would go dark forty-six years later for Willie's son, Oscar.

With the death of the three Hammerstein boys within a single year only Arthur remained to carry on the theatrical tradition of the family—that is, besides "the old man." But old Oscar was now totally gone on opera and was soon to borrow heavily against and eventually to lose the last remaining pin of his empire, the Victoria.

Although Willie's death weighed heavily on the sensitive young Oscar, it eventually came to be salvation and liberation, for soon he came to realize that their only real rapprochement had come in the last year or so of his father's life. Over the objections of Uncle Arthur, who said he had promised Willie on his death-bed to be sure that Oscar would have nothing to do with the theater, the nineteen-year-old announced that he had decided to join the Columbia Players.

The Columbia Players, like Harvard's Hasty Pudding or Yale's Whiffenpoofs, was a respected all-male collegiate troupe devoted to an annual musical. At Columbia the show ran for a full week and was even accorded the professional status of a review in New York's commercial newspapers.

In his first venture, the 1915 college show called *On Your Way,* Oscar played a poet, Clarence Montague. He sang and danced to two songs. The whole show was written by rather untalented upperclassmen, the brothers Kenneth and Roy Webb. Their macho and rather juvenile concept of Oscar's first number, "Looking for a Dear Old Lady" was of a young man who's looking for a rich old lady to keep him. Some of the lyrics of Oscar's second song, "In Kewpie Land," on which he collaborated are printed below:

In kewpie land they need no cash,   They never saw a subway train,
There are no bills to pay.   They never took the El.
There are no gentle grafters there   They never rode in taxicabs
Like we have here today.   That tear around like the deuce.

That taste of show business whetted Oscar's appetite, and by time next year's annual rolled around he involved himself more heavily. This show, *The Peace Pirates*, was a take-off on the "Peace Ship" Henry Ford financed to try to bring World War I to an end. Ford's venture was as ludicrous as it was unsuccessful but the satire written on it hit its mark dead center. The loose libretto was largely the work of Herman J. Mankiewicz, who after his undergraduate Columbia days was to turn out Academy Award-winning screenplays of the caliber of *Citizen Kane*. He allowed Oscar to add an interpolated blackface scene. Oscar named his character Washington Snow, in vague reminiscence of the little black boy whose soul turned white.

Gangly and dressed in britches that made his body seem all legs, Oscar's performance was reportedly "thoroughly original and distinctly funny and demonstrated his ability to put over a song." So said the critic for Columbia's *Daily Spectator*, the writer being none other than classmate Lorenz Hart. Hart, the paper's dramatic critic, was a valued cast member of *The Peace Pirates*. His diminutive frame and saucer eyes while doing a take-off of Mary Pickford reminded Oscar of "an electrified gnome."

But perhaps the most prophetic meeting during his Columbia days was to happen the following year when the Columbia Players presented *Home, James*, which Oscar co-wrote with Herman Axelrod. After a performance in March 1917, Morty Rodgers, Oscar's fraternity brother, brought his kid brother Richard backstage. Here, in an excerpt from *Musical Stages*, his autobiography, is how the composer remembers their first encounter:

Going backstage at a Varsity Show was heady stuff for a fourteen-year-old stage-struck kid, and I was overawed when I was introduced to the worldly upperclassman who had not only acted in a Varsity Show but had also written its lyrics. Hammerstein was a very tall, skinny fellow with a sweet smile, clear blue eyes and an unfortunately mottled complexion. He accepted my awkward praise with unaffected graciousness and made me feel that my approval was the greatest compliment he could receive.

Oscar, too, in later years was to recall that first handshake with humor and remark that Richard Rodgers was then still wearing short pants. The composer vehemently denied this, saying he had already graduated to "longies."

In his book, Rodgers goes on to talk of how he aspired to emulate Hammerstein. "That afternoon I went home with one irrevocable decision: I would also go to Columbia and I would also write the Varsity Show. I also decided that I couldn't waste much time before starting out in my chosen profession." The young composer contacted a lyricist, the team went to work and in three months time had copyrighted their first song.[2]

Beyond this fateful meeting that was to bear the fruit of a golden collaboration some thirty-six years later, one can see in *Home, James* the emerging pattern of what would become a typical early Hammerstein libretto—entertaining songs with good rhymes which didn't always fit the story and "a light romance with complications."

In a section he calls "Notes on Lyrics," of his *Collected Lyrics*, Hammerstein is unduly hard on his "showstopper" in the second act, "Annie McGinnis Pavlova." Although the execution is not flawless, the concept is fresh and original.

VERSE	REFRAIN
Clancey was fond of a show,	Annie McGinnis Pavlova,
From opera to movies he'd go.	I'll stop you from puttin' one over.
On dancin' and prancin' Clancey was keen,	'Twas in Hogan's back alley
But the Ballet Russe he never had seen.	You learned the bacchinale
For five dollars a throw he decided to go	And now you're the pride of the Ballet Russe.
And he bought him a seat in the very first	They call you zephyr, a fairy, and elf,
row.	Put on your flannels, take care of yourself!
Annie came out. Clancey yelled, "Stop her!	For the costume you're wearin is a shame to
She ain't no Roosian, for I knew her poppa."	old Erin,
	Oh, Annie, you'd better go home.

The lyric of "Annie McGinnis Pavlova" tells a great deal about the teenage Hammerstein's character and provides perhaps a clue to his forthright, some might say his goody-goody, lyrics. To understand its relevance one must return to what Oscar termed "a black mark in my history." The incident occurred in his childhood. Because of danger from passing horsecars both he and Reggie were forbidden to leave the Harlem sidewalks. One day when the brothers were playing, Oscar threw the ball so hard that it had to be retrieved from the street. "Reggie started out to get the ball and I called to him a warning that he was not supposed to leave the pavement. He did anyway and he retrieved the ball.

[2] Only two years later in March 1919, seventeen-year-old Richard C. (Charles) Rodgers made good on his vow when he wrote the music and some of the lyrics for the annual Columbia University Players Varsity show, *Up Stage and Down*.

Later I told my mother on him. She let him off with a mild rebuke for breaking the rule, but at the same time she let me know that she did not admire me for having informed. . . . I was ashamed of myself then and [at fifty-eight] I am still ashamed." Obviously. And as an attempt to write out his shame, Clancey in the song might be the same kind of whistle-blowing informer.

Part of the natural, openhearted effect of *Home, James* was achieved simply because none of the three major creators pushed too hard, as each expected to make his living in another calling. Robert Lippmann, the composer, was on his way to becoming a successful orthopedist; Herman Axelrod turned to real estate; and Oscar, although he hated it, was still heading into law.

Contacting the firm of Blumenstiel and Blumenstiel, because he thought a job might make his studies more interesting, he was surprised to be taken on. Perhaps they gave him the job because Blumenstiel and Blumenstiel managed the securities Willie had left that supported Mousie, Reggie, and himself; more likely it was because he came so cheaply. Oscar was paid five dollars for his week's work.

The young clerk enjoyed his work, looking up precedents, copying depositions and the like, and when the spring term at Columbia ended in June 1917, he was offered full time employment as a summons server because he was so tall and imposing. Yet when he was bluntly told "the fellow you're looking for is out!" and the first doors were slammed in his face, he was shifted back to indoor work. Now he knew he was not long for Blumenstiel and Blumenstiel, even Columbia, and certainly "the law."

When the United States entered the war in 1917 and the draft began in early June, army service seemed to Oscar like an honorable way out of his dilemma. Informed by friends and family that he would be certainly be rejected for being underweight, he stuffed himself with bananas and drank gallons of water before his physical examination. But still he was turned down.

To ease the disappointment of his rejection, he went away for the weekend to Deal, an elegant New Jersey beach resort. There at a house party he met Myra Finn, and after a game of spin-the-bottle realized he had fallen in love with and was determined to marry her. The diminutive Miss Finn, just a year younger than Hammerstein, was a distant cousin of Richard Rodgers's. She stood but four feet, eleven inches tall and was often described as "cute." She and the six foot two Oscar certainly made an odd couple.

Over the objections of Myra's parents and Mousie, Oscar and Myra announced they would be wed at summer's end. He entreated his Uncle Arthur for a job, any kind of job, boldly announcing that he did not intend to return to Columbia Law and that he intended to make writing for the theater his life's work.

Arthur brought out all the stock objections: his deathbed promise to his late brother, Oscar's lack of experience in professional theater, and his final statement that two generations of Hammersteins in show business were enough. Oscar trumped his uncle's last pronouncement with his declaration that two generations proved that theater was in his blood. Finally Oscar blurted out that he needed the money now because he was getting married.

Arthur, forced to relent, said he would give his nephew a trial at twenty dollars a week. Oscar would be assistant stage manager and general gofer on his current Broadway hit, *You're in Love*. The surrogate father warned his young protégé that he was to absorb the technique of playwriting from stage managing and attending the theater and was not to try his hand at writing a script for at least a year. Oscar acquiesced, and Arthur in turn, soon after his current production closed, promoted his nephew to a permanent post as production stage manager.

Oscar and Myra were married on August 22, 1917, in Myra's parents' apartment on West End Avenue, and after a brief Canadian honeymoon Oscar settled down as stage manager for Arthur's next productions. Quick to learn, he had easily made the leap from the Columbia Varsity into the professional theater. During an out-of-town crisis in the try-out of *Furs and Frills*, the harried authors, knowing Oscar's words would be drowned out by a tap dance number, permitted him to write lyrics for the second act opening. "This was my first professional work," he commented. "I was writing a lyric for a hostess to greet her guest shortly after the curtain rose." A sample is printed below:

> Make yourselves at home.
> 'Neath our spacious dome.
> Do just as you please
> In twos or threes if you'd rather—
> But rest assured you'll be no bother.

At the time, Oscar must have thought his lyric passable, but later the mature Hammerstein went on to coruscate his lines. "How did she know they would be no bother?" he asks. "As I remember a bizarre assemblage coming on the stage carrying tennis rackets, wearing riding clothes and sport costumes of outrageous color, it seemed almost certain that they would raise hell with her house before the weekend was over."

Yet Uncle Arthur, whose taste was for the pretentious, must have believed in his nephew's ability, for far before the year of apprenticeship had elapsed he was to suggest Oscar adapt a melodramatic story he liked into a serious, nonmusical play; that is, if Oscar liked it too.

"I would have liked the telephone book," his nephew was to say as he immediately began work on this story about a girl who agrees to marry a man she does not love in order to escape her unpleasant home life. The artificial plot then moves the heroine to work in a gambling resort where, through the intervention of a benevolent grandfather, she is eventually reunited with her earlier, childhood sweetheart.

The play, called *The Light*, was put into rehearsal in late April 1919. After Oscar I, whose name was often headline news, attended the dress rehearsal prior to out-of-town performances, the *New York Telegraph* reported, "He patted his grandson on the back and assured him he had written a good play. Whereupon the younger Hammerstein promptly resigned his position of stage manager of *Tumble Inn* at the Selwyn, notifying Uncle Arthur that henceforth he intended to devote himself entirely to playwriting and thus give further lustre to the family name."

Perhaps it was because the deus ex machina was a magnanimous grandfather, or simply because Oscar I's critical faculties were failing, but his critique was a heavily inflated opinion of *The Light*.[3] Comments in the New England papers were devastating: "Serious moments are absurd. . . . wretched material. . . . worst is a scene where the heroine's former sweetheart throws her to the floor."

On opening night in New Haven, Oscar, dismayed because he knew he "had a big flop . . . just ran out of the theater, went into a park and sat on a bench." But while he was sitting there, an idea came to him for a new show and he "started writing it."

This was to be a familiar Hammerstein pattern, he wrote, years later, that he was never discouraged by failure. Even though he was to call his first play ever after "*The Light* that failed," he could take comfort that two days before, on May 18, *Up Stage and Down*, the 1919 Columbia varsity show, had just been mounted on Broadway under the title *Twinkling Eyes*. Directed by Lorenz Hart, the revue included "Weaknesses," "Can It" and "There's Always Room for One More," all with wisecracking Hammerstein lyrics. The composer? Seventeen-year-old Richard Rodgers.

---

[3] The play eked out seven performances: three in Springfield, Massachusetts, and four in New Haven, Connecticut; the production never reached New York.

Hammerstein

1919–1924

J UST TWO MONTHS AFTER *The Light* failed, the light failed for Oscar I.
His unhappy last years and the violence that led up to his death were as
dramatic as any opera he ever produced.

During an ocean crossing four years earlier, when he was sixty-eight, he had
met and fallen deeply in love with Emma Swift, a big-boned, imposing woman
with great lust for life, who was thirty five years his junior. Hammerstein's
reasons for marrying the volatile Miss Swift are obvious, and hers, too, are
understandable. He was famous, and she evidently thought he was very
rich. Her suitor talked of producing operas, building theaters, and royalties
from old and new inventions. Instead, in the course of several months she discovered he had diabetes, often fainted, and needed hospitalization. Deceived,
irritated at his braggadocio and well-known irascible temper, and resenting that
she was transformed into a wife and nurse for a dying man—and a penniless
one at that—when the diminutive Oscar struck her, she returned his blows.
Because of Emma's robust strength and size, Oscar always got more than he
gave.

Arthur, his only surviving son, had seen to it that some of the proceeds from
the sale of Oscar I's last remaining theater went to buy a house for his father
and Emma in suburban Atlantic Highlands, New Jersey. He hoped his father
could live out his life there in relative tranquility. The house was far enough
out of New York to keep "the old man" out of trouble and the deed, lest his
father put up his only remaining asset as collateral on another operatic venture,
was kept in Arthur's safe.

But in July 1919, Oscar I was found lying unconscious at the local railroad
station and immediately taken to the hospital. When he regained consciousness
he recounted the harrowing story of how Emma had taken a pail of cold water
and thrown it over him as he lay in bed. The startled old man immediately

dressed (in his usual striped trousers, tailcoat, and top hat) and with the aid of his cane hobbled down to the railroad station, intending never to return to that house of torture. Sitting on a bench, awaiting the New York train, he fainted.

Fearful of another confrontation with Emma, before he went back into unconsciousness from which he was to wake intermittently until he died at 7:30 P.M. on August 1, 1919, he gave orders that no one was to be admitted to his hospital room. But that very afternoon, even though Oscar I could not have been aware of it, his grandson Oscar 2nd, having heard that the patriarch was dying, had quietly entered the room.

The young man was to write of his feelings during and after the deathbed scene and to observe that those five minutes listening to the patient gasping for breath before the nurses shooed him out was the longest time they had ever shared together.

"I walked down Park Avenue feeling lost and unclassified. My grandfather was dying and I didn't know how I felt about it. I had a deep sorrow to give way to or to resist stoically." Certainly Oscar was thinking of the resentments he felt toward his own father for forbidding him a career in theater when he wrote the next sentence: "I had no resentful memories of a domination from which I could not feel free." And then returning to his feelings about the demise of his grandfather he was to continue: "I could make no crass speculations concerning my probable inheritance in his will—I knew he was broke. I had none of the conventional thoughts of a bereaved grandson. It was an uncomfortable feeling, the more uncomfortable because in some vague way my heart had been touched and I didn't know why."

All the papers devoted columns detailing the highlights of Oscar Hammerstein's career. Perhaps the most human tribute was Carl Van Vechten's which concluded:

> It was not in Oscar Hammerstein, I think, to inspire affection. His way was too big, his egoism too colossal, his genius too evident. These qualities made men stand a little away from him . . . he could and did command admiration for the things he accomplished, more than that, admiration for the way he failed.
> He was not, as a matter of fact, what is called a good loser. He groaned and moaned over loss, but in a few days the board was erased and with a clean piece of chalk he was drawing a new diagram, making a new plan. He was an artist, he was a genius.

The funeral, like his son Willie's, was attended by all the notables of the show business fraternity. Oscar I, who coined the phrase "it is incredible how many people can stay away from a bad show," would have been proud to know that Fifth Avenue's massive Temple Emanu-El was filled to overflowing. After a simple Hebrew prayer, the great Irish tenor John McCormack, representative

of the many European artists who had made their debuts in America under
Hammerstein's aegis, sang "The Lost Chord."

Years later, when he was in Hollywood and finally got in touch with his
feelings about "the old man," young Oscar outlined, as mentioned before, a
biography of his grandfather. "He couldn't hurt me now. He couldn't humiliate
me," he was to explain. "The fears and resentments developed in my childhood
were no longer a block to our union. It is ironic and sad and strange that I did
not begin to understand or like my grandfather until the day of his death. But
he was a strange man and so, perhaps, am I."

At the time of his grandfather's death, Oscar and Myra were living at 122nd
Street and West End Avenue. It was there on October 26, 1918, that their son
William (named after Oscar's father), was born. Their second child, Alice
(named after Oscar's mother), was born two and a half years later on May
17, 1921. Richard Rodgers's father William was the obstetrician who delivered
both babies.

When he was puffed with what looked like the certain success of *The Light*,
Oscar had boasted that he was leaving his uncle's employ to strike out on his
own as a playwright. After the demise of his first play and during the early
years of the "roaring" decade, Oscar was frequently on the road tinkering with
one of Uncle Arthur's musicals. When he was at home, Myra, who at first had
been serious about her husband's career, typing additions to his scripts and
willingly accompanying him to all the latest plays, turned argumentative.
Within the next few years their marriage would become a rocky road. "She was
the kind of woman who, when she entered a room, if the window was open,
wanted it shut," reported her daughter Alice of life with mother, "and if it was
shut, she wanted it opened."

"Myra Hammerstein was selfish rather than supportive. Her interest in her
husband's work was soon replaced by her interest in her social life," Brahms
and Sherrin's book, *Song by Song,* recounts. Yet in sum, Oscar's determination
to master every aspect of his craft left him little time to develop the craft
of marriage.

It does not seem as though the young man tried, for Oscar soon began sleeping in the other twin bed in his five-year-old son William's room. Yet William,
who recalled his joy at waking up and finding his father in the adjoining bed,
because as a toddler he could slide down his long-legged father's knees, recently
recalled how his father cruelly baited his mother. "I can remember things that
went on between them. . . . He was mean to her, he used to play tricks on
her . . . to aggravate her. He thought it would be funny—he learned it from
someone—you leave a door partially closed, and you put a telephone book on
top of the door so whoever opens the door gets the phone book on the head.
So he did that to my mother, he thought that was funny as hell, and it's not

at all funny and she got terribly upset. She was a kind of hysterical person. She overplayed everything, and I think that just got on his nerves."

Oscar not only had a deteriorating marriage to contend with, he had Uncle Arthur's prudishness and lack of adventurousness to face. Arthur's taste for drama and sentimentality made him an atypical producer in a decade that was avidly in search of escape and of veiled sexuality. After the deprivation of the war and the freedom brought by the urge for woman's suffrage, most Americans went to theater, especially musical theater, looking for escape. But the popular *Follies, Vanities, Scandals* were not for Arthur. Known as "the Ziegfeld of Operetta," his agendas often included light comedy or even, as in the ill-fated *The Light,* tragi-comedy. An Arthur Hammerstein production usually included a dash of burlesque and horseplay—invariable featuring a "shtick" having nothing to do with the story performed by a featured comedian in the middle of the second act.

His success was almost an accident, for early in his career, after working closely with "the old man," Arthur had been fortunate enough to involve himself in Victor Herbert's *Naughty Marietta.* When he and Herbert had a falling-out, the producer was prescient enough to hire young Rudolf Friml to write an enormously successful operetta, *The Firefly.* He was also clever enough to have hired Herbert Stothart as musical arranger, director, and conductor of his last five shows. Stothart, an expert at sensing when tempos needed reviving or when songs needed rearranging, was eager to try his hand at composing.

It was only natural that stage manager Oscar Hammerstein, eager to learn about musicals, would want to collaborate with Stothart, and that Stothart would have the experience and knowledge as to what kind of musical Arthur would be likely to commission. Oscar concocted a story that concerned an American doughboy who has been nursed by Toinette, a French girl, and promptly falls in love with her. Once back in the States the soldier returns to Joan, his Arkansas sweetheart. The history and the genesis of the musical can be found in its search for a title. First called *Toinette,* and then *Joan of Arkansaw* (sic), the true blue element finally surfaced in its eventual final title, *Always You.*

Once they were given the go-ahead, Oscar wrote the play, leaving room for the songs. Then, after the collaborators discussed whether the song needed should be a waltz, polka, drinking song, ragtime—or, less precisely, after pinpointing the needed mood—Stothart wrote the music. This was the usual practice by which musical comedy or operetta was concocted in early twentieth century, since English was not native to most of the leading composers for the lyric theater. Kalman, Romberg, Friml, and all the leading lights of operetta were of German descent and would generally mangle accents and misplace prosody if they were handed a lyric to set to music.

But there were two more important reasons for writing the music first. Most

often it was to allow the musician the freedom to compose his melodic line without being imprisoned in a rigid form that a set number of lines imposes. The other was ragtime. A syncopated tune intending to become a dance, hopefully a dance craze, demanded a lyric which could only be written after.

Throughout his career, not only at its onset, Hammerstein was to collaborate with Rudolph Friml, Sigmund Romberg, Jerome Kern, George Gershwin, Vincent Youmans, Arthur Schwartz, and Ben Oakland among others, in addition to Stothart. With each of these he was write his lyric *after* the composer had handed him the music. The outstanding exception was Richard Rodgers, who, when he worked with Hart, always followed the musical comedy tradition of writing the music first. By the time Rodgers collaborated with Hammerstein, however, he had come to feel that the songs were a logical outgrowth, an extension, of the libretto, whose words had to be *set to music.*

Besides, Rodgers sensed that in Hammerstein he had found a poet, and the music he wrote rose to the occasion that art demanded. Except for Gershwin, Rodgers was probably the best trained of all the Broadway composing fraternity and had acquired enough technique and experience in writing *both music and lyrics* to be able to follow the pattern of his forebears. In preceding centuries, Mozart had set Da Ponte's librettos, Beethoven had set Goethe's poems, and Debussy had musicalized Verlaine's *poèmes,* not to mention Sir Arthur Sullivan, who would not think of setting pen to music paper until he had Gilbert's completed lyric in hand.

Unfortunately Stothart, never a very inspired composer, was to write a very clichéd score which likewise brought out only cliché lyrics from Oscar. The title song, "Always You," has perhaps even for 1920 one of the tritest quatrains of the time:

Now that fate has called me from your side,   Thru the days of darkness I'll abide,
Dear little girl, little girl of mine,   Heart of my heart, I will wait and pine.

The score was replete with good concepts badly executed: "Pousse Café," a song in pretentious (and faulty) French; "Misterioso," listed as an eccentric and containing lines like "we are as vicious as you could wish us"; and "A Wonderful War" ("For we had oodles of heroes handsome / Lieutenants and Captains too, / When with a glance you'd always land some / What else was a poor girl to do?").

A bored Oscar and his first wife, née Myra Finn, set sail for Europe. They were divorced in 1929. Contrast this with Hammerstein's smiling expression as pictured here late in his life beside his second wife, Dorothy Blanchard. Rodgers & Hammerstein Organization (R & H).

Perhaps the only hint of the charm and honesty that was to surface in the mature Hammerstein is to be found in a revue-type song, published as "A String of Girls" but known as "The Tired Business Man." It was cut before the show opened in New York. Although it doesn't always scan and its rhymes are often forced, its lyric pretty well describes an Arthur Hammerstein show. Was it cut because the message struck too close to home?

VERSE
It's a cinch to put across a show
If you try, 'd'ever try?
You can keep them laughing if you throw
Lots o' pie in the eye.
Pretty costumes and scenery
These of course, you must buy.

VERSE 2
You've omitted one important thing,
Here's a hunch, here's a hunch.
Is it of the pretty girls you sing?
That's the bunch, that's the bunch.
It's the chorus that does the trick
They're the crowd with the punch.

REFRAIN
Start with a little plot,
Cook it, but not too hot,
Throw in a heroine,
Maiden so simple and ingenuish.
Then let your tenor shine
With his high C;
Write in a well-known joke,
Use all the old-time "hoke"
For this is the surest plan
To entertain the tired business man.

REFRAIN 2
Bring on a string of girls,
Each with her string of pearls,
Teach them some tricky twirls,
Dress them in costumes
A trifle naughty.
Sing of a summer night under the stars;
Light up the yellow moon,
Write up a jazzy tune,
For this is the surest plan
To entertain the tired business man.

The show received so-so reviews. Most critics found it "traditional and conventional," one even took his clue from Oscar's lyric and wrote that *Always You* was "for the tired business man with the prettiest chorines in town." The score, as most critics noted, was not very exciting, but they all added that Arthur "has actually managed to cut out jazz, to eliminate any reference to prohibition, to cater to the unsyncopated mind." ·

Arthur, who was to gain much publicity from the fact that he was bringing forth his nephew's Broadway debut and that the show featured French star Julia Kelety in her American debut, came in for the lion's share of the kudos. Somehow he was able to keep the show running for eight weeks, during which time Oscar, in the wings nightly, learned much about the craft of writing a musical. He was surprised when laughter came on unexpected lines or when his best jokes were greeted with stony silence or groans. Now he began to realize that an audience's involvement, interest, and laughter depended on motivation, and their empathy with the character's predicament.

By the time *Always You* was ready to leave New York to tour, Arthur was preparing his next production, which was to be a vehicle for comedian Frank Tinney. Of course Oscar wanted to be involved in it, and Uncle Arthur, realizing his nephew had done a professional, if not inspired job with *Always You,* consented. It was obvious that Oscar needed to learn more about his craft, and Arthur suggested he collaborate with Otto Harbach, the acknowledged master of the lyric-libretto, who had written twelve shows for Arthur before this meeting. As further insurance and aiming for a massive hit, Arthur turned the collaboration into a triumvirate by adding Frank Mandel. Mandel, a clever humorist who had worked with Harbach before, was eleven years older than Oscar and not a lyricist, but a script-writer. He would bridge the gap between the intense, raring-to-go Hammerstein and the doyen, professorial Harbach. But Arthur was not as altruistic as he sounds. His next project was in dire need of some levity and youthful ideas.

Harbach, Hammerstein, and Mandel got along famously and were frequently to work as an ensemble or a tandem in years to come, but Oscar's closest association would be with Otto Harbach.

Twenty-two years Oscar's senior, Harbach had been successful since he had written *Three Twins* in 1908, which produced the hit song "Cuddle Up a Little Closer, Lovey Mine." He had done lyrics and libretto for Arthur's first massive hit, *The Firefly,* as well many of his later productions like *High Jinx, Katinka,* and *Tumble Inn.* Now he had already written this latest musical, *Tickle Me,* and inscribed the title page as "a musical laugh in two hysterics and ten screams with a book as light as a feather and score of tickly tunes." A spoof on the movie cliffhanger theme, *Tickle Me* concerned the misadventures of a crew sent to make a film in Tibet. There they are put upon by "Indian Indians," a "Hotcha-cha," and an evil "Dardanella" (the names of each of these villains were supposed to produce laughs). The thrust of the plot involves comedian Frank Tinney, a Harold Lloyd look-alike and indeed the star of the show, who is sentenced to be tickled to death with a peacock feather.

The show already having been written, Oscar was able to contribute only a sentimental ballad, "If the Wish Could Make It So," near the end of the second act and hold book (assist at rehearsals). Certainly there was not much he and Mandel could do to improve the hodgepodge of mistaken identities, absurdities, non sequiturs, bad jokes, and double takes that pervaded *Tickle Me,* and Oscar, eager to use his recently acquired backstage knowledge, hoped never again to be trapped in a star vehicle. Although he was forced in his very next show to tailor his lines around another well-known comic, later in his life he was to write:

> That was very early in my career. After that, I don't believe I ever wrote another comedian vehicle. I found out early in my career that I preferred to write musical plays that did not depend on stars, because then you owned a piece of musical and literary property that was detached from personalities and you could play it anywhere.

In spite of Oscar's imprecation against the star vehicle (which, it will be noted, was not written out of artistic integrity but from a financial point of view), there is no doubt about what turned *Tickle Me* into a hit and allowed it to chalk up 207 Broadway performances—Frank Tinney. The *New York Journal* wrote, "Frank Tinney, always good, is at his best." Robert Benchley commented in *Life* that "Mr. Tinney remarked he had never dared to have more than three children because statistics show that every fourth child is a Chinaman."

Arthur, swelling with the success of his show, immediately set out to repeat the formula. This time he would feature Jewish comic Ben Welsh. The libretto would again be written by the threesome Harbach, Hammerstein, and Mandel to a score by Herbert Stothart; Bert Finch, who had choreographed the hit show, and set designer Joseph Physioc also would be retained. Coming in at the outset, this time Oscar would have real input. Harbach and he talked out the story line and worked individually and collectively on the songs. This time he earned his credit as collaborator on lyrics and book—and, of course, his share of the royalties. Unfortunately, although it was a far better show than *Tickle Me,* the new show, *Jimmie,* which opened barely twelve weeks later, was a flop, barely eking out 71 performances. But while it lasted, twenty-five-year old Oscar had the heady distinction of having two shows running simultaneously on Broadway.

Oscar and Otto developed a unique way of working together, one that seems amazingly without ego for creative people. Harbach, the teacher (in showbiz circles he was called "the professor"), would generally suggest the idea or concept of how the song would fit into the play. Then he would usually give Oscar the title—as if it were an assignment. Oscar would work on the lyric, returning it to his mentor to polish. Since each man was capable of turning out a song on his own, they were easily able to combine in one song the best each could offer. Once the lyric was polished, neither man chose to remember who wrote which line, so deep was their collaboration and trust.

Hammerstein was careful to exclude from the publication of his collected *Lyrics* most songs that were not entirely of his own creation, even the many where it was reported he had contributed the majority of the lines. Only two, "Who?" and "D'ye Love Me?" (both from *Sunny*), are listed as a collaborative effort with Otto Harbach. (Although Harbach never published a book of solo lyrics, every song that came out under the collaboration always listed both names.)

One gets an idea of the way the duo, actually all the librettists of the era, manipulated their story line in "Below the Macy-Gimbel Line," a song that was introduced early in the first act. The scene is Carlotti's restaurant in Greenwich Village, and the main story thrust is how Jimmie's true parentage is being kept from her. (Eventually she discovers she is an heiress and a Jewess, the daughter of millionaire Jacob Blum, not the Carlottis who own the seedy restaurant.)

The song, a play on words, has nothing beyond its setting to do with the plot; rather, as a take off on the Mason-Dixon line and Greenwich Village bohemianism, it is reprised as a delightful soft-shoe by the gamine, Jimmie. Its technique, especially a rhyme such as "train" and "again," would never have gotten by a mature Hammerstein. Today one might consider the lyric ethnically offensive, but much of what passed for humor in the '20s jibed Jews, Blacks, Italians, and especially Asians. Each of these ethnic groups came in for its share in the libretto and lyrics of *Jimmie*. The lyric goes:

VERSE
Honey, I'm a-going to the South
Where the waffles melt in your mouth.
Down in Greenwich Village,
That's the place I want to be.
I'm gonna hop upon the subway train,
You'll never see me in Harlem again.
For I'm a-goin home,
No more to roam
From the folks I long to see.

REFRAIN
Down below the Macy-Gimbel line
Fine old Southern people have their home
No one ever uses soap or comb
In the land of cotton stockings,
Bobbed-haired ladies shine,
Never see no mammies, sakes alive!
But there's an Uncle Joe to lend you five.
Though we have no levees here,
Our Levy's trust us fine.
The worst families of Old Virginia,
Live down below the Macy-Gimbel line.

The unquestionable hits, favorites with all the critics, were a sentimental lullaby called "Baby Dreams" and the title song, "Jimmie" (She's a chip of old New York / Vagrant and whammy / A dear little bluffin' / Ragamuffin), but the reviewers overlooked the far superior "I Wisht I Was a Queen," which is not unlike "Just You Wait," the diatribe Alan Lerner would write a quarter century later for an irate Eliza in *My Fair Lady*.

Famous queens of history
I read a book that tells about 'em all
One of them I'd like to be
But I fear that I was born too small.
Gee, I wisht I was a queen,
With a crown upon my bean
I'd like to hold
A golden scepter in my hand.

Nurses and teachers,
I'd kick every one of them off my land.
In my court I'd be so gay,
I could play in the mud all day.
No one to cross me,
No one to boss me,
No one to make me keep clean
That's why I wisht I was a queen.

If the number was overlooked by the critics it was certainly due to misplacement. Obviously the collaborators had not observed a cardinal rule of show business, for they gave their diminutive star, Frances White, two other comic numbers preceding the above. The first, "Do, Re, Mi," which Oscar would resurrect to better advantage in *The Sound of Music*, incorporates the quasi-joke line, "so fa, so good," but the second, "Some People Make Me Sick," an incipient cabaret song, contains by far the most humorous and least forced lyrics in the show.

I just got a terrible licking
And I didn't do nothing at all.
My whole blamed family picks on me
I suppose it's because I'm so small.
Nothing I do seems right to them
And nothing I say is so.
And who told them that they were so great?
That's what I'd like to know.

Some people make me sick
For instance, my big sister
She puts on such airs
When her boyfriend calls
And insists that I call him Mister.
He calls on her ev'ry night
And he combs his hair so slick.
But whenever he leaves her,
His hair is all mussed,
Ah, some people make me sick.

My cousin Maud is terribly jealous
And she's just as mean as can be.
She got so mad the other night
Cause her beau was talking to me.
We was having a quiet little chat,
He was falling in love, I suppose,
When in comes Maud and she says to me,
"Why don't you blow your nose."
(I was so humiliated)

Some people make me sick.
Now can you imagine that jealous dame
Why she jumped at the chance
When that guy proposed
Tho he haint got a cent to his name.
He says to her "Will you marry muh?"
We'll do some light-house keeping
Who wants to live in a lighthouse, huh?
Ah, some people make me sick.

Although *Jimmie* was not a triumph, Oscar was still riding on his success with *Tickle Me,* and during the years between 1920 and 1924 when he and Otto Harbach would write one of the classic operettas, *Rose-Marie,* he had irrepressible industry and an urge to succeed in all forms of theater. Almost everything he wrote failed dismally, perhaps because he tried so hard for success.

In 1921 he and Frank Mandel turned out a music-less comedy, *Pop,* which never even opened in New York, and the next year, collaborating with British librettist Guy Bolton on a Cinderella story, they produced a resounding flop called *Daffy-Dill.* A year later, again with Mandel, Oscar wrote a musical for Nora Bayes that featured the popular Norma Terris and Harry Richman. But even that stellar threesome couldn't keep *Queen O' Hearts* afloat for more than a month.

Then in the spring of 1924 Oscar wrote two straight plays, one produced by Uncle Arthur, the second by Sam Harris. Both of these moralistic serio-comedies were collaborations with Milton Gropper, a former Columbia University chum. Gropper, a playwright who had written only one mildly successful play, *The Charwoman,* was to be in and out of Oscar's life as friend, idea man, and later litigious adversary.

Their first fiasco, *Gypsy Jim,* dealt with a disguised eccentric millionaire who restores a family's faith in themselves; their second, *New Toys,* bore the subtitle: "A comic tragedy of married life after the baby arrives." Doyen critic Alexander Woollcott aimed a pointed barb that skewered both plays with a single line: "The process of disintegration proceeds so far that the final scene of *New Toys* reaches a level below any explored even by *Gypsy Jim.*"

Gropper and Hammerstein never collaborated again but did remain friendly, for Oscar—who wore his heart on his sleeve—often discussed plans and took advice from his many acquaintances, Gropper among them. Realizing that Gropper was barely able to eke out a living, once Oscar became successful, he employed his former classmate to research stories or synopsize plays and gave him a weekly salary. They had a disastrous falling-out in 1951 after the opening of *The King and I,* when Gropper sued Oscar for a quarter-million dollars, claiming he had supplied the idea for the Jerome Robbins ballet "The Small House of Uncle Thomas." Although the suit was patently without proof and Oscar, seeing red and feeling betrayed, wished to fight it, the R & H lawyer pointed out that it would be cheaper and more expedient to settle out of court. Gropper was finally paid $50,000, but, in a final ironic twist, when he died two years later he willed the $50,000 back to Oscar.

The Gropper and Mandel failures may make the reader feel that Oscar produced nothing but disasters in the early 1920s, but all was not as bad as it might appear in the preceding paragraphs. In between these calamities Oscar and Harbach produced *Wildflower*, a hit whose magnitude was disproportionate to its merit. It was based on Arthur Hammerstein's idea of the consequences of uncontrollable anger. Then in 1923 they wrote the first truly woman's lib musical, *Mary Jane McKane*. The latter was a product of a first (and last) collaboration with lyric-librettist William Cary Duncan. Both had songs by the young Vincent Youmans.

Oscar's collaboration with Vincent Youmans was to be the first of many with "world-class" composers. Even though *Wildflower* was only Youmans's second show, he had already scored two years earlier with *Two Little Girls in Blue,* which had lyrics by Ira Gershwin (then still writing as Arthur Francis).

Youmans was born on September 27, 1898—a mere day after George Gershwin—into a wealthy family of hat-store owners, educated at private schools, and given piano lessons from boyhood. His physical appearance, indeed his career, although his output was smaller, was to be not unlike Gershwin's. Both were acclaimed while they were still very young, both were possessed of an angular, slight frame, and both died young. Each man left a legacy of songs with jagged rhythmic patterns usually built on a stunning melodic motif—repeated and reharmonized until it had rendered its last ounce of emotion. A comparative mental replaying of George's "The Man I Love" or "I Got Rhythm" and Vincent's "More Than You Know" and "Tea for Two" should prove the point.

Musical historian Stanley Green summed up the facts of Vincent Youmans's life when he wrote: "His Broadway output consisted of twelve scores and his Hollywood contribution comprised two original film scores. He became a professional composer at twenty-two, an internationally acclaimed success at twenty-seven and an incurable invalid at thirty-five."

And Otto Harbach, who developed an abiding friendship with Youmans during the writing of *Wildflower*, was to comment after the composer's death at forty-seven: "He killed himself by not being satisfied with just being a great composer. He wanted to be the writer of the book, own the theater, buy the costumes, supervise everything. He got mixed up in so many expensive propositions that were out of his area. He took a lease on the Circle Theater and had to pay rent whether it was playing or not. It broke him and he simply went to pieces."

Youmans developed tuberculosis in the early '30s and went to Colorado to live, returning intermittently to Broadway. Harbach described him as "slight—but with the deepest bass voice I've ever heard. He whistled the songs as he played."

Vincent Youmans and Oscar alone might have brought a freshness to his worn operetta concept of *Wildflower*—a tempestuous ingenue who must constrain her anger in order to inherit a large legacy—a story not unlike that of the unsuccessful *Jimmie*, but Arthur insisted that Harbach and Stothart be there to bring their respective formulaic professionalism to the production. Although libretto and lyric credits were shared equally by Hammerstein and Harbach, Stothart was given short shrift when most of the better lyrics were taken to be set to music by Vincent Youmans.

Theatergoers left humming "Bambalina," a sort of "musical chairs" song-dance led by an old fiddler and the show's only hit. It certainly was the prototype for a later hit, "The Carioca." Its refrain follows:

When we're dancing at the fair	Times when he may choose to stop
We have to watch and keep aware,	Give me a good excuse to prop
When good old Bambalina calls a stop—	My little head against my partner's chest.
That means I must stand still	So you see the reason why
In your arms and hold your hand still,	Though other dances I may try,
For we dare not dance or skip or kick or hop.	I always like the Bambalina best!

Most of Vincent Youmans's attractive melodies were set to old-fashioned stagy lines like "When I see you swaying while the breezes all caress you / Then my heart is praying that I too may hold and press you." *Wildflower* had few lyric gems. Only "I Love You, I Love You, I Love You," rising above its often-repeated title, displays some of Hammerstein's simple naturalness. (That's the general meaning of it all / Do write soon, miss you so / Long to hug and kiss you so / Dearest can you read my silly scrawl? . . .)

Despite being dreadfully derrière-garde (or perhaps because of it), *Wildflower* received splendid reviews from the Broadway fraternity and played 477 performances. It was the first Hammerstein musical to open in Britain, although there it managed only 114 performances. St. John Ervine, writing in the *London Observer* in February 1926, gave it perhaps its most honest and cogent critique:

> It may be that English authors and composers are tired or simply no good, but I cannot help thinking that even the least capable of them could have concocted a play as good as this. It has some lively music but none of it is especially notable. . . . The story of the play cannot have caused the authors any brain-fog, and what there is of it was badly told.

Between the New York and London debuts of *Wildflower*, the same group, minus Otto Harbach, adapted a story by William Cary Duncan that truly tried to break away from conventions. The plot, more like the smart contemporary ones that were being written by Rodgers and Hart, centered around the pretty secretary of the title, *Mary Jane McKane*, who didn't want to be loved by the young handsome boss for her looks alone. She tries to disguise her beauty with thick glasses and crimped hair. Of course this leads to typical musical comedy mistaken identity.

The show begins quite promisingly with a musical number which incorporates the rumble of a subway train and the intermittent comments of its work bound passengers. The hero, Joe, sets the pace at the outset with "Speed":

> Never slow
> And you'll never go wrong.
> Always keep moving, moving along.
> Makes no difference what you do
> As long as you
> Have speed.

All seems well through the obligatory love duet, "Down Where the Mortgages Grow," with lines like: "I'd like a lamb with some nice tender chops in him / I'd keep a toad and make beer from the hops in him / Oh, what joy and what delight / Being a suburbanite." But before we get to a threnody on the tribulations of working in an office, called "Time Clock Slaves," which opens the second act, *Mary Jane McKane* gets bogged down in convention. Songs seem to be inserted arbitrarily and with little concern for the character who is to sing them. Soon the mood of freshness—not to say believability—dissolves with song lead-ins as bad as:

> Jane: My uncle made everyone work hard. I started a pair of mittens for him once. I never would have gotten them done if he hadn't kept after me.

SONG: "STICK TO YOUR KNITTING"

Just you stick to your knitting, dear
And the hours will run.
Just you stay where you're sitting, dear,
And pretend it's fun.

Little fingers may ache
But by and by comes chocolate cake.
And then a great big slice
Tastes twice as nice
When you know your knitting's done.

And the old operetta conventions and second-rate rhymes allayed at the beginning of the show rise again towards the end with the title song.

. . . JANE, how came you so fair to see?
We proclaim you the kitten's knee.

You're just as sweet as sugar cane, Jane,
And you're the main Jane for me.

One of the best tunes in the show, which posterity has come to know as "Sometimes I'm Happy," was given a title dreadfully inappropriate to its zippy melody and called "Come On and Pet Me." We will never know whether Oscar or his co-lyricist William Cary Duncan was responsible for the travesty. We do know that when the song was excised from the show before the New York opening, Youmans snatched it back; with a new lyric by Irving Caesar and Clifford Grey it became the hit of *Hit The Deck!* in 1927.[1] Below, for the record, is the original version:

Come on and pet me
Why don't you pet me?
Why don't you get me
To let you pet me?
You never ask me out for a spoon,
For all I know, there ain't any moon.

I'd like to bask in
Your fond caressin'
You do the askin'
I'll do the yessin'
Within your arms I'd stay for a year
Come on and pet me, dear.

In spite of its obvious capitulation, *Mary Jane McKane* proved to be more contemporary and adventuresome than any musical Oscar had yet attempted. The critics, seeing this as a departure, wrote, "Capsizes many of the current notions . . . few musical comedies have broken so many rules," which permitted it to run for more than three months (a considerable achievement in 1923) and to tour.

Shortly after *New Toys*, the second of Oscar's plays with Milton Gropper, closed in April 1924, Uncle Arthur had an idea for a stage effect so breath-

[1] Youmans, who characteristically used and reused original material, also recycled "My Boy and I," a soupy waltz. With the rhythm changed from 3/4 to a sprightly 2/4 and with a bright lyric by Otto Harbach, it became the title song of *No, No, Nanette*.

taking that he wanted to build an entire musical around it. He had been told about an ice palace built during the annual carnival in Quebec which was melted down as a climax to the week's celebration. He reported to Oscar and Otto, whom he planned to send on a research trip to Montreal, that his musical would end as the natives on snowshoes climbed the hill overlooking the city of Montreal and with fiery flambeaux melted the palace into a glittering river. Neither Otto nor Oscar had the temerity to ask Arthur how he planned to do this on stage nightly and twice on matinee days.

Staying at the Chateau Frontenac and investigating Montreal from there, Hammerstein and Harbach soon found out the ice palace story was bogus. But they did come up with a story that involved a prospector, a true-blue heroine, an Indian, a half-breed, a low comic, a titled Englishwoman, and a whole troupe of Mounties. All this was set against the lush background of the Canadian Rockies and featured a crime that had never been seen in operetta before: a murder.

Harbach reported it took a long time to get the story on its feet, but once it was worked out, Rudolph Friml, who had left Arthur Hammerstein's aegis to write an unsuccessful operetta called *Cinders*, begged to be allowed to return to the fold and write the score with his old friend Harbach. "Only if you work with Herbert Stothart," was Arthur's reply. Friml was forced to capitulate, but, in truth, the best songs in *Rose-Marie* had melodies by Friml.

Otto Harbach recalled how he, Oscar, and Friml worked on the score:

He had what he called his little sketch-book—a book as thick as The Bible. In it he had just the beginnings of melodies. He knew how to develop these. . . . He'd sit at the piano and play his little phrases, and every once in a while there would come one that would attract our attention, and I would say, "let's see what that might lead to." And then he would go all over it, and we'd have a melody. Sometimes a composer likes to set words to music, like Victor Herbert, but Friml couldn't do that because being a foreigner, he would always get the accents wrong. So it was safer for him to write the music first and then I would put words to it that would sound as though the words had been written first.

Nobody could improvise like Friml. Friml filled the music so full of furbelows and frills that it was difficult to tell what the tune was. I would play his tune on the violin. Friml said, "This is the acid test, if it sounds good with Otto playing the fiddle it's bound to be good!"

Harbach was very persuasive in not allowing Arthur with his penchant for the sentimental to ruin the show. Friml had written music for a bridal procession, a waltz which began "when the doors of my dream were opening wide," and Arthur and he had both said that this was going to be the theme of the whole show. "I had planned another song from the lead sheet [a melodic line]

Sketches of the principals of Rose-Marie. Clockwise from "heart-throb" Arthur Deagon's splendid profile: Dennis King, Dorothy Mackaye, Mary Ellis, Frank Greene, Lela Bliss. MOCNY.

44

Friml had given me. There was quite a row, because Arthur thought this was a surefire hit. I said I wanted 'Rose Marie I love you.' "

Of course he was right. The bright title song is typical Harbach:

Oh, Rose-Marie, I love you!
I'm always dreaming of you.
No matter what I do I can't forget you
Sometimes I wish that I had never met you!

And yet if I should lose you,
'Twould mean my very life to me.
Of all the queens that ever lived I'd choose
   you
To rule me my Rose-Marie.

Rose-Marie contains the first truly generic song written as a plot device in the famous "Indian Love Call," when Marie uses the yodel-like theme to warn her lover, as Harbach put it, "to get the hell out of here because they're after you, under the pretense of just singing a song." Recalling its genesis, Otto explained to Friml what the melodic construction should be like. "I wanted a song in which there would be a phrase, like a call in the mountains. In trying to tell him about that call, I *gave* him that call. We were in his car and he said 'what do you mean by a mountain call?' and I said 'like you hear in the Swiss mountains—I want something where the voice is coming down.'"

There is no doubt that the "Indian Love Call" became one of the operetta-dom's biggest hits. When Brooks Atkinson in his book *Broadway* described its effect as "hokum, but it drenched the town in banal musical mystery," he must have been thinking of its quasi-poetic words. The entire song, by now an accepted classic, shows no sign of diminishing popularity. It is Friml and Harbach with a strong assist from Hammerstein in top form, with all the conventions of the times. I have printed the lyrics of the verse below.

Ooh! Ooh!
So echoes of sweet lovenotes gently fall
Thru the forest stillness,
As fond waiting Indian lovers call!

When the lone lagoon
Stirs in the Spring
Welcoming home some swany white wing,

When the maiden moon
Riding the sky,
Gathers her star-eyed dream children nigh.
That is the time of the moon and the year
When love-dreams to Indian maidens appear
And this is the song that they hear:

Since *Rose-Marie* was pure operetta, it demanded a subplot of droll lovers, and so besides these well-known romantic numbers illustrated above, the score contained much light music for the comics. "Hard-Boiled Herman," "The Minuet of the Minute," and "One Man Woman" all seem to have the Hammerstein touch. And next to "Indian Love Call," the hit of the evening was "Totem Tom Tom."

First-nighters were faced with a rather pretentious program note explaining

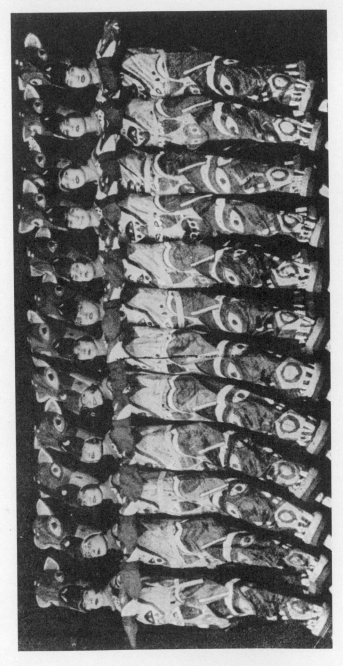

Only a producer like Arthur Hammerstein, who abhorred nudity, could conceive of costuming his chorus line in unwieldy totem-pole dresses. MOCNY.

why the songs were not listed in usual order. It ran: "The musical numbers of this play are such an integral part of the action that we do not think we should list them as separate episodes.[2] The songs which stand out independent of their dramatic associations are 'Rose-Marie,' 'Indian Love Call,' 'Totem Tom Tom' and 'Why Shouldn't I' in the first act and 'Door of Their Dreams' in the second act."

Before the preview performances the show ran into problems when Arthur could not come to an agreement with stagehands' unions. Once he passed that crisis, because of the delays and the tremendous size of the production he ran out of funds and was forced to invest $85,000 of his own money. But by the time the curtain came down on the first Broadway performance, September 2, 1924, he knew he had a hit of massive proportions. Harbach says, "I made Arthur a millionaire." Actually, *Rose-Marie* made all of them millionaires. At the latest count the show has grossed close to thirty million dollars.

After 557 New York performances, *Rose-Marie* toured for three solid years, afterwards swinging back to Broadway for another long engagement. Arthur mounted five road companies that toured all over America. The show ran for over two years at the Drury Lane Theatre in London and had extensive engagements in Paris, Berlin, and Sidney. It has been filmed by MGM in three versions starring, successively, Joan Crawford,[3] Jeanette MacDonald, and Ann Blyth.

After the opening of *Rose-Marie*, Oscar wrote that *Wildflower* had been a timid attempt to bring back the operetta. Dancers and choruses were obvious compromises to pander to a public who demanded those elements. By that time he was already searching for a term to replace the unstylish "operetta" and had found the term "musical play," a name often employed for operettas ashamed of their genre.

And Oscar, carrying the hubris of the newly successful, believed that in *Rose-Marie* he had mastered his craft. He felt so strongly that libretto writing is a misunderstood craft that he went so far as to send a note to drama critics in all the principal cities where the show would appear, asking them *not* to review the work.

His caustic remark stating that "the critics' problem is that they do not recognize what a good libretto is, and do not realize that a good musical comedy must not necessarily be a good play," may have angered many of the reviewers,

[2] Not until Stephen Sondheim's *Passion* opened in May 1994 was the public to be deprived again of a listing of musical numbers in the program.

[3] Crawford and James Murray starred in a silent version, but an accompanying score for piano, organ, or even full orchestra was available to all movie theaters. In the larger cinema palaces an off-stage singer positioned in the wings sang the "Indian Love Call" during that sequence.

but his message was essentially true. Unfortunately Oscar did not yet possess enough technique to carry his message beyond paper and into his scripts, and he was far from altruistic enough to starve for his ideals. Thus, his diatribe was mere bluster.

He was still dazzled by Harbach, Stothart, Friml, and the older generation with whom he felt it had been his privilege to collaborate. But he was learning fast, and although he still aimed for success at any cost, *Rose-Marie* had made him rich and famous and allowed him to move on to work with the composer he most admired, Jerome Kern. Perhaps Kern's fame coupled with Oscar's new-found success would allow a producer to take a chance on his writing the perfect libretto—the kind he had in his head?

Hammerstein

1924–1927

WITH *ROSE-MARIE*'S ACCLAIM IN HIS POCKET, the autumn of 1924 was a heady time for Oscar. No sooner had the New Year dawned than he decided to take his family to Britain while he supervised the London production. It promised to be a holiday in work disguise, but actually it turned out to be a grueling month of intensive rehearsal.

From Oscar's difficulty in replacing highborn British accents with Canadian half-breed speech to the apocryphal story of a chorine auditioning with the "Indian Love Call" who sang, "When I'm calling you, double o, double o," nearly everything had to be restaged and redone. But the effort was to pay off, for *Rose-Marie* became the darling of the British public and was to hold the longevity record at the Drury Lane until the advent of *Oklahoma!* more than twenty years later.

Once the show was running smoothly, Oscar sent Myra and the children back to the States and, planning to get on with his much delayed vacation, spent six soul-searching weeks alone in Paris. Never idle, he plotted a new operetta wherein the City of Light would be the meeting place for a love story concerning a deposed White Russian prince and a tempestuous revolutionary called "The Flame."[1]

As soon as he was back in New York he asked Otto to come in with him on the project. Harbach answered that he liked the idea but was committed to a show for producer Charles Dillingham, hopefully a vehicle for superstar Marilyn Miller. If Oscar would collaborate with him on this one, they could do the Russian-Parisian opus immediately after. Otto saved the clincher in his invita-

---

[1] Chosen either for insurance or good luck, for *Rose-Marie*'s heroine's full name was Rose-Marie La Flamme.

tion until the end of the conversation, when he announced that Jerome Kern was to do the music.

Kern had long been the idol of everyone in musical theater and was recognized as the father of the modern musical comedy. His Princess Theatre shows, written between 1915 and 1920 mostly with P. G. Wodehouse and Guy Bolton, had pioneered the intimate, integrated musical. But he could write big shows as well. *Sally*, his 1920 hit starring Marilyn Miller and produced by Ziegfeld, was one of the longest-running shows in a decade.

Now he was contracted to write three shows for Dillingham, and since Miller had had a falling-out with the mercurial Flo Ziegfeld, Kern's idea of building a show around the diminutive Miss Miller seemed as if it would result in a sure-fire hit of *Sally* proportions. Although Kern's least favorite show, *Sunny*, which Harbach, Hammerstein, and Kern wrote to Marilyn Miller's specifications, turned out to be a *Sally* clone. Harbach recalled that "although they looked somewhat like Mutt and Jeff, from their first meeting Hammerstein and Kern had a terrific affinity. Those two just grabbed each other like electricity and it was a wonderful team." Somewhat chagrined and relegated to the background in this triumvirate, the retiring Harbach reported, "My relationship with Jerry was a business one because I never felt comfortable with him." Although Harbach and Kern together were to produce splendid artistic and commercial successes—*The Cat and the Fiddle* and *Roberta*—Harbach adds, "I never felt that I was his equal because he was a bit ritzy compared to me. When I worked with him I always felt that he wished it was Oscar he was working with."

There is no doubt that Oscar Hammerstein's admiration for Jerome Kern was boundless. Here, in contrast to the composers he had collaborated with—Stothart, Youmans, and certainly Friml—was a musician who not only believed in the integrated book and story but was erudite and literate enough to be a true collaborator involved in every aspect of a musical's staging. Harbach, as the senior member of the trio and always the mentor, certainly must have been irked by Hammerstein's dazzlement with Kern. The original champion of songs and book being welded into a whole, Otto was accustomed to being in charge of the entire project. Besides, unlike his younger colleagues he was a realist and knew that although Kern and Hammerstein talked in high-flown rhetoric of fashioning the integrated book, it would be an impossibility in *Sunny*, a show that had been precast even before its first rehearsal.

All the featured players would be given time to do their specialties. Cliff Edwards, also known as Ukelele Ike, one of the stars, had an ironclad contract with Dillingham that called for him to do his own numbers (written by others) at precisely 10:15, and this had to be worked into the plot. Vaudeville favorite

Pert Kelton also interpolated her own songs.[2] Even Marilyn Miller, notoriously simple-minded, whose involvement in *Sunny* was contingent on her approval of the story, would do her specialty dances in the second act.

Oscar painstakingly told Miss Miller the story he and Otto had created from an idea of Kern's about a British circus bareback rider (her role) who is in love with handsome Tom Warren. She avoids the advances of her ex-husband, the circus's owner, by stowing away on a ship bound for New York. En route she meets and marries Jim Deming (Jack Donahue), a rich passenger, in order to be allowed to land. Once in New York, she divorces Jim to marry Tom, but as soon as the divorce is final she discovers she really loves Jim. As the final curtain falls, Sunny and Jim decide to remarry.

Marilyn Miller is reported to have listened attentively to this complicated saga and is said *not* to have asked the obvious question, as to why she had to get married so many times in this story. But her only query before she gave the trio the go-ahead, was reportedly, "When do I get to do my tap specialty?"

Perhaps because the libretto of *Sunny* was full of the same old contrivances, it received hearty approval from its star. The score contained few gems, and only one that has since become a standard—the most interesting number and the hit of the show, "Who?"

The first word, a monosyllable, must be held for more than two measures. We get an insight into Hammerstein's lyric sensitivity by noticing that the melodic pitch is the sixth of the scale, itself a questioning tone and one that wants to pull downward to an answer. Of the inevitability of the title there is no doubt, since a long-held note must be sung on a vowel.[3] The only non-nasal open sounds that could possible be held that long are "ah" (for which there are few interesting rhymes), "ooh" (as in true),[4] and "oh" (which Oscar would utilize in the ultimate long-holder some twenty years later as the opening of "O———klahoma!"). "Who" also allows for some terrific rhymes like "true," "blue," "to," and the punch line answer to a lover's question of "who?" "No one but you."

[2] This tradition, born around the turn of the century, continued until the late '20s in the United States. Al Jolson was perhaps its most lavish practitioner, always stopping his shows and singing requested repertoire sometimes for a full hour. The tradition survives still in the British pantomime and the French music hall.

[3] Hammerstein was convinced that the lack of success of his moving song from *Carousel*, "What's the Use of Wond'rin'," was due to his use of the final word "talk" to end the lyric rather than an open vowel sound. "Talk," he said, "cuts off the melodic line too abruptly."

[4] Of course the "ooh" sound had been used to good advantage in the "Indian Love Call" conceived by Harbach and Hammerstein for *Rose-Marie* only the year before.

Marilyn Miller as Sunny *the circus queen, shown here with Jack Donahue.* MOCNY.

Because he had created this lyric without collaboration, Oscar's talent was recognized by his colleagues in the songwriting profession. But it must be noted that with the exception of "Who?," discussed above, "D'ye Love Me?" (printed below) whose verse is truly sophomoric but whose chorus has a modicum of charm, and the title song, *Sunny* contains no hidden lyric gems.

VERSE

When a man begins to angle
And a heart he tries to entangle,
He'll sing love's lullaby.
Like a sandman while he's napping
He'll begin his tender love tapping,
Proving but a gay romancer,
Unless this question he can answer.

REFRAIN

D'ye love me? (Um - hu!)
D'ye mean it? (Um - hu!)
D'ye promise to love me always?
Forever? (Um - hu!)
And ever? (um - hu!)
And not for a week and a day.

For there are men who love
Now and then,
But they vanish when
They hear wedding bells playing.

So you must agree
It's going to be
Forever or *never* with me.

Certainly it was Marilyn Miller's charm that made *Sunny* the most successful show of the season and allowed it to chalk up 517 performances on Broadway. Burns Mantle, critic for the *New York Daily News*, wrote: "Dillingham has put everything into it but the Hippodrome elephants. . . . It is a little like *Sally* in size and also in quality . . . past Kern's score and past the beauty of the decorations, the personnel is the big note . . . an entertaining show. In addition it has Marilyn in person. You might ask for more, but would you deserve it?"

Otto had been right. This was a star vehicle. Not a single reviewer complimented the libretto or commended the lyrics. And as for the songs being integrated into the libretto, none was.

True to his word, Otto suggested they begin work immediately on the project that appealed to Oscar during the Paris holidays and which he called *Song of the Flame*. It would be brought in under the banner of Arthur Hammerstein, who had hired Rudolph Friml in hopes of producing a Russian *Rose-Marie*. But Friml, dissatisfied with Arthur's contractual terms, bowed out. Since Oscar was now so successful, he and Harbach could name their own composer. They went right to the top—George Gershwin.

Gershwin was every bit as ambitious as Hammerstein. They both had a tendency to bite off more projects than they could chew comfortably. In the busy month of December 1925 alone, Gershwin premiered his Piano Concerto in F on the third and opened his big new musical, *Tip-Toes*, on the twenty-eighth, two days before the premiere of *Song of the Flame*.

Certainly *Song of the Flame*, their only collaboration, sounds like it was shuffled to the bottom of the Gershwin pack and bears no mark of the composer who only a year before had produced the "Rhapsody in Blue" and had already written such monumental songs as "The Man I Love" and "Fascinatin' Rhythm." How could it? George was overwhelmed, first because this was blueprinted as an Arthur Hammerstein production, meaning full-blown operetta rather than the sassy musical comedies Gershwin had been involved in under the aegis of Aarons and Freedley. That meant old-fashioned hokum instead of jazz. More than that, he was saddled with Stothart, Arthur's perpetual ally, as co-composer. Instead of brother Ira, the catalyst for many of his most successful songs,[5] the Harbach-Hammerstein team, who he felt were totally "square," were contracted to write book and lyrics.

The story they created including double identity would be better realized in Hammerstein's very next project, *The Desert Song*. In *Song of the Flame*, the first act, which occurs before the revolution, concerns a Russian noblewoman, (known as "The Flame" because of the red dress she wears) who incites the peasants to revolt. She falls in love with a nobleman, but because of their differing ideologies they are forced to part. In the second act, they meet again in Paris and are able to accept each other for what they are.

The book and lyrics were hopelessly old-fashioned even for 1925, with most critics referring to Hammerstein and Harbach's work as staunch and serviceable. But what the show lacked in inventiveness was made up for by the sheer size of the production. Arthur hired a cast of two hundred, which included the Russian Art Choir and the American Ballet. All this hullabaloo plus the Gershwin name (although most of the music was created by Stothart) allowed *Song of the Flame* to make it to 219 performances in New York.

It must be admitted that the show's title song, whose words are credited to Harbach and Hammerstein and whose music was written by Stothart and Gershwin (as Cole Porter commented: *Four* men to write *one* song?), did attain a modicum of success, perhaps because its opening harmony moved freshly from E minor to C7 to B7. But its lyric ("What's that light that is beckoning? / Thru the night it is beckoning / Come, come, come come / Take your new day of reckoning") shows the two OH's at their worst.

Because of its elephantine proportions, *Song of the Flame* would seem an unlikely show to tour, but tour it did. Then Arthur sold it to Warner Brothers and it was filmed in 1930.

In the summer of 1926 Oscar and Otto signed for two shows: *The Wild Rose*, which would have music by Rudolph Friml, and *The Desert Song*, which would be produced by Schwab and Mandel and for which Sigmund Romberg would write the score. The former was to be another Arthur Hammerstein operetta, this time based on a play that had starred Douglas Fairbanks. Otto Harbach recalled:

> It was a swashbuckling role, jumping off balconies and swinging from chandeliers. So I started it and I wrote it as a musical play. I had to frame the whole thing again using just the basic idea. The situation was that there was a little English girl that Arthur got crazy about. . . . The night before we opened she came on stage with her hair chopped off just like a boy. The bob . . . she had lost her sex

---

[5] Ira often sparked George's inspiration. In 1924 he handed his brother a jotting: "Do do do what you done done done before." Ira recalls that half an hour later they had the completed song.

appeal and the play got nowhere. I said to Oscar, "you bring in *Desert Song* and I'll bring in this show."

Unfortunately, *The Wild Rose* was a flop. It is out of the domain of this book, for by Harbach's own admission, although both names appear on the work, its book and lyrics are almost exclusively his. But *The Desert Song*, was a true collaboration and a gigantic hit to boot. It has never been out of the repertory of light-opera troupes all over the world.

Otto and Oscar were inspired to write the operetta because of headline newspaper accounts of the exploits of Berber chieftain Abd-el-Krim, who battled both the French and Spanish in the early 1920s as leader of the Riffs in their revolt in Morocco.

Their hero, always garbed in red and known as "the Red Shadow," for some inexplicable reason was changed from Arab to Frenchman fighting his own countrymen. He bears a close affinity to *Song of the Flame*'s heroine, not only because of their similar red costumes but because the role is a double one. The dreaded Red Shadow is really Pierre Barbeau, *pretending* to be the dim-witted son of the commanding general. Only the audience is in on the deception, and when the masked Shadow abducts the heroine, she falls in love with him. The climax comes in the second act when the Shadow's hiding place is discovered and General Barbeau challenges him to a duel. Rather than combat his own father, the Shadow lays down his sword. Realizing the ruse can no longer be continued, Pierre announces he has slain the Shadow—much to the dismay of Margot. But at the final curtain when Pierre dons his Shadow mask, Margot rationalizes his deception and together they look forward to helping achieve a Riff-French reconciliation.

The success of *The Desert Song*, which, by the way, has some of the psychological overtones of *The Phantom of the Opera* and to which Clark Kent and "Superman" owe their very existence, was certainly due to Romberg's lush and frequently stirring score. But as in all enduring works there is always something in the *air du temps*. At that time American women were fascinated with the idea of abduction by a sheik. The cinema's Rudolph Valentino's recent premature death had no way diminished that fantasy but had made him even more enigmatically heroic. Romberg was certainly on that wavelength when he wrote the exotic melody he gave to Hammerstein while the show, called *My Fair Lady*, was still working out its kinks in Boston. And Hammerstein, sans collaborator, had never set a melody more sensitively than with the lyric he produced on that long train ride to New York. It clinched the show's title. That ballad insisted the musical be called *The Desert Song*.

An examination of the first dozen bars of both *music and* lyrics below eluci-

dates the lyricist's craft. For example, starting with Bar 1 we find Hammerstein's signature "oo" sound; in Bar 2 the lyric swoops up on "heaven" as it
should; Bar 3 has another "oo"; in bars 5, 6, 7, and 8, the words "sand,"
"kissing" and "sky" are most appropriate, creating an image of dunes meeting
the horizon. At the end of that bar we are wafted up by "a desert breeze,"
which whispers a lullaby and there is an almost seductive intimacy in the line
"Only stars above you to see I love you."

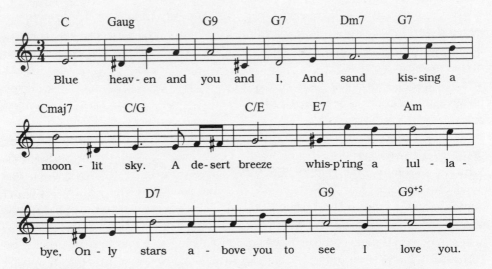

Coupled with one of Romberg's most exotic musical lines, which relies heavily on ninths, thirteenths, and augmented chords to create tension as well as
sweeping appogiaturas, "The Desert Song" is one of operetta's most successful
achievements. The balance of the lyric, except perhaps for the rhyme of "calling" with the forced word "enthralling," is hardly less exotic.

> Oh, give me that night divine,
> And let my arms in yours entwine,
> The desert song calling,
> Its voice enthralling
> Will make you mine.

Additional cinematic timeliness was in evidence in the up-to-date lyrics Otto
and Oscar supplied when neatly balancing their exotic romance with a typical
operetta subplot. This introduced a bumbling American reporter and his girl.
Clara Bow, the movies' "It" girl, was not overlooked either.

VERSE

There was a time when sex appeal   Mister Freud
Had quite the most complex appeal.   Then employed

Words we never heard of.	In one word
But the seed of sin	She defines
Now at last has been	The indefinable thing,
Found by Elinor Glynn.	She calls it "It."

The Desert Song was ecstatically received by press and public. It proved to be a bonanza not only for its producer but for Oscar who, for once, had invested heavily in his own show. Most critics felt it surpassed anything heretofore in Romberg's considerable oeuvre, which included *Maytime, Blossom Time,* and *The Student Prince,* and even outclassed Friml's reigning hit, *The Vagabond King.* It gave 471 performances in New York and 432 in London and has thrice been filmed by Warner Brothers. In the 1943 version most of the score was replaced while the Riffs foes became Nazis. But still its indestructibility prevailed. Although productions these days are liberally sprinkled with "camp," *The Desert Song* has been a staple of light-opera troupes the world over for more than seven decades.

Perhaps the very success of the project sounded the death knell of the Harbach-Hammerstein collaboration. Although they did three more shows together—*Golden Dawn* in 1927, *Good Boy* the following year, and a show that never opened in New York more than a decade later—their golden alliance had tarnished. Both men were to remain friendly and hold each other in high esteem, but ever afterward Otto felt Oscar had left him left stranded on the road, holding the withering bag of *The Wild Rose.*

Looking back, he complained that he had opened the door for Hammerstein, who had gone through alone to work with Friml, Youmans, and Kern, to all of whom Otto had introduced him. But it was more than that. While Otto had been on the road furiously and hopelessly rewriting *The Wild Rose,* on November 12, 1926, *Criss Cross,* a light comedy Otto had co-authored with Kern and Anne Caldwell, opened. It was to be a fateful night, for before the second act was through, Kern wangled an introduction and proposed to award-winning[6] novelist Edna Ferber that he adapt her raging best-seller, *Show Boat,* into a musical. Ferber believed it would be impossible to squeeze her multigenerational, sprawling novel into a single musical evening, but Kern had a vision as to where cuts could be made and within a month had optioned the property from Ferber. Even before that he telephoned Oscar, asking him to read the book. Hammerstein was as excited about the project as Kern, and, realizing they had something of magnitude and magnificence, they began working more devotedly than either had ever done before. Within a few weeks Florenz Ziegfeld had been enlisted as producer. Otto soon found out about this important—and, at first, secret—collaboration, and for almost a year while he worked with

[6] Miss Ferber had won the Pulitzer Prize in 1925 for *So Big.*

Hammerstein and even with Kern on other projects, his exclusion from *Show Boat* rankled.

"The next thing they did was one of their biggest hits and I was not invited to join with them on the project, *Show Boat*," Harbach was to complain. "Dreyfus[7] thought it was a lousy trick after I'd brought Oscar in. After that I came second when it came between Oscar and Jerry."

But the fault, if there was any, was not Hammerstein's but Kern's, who realized the time was right for the triumvirate to become a duo. His vision enabled him to elicit from either man his most inspired work. Just as Kern was later able to mastermind his and Harbach's hit modern operetta, *The Cat and the Fiddle*, or *Roberta*, which produced "The Touch of Your Hand," "Yesterdays," and the immortal "Smoke Gets in Your Eyes," he understood that *Show Boat* needed a younger, less operetta-bound libretto-lyricist like Hammerstein.

Still, Otto's pique was understandable, for only two weeks after Ferber had met Kern, Kern announced that he had contracted with the novelist that "the new team of Kern and Hammerstein" would write a musicalization of *Show Boat*. On top of that, *The Desert Song* opened and through an oversight there was no reserved seat for Otto Harbach. "Oscar gave a party afterward at his place in the country," he stated angrily. "I wasn't even invited."

But neither man had time to dwell on trivialities, even if Oscar seemed to be having all the hits. Hammerstein shouldered his responsibilities and left in early 1927 to supervise the British production of *The Desert Song*. The enforced separation obliged Otto to stay in the United States and tinker with a frothy libretto titled *Kitty's Kisses*, a failure he had been contracted to supervise. Once that closed he involved himself with the libretto for one of Kern's worst fiascos, *Lucky*.

More bad news was in store for Otto with the opening of their next show, *Golden Dawn*, on November 27, 1927, while Oscar was shuttling back and forth and readying *Show Boat* for its New York opening a month later.

Golden Dawn was a curious attempt at an African quasi-opera (pretentiously listed in the program as "a music drama"), which premiered at the brand new Hammerstein Theatre on Broadway. Arthur had built the handsome edifice and dedicated it to the memory of Oscar I. Before the curtain was allowed to rise on opening night, the black-tie audience was forced to listen to forty minutes of eulogies to the "old man." By that time, Otto was surfeited and totally outnumbered by Hammersteins—including, besides his collaborator Oscar, Arthur who produced it and brother Reggie who directed. Otto proclaimed that "when

---

[7] Max Dreyfus, who established Harms as a major music publishing house, published the works of Kern, Romberg, Friml, Youmans, and Gershwin—later Kurt Weill, Richard Rodgers, and Cole Porter. For half a century his vision directed the course of popular music in the United States.

the show started out the funereal orations put everybody down in the depths. It put a pall over the whole evening."

Winchell agreed, dubbing the show *The Golden Yawn,* while most other critics termed it pretentious, dreadful, and interminable. The book, a combination of *King Kong* and *The Emperor Jones* with a bit of Rider Haggard thrown in, was full of violence but contained little excitement. The music by Viennese operetta composers Emmerich Kalman and Robert Stolz was Mittel-European *schlag* and totally wrong, while the decision to cast white members of the cast made up to portray native Africans was ludicrous. Trapped in a subsidiary role in this fiasco, fortunately unrecognizable in his blackface, was Archie Leach, who would move on to Hollywood as Cary Grant.

The melodramatic, predictable, and eminently racist story concerns Dawn, an English girl who has been captured by an African tribe and is brought up by her native nurse, Moodah, to believe she is destined to be its princess.

At the end of the play, when we are told the tribe is desperately waiting for a rainfall, Moodah warns the villain, Shep Keyes, saying: "Don't you put your black hands on her. You know she's white." An astonished Dawn says, "White? I'm white? Oh Moodah, what do you mean?"

MOODAH: "Oh, my baby, for what I've done to you—I ought to die."
SHEP KEYES: "There's a crowd of niggers coming down this path."
MOODAH: "The nuns! They will help me."

At once Dawn's black father, mysteriously a Catholic, bashes the head of villain Shep Keyes with a cross and the long-awaited rain falls. The nuns say the miracle occurred because the blacks have gotten religion and because they suddenly realized the white man's God is more powerful than their belief in idols. At the final curtain, before Dawn falls into her lover Steve Allen's arms, the sisters announce they will benevolently send her to have education in the land of her fathers.

Once again the program pretentiously announced: "The musical numbers are an integral part of the story as it evolves and therefore are not listed as individual songs. The titles of the principal themes are 'When I Crack My Whip' and 'Dawn,' " etc. This was followed by a long list of credits.

Prominent in the score, which incomprehensively is written in dialect reminiscent of the American South, an impossibility in Africa, were "My Bwana" and "Jungle Shadows" (Jungle shadows falling / Like a leopard calling / Black jaws open wide / You better run and hide / White man, when they've found you / Black arms will surround you / When they hold you so / They never let you go . . .). The only song the critics even remarked on was "When I Crack My Whip," a stagy, sadistic effort.

The finale of the jungle potboiler Golden Dawn, 1927. *Princess Dawn (Louise Hunter) is rescued by hero Paul Gregory, while Marguerite Sylva as her nurse Mooda smiles beatifically.* MOCNY.

VERSE

Listen little whip
While I got my grip,
While I got an arm
With strength left to sling ya.
You're a friend to me,

You can make me be
Way above de heads
Of frails while I sling ya.
I don't need no brains,
I can hold de reins,
Long as I have you to give dem pains.

REFRAIN:

Dey heah me crack my whip
An dey crawl.
Cause when I crack my whip
Dat's my call.
Fo in dis ole hell-hole down in Africa
I am de big boss of dem all!

You muh bow down to me,
White or black.
You got to bend yo' knee.
Break yo back.
An when I wants a thing
Den I gets dat thing.
Fo in dis ole hell-hole
I'm de king!

Hammerstein and Harbach had reached their nadir in this misguided opera, yet the times were such and Arthur a clever enough publicist to ballyhoo *Golden Dawn* into 184 performances. He even sold it to the Warner Brothers who turned this tale of Dawn, the white princess of a primitive tribe, into a film. It should have starred Pearl White; unfortunately, stage luminary Vivienne Segal was trapped in the disaster.

Hammerstein

1927–1928

THE YEAR 1927 CHANGED OSCAR HAMMERSTEIN'S LIFE. By its close Arthur's glorious new Hammerstein Theatre where *Golden Dawn* was running (not a hit but a contender) would bring added glory to the family name. *The New Moon* would have its first performances in Philadelphia, and *Show Boat* would open the next night at the auditorium Ziegfeld had built and named after himself. Oscar was not to come up with as big a hit as the latter again until he struck gold in *Oklahoma!* sixteen years later, for *Show Boat* was to alter the face of the American musical forever. But there was much more.

In March, aboard the *Olympic* traveling alone to England to supervise the London production of *The Desert Song,* he met and fell in love with Dorothy Jacobson, née Blanchard. The Jacobsons, traveling to Europe on holiday, were friends of Oscar's lawyer, Howard Reinheimer, who had come to the dock to wish his client bon voyage. Reinheimer presented his friends to the solitary traveler whose wife, looking after their children, could not join him abroad until the school year ended in June.

Dorothy was the antithesis of Myra. Tall, auburn-haired, blue-eyed, with a strikingly patrician beauty, she was born in Tasmania and had grown up in Melbourne before moving to London to seek a modeling and acting career. She was only seventeen when she met William McKinley Nicikle and married him. When both the marriage and her career fizzled, she divorced Nicikle and came to New York, hoping her height and beauty would serve as entré into the "profession" of showgirl, but met with little success. Eventually, though, she did land a part. Dressed in gray chiffon she undulated, portraying smoke, while Jack Buchanan sang "Smoke Rings" in *Charlot's Revue of 1924.* But she soon realized the stage was not for her and married wealthy diamond merchant Henry Jacobson because, she said, "he was so nice."

On deck, Dorothy and Oscar soon began talking about theater, and shortly after she mentioned how disillusioning her own stage experiences had been, he brought out a lyric that corroborated her words. It was "Life On the Wicked Stage," a light piece for his and Jerome Kern's upcoming *Show Boat*. They laughed about this and actually found so many things in common to laugh and talk about that by the time the liner docked in Southampton they knew they had fallen deeply in love.

They continued to meet throughout the spring, and when Myra finally arrived in London and met Dorothy for the first time in the lobby of the Carlton Hotel, it has been reported she asked Dorothy if she had a lover. When Dorothy replied in the negative, Myra is reputed to have confessed to having left a "divine lover" back home. The conversation is recounted in *Getting to Know Him*, Hugh Fordin's posthumous biography of Hammerstein, commissioned and underwritten by the second Mrs. Hammerstein in 1975.[1]

By the time the Hammersteins returned home in early summer so that Oscar could polish the forthcoming production of *Show Boat*, the two couples had outwardly become good friends. Henry Jacobson was aware of his wife's feelings and suggested they rent a summer house nearby the libretto-lyricist so "the infatuation might run its course." Myra must certainly have known about Oscar's deep feelings for Dorothy because their children remarked that besides having separate sleeping arrangements, their parents only rarely spoke to each other now.

Part of Oscar's pique did not concern Myra at all. It was annoyance at Ziegfeld for what seemed like insensitive interference in his, Oscar's, *Show Boat*. The "Glorifier of the American Girl" felt that *Show Boat* was nowhere near ready. Most of the libretto and score for the first act had been written by the time Oscar had gone abroad in March, and Ziegfeld sent a bitter wire to Kern complaining that Hammerstein's work was far too serious.

"In its present shape it hasn't got a chance except with the critics and I'm not producing for critics and empty houses," he wrote, adding that since Hammerstein never worked alone he ought to accept the advice of Dorothy Donnelly (actually a hack writer of operettas, but because of her numerous hits, one Ziegfeld admired). He ended a long diatribe by finding "Hammerstein's present layout too serious, not enough comedy."

It must be remembered that Ziegfeld, who had very little taste, had had no

---

[1] *Getting to Know Him* understandably seems to lean heavily in favor of Dorothy Blanchard Hammerstein's involvement in every aspect of her husband's life and career. It even features her picture along with her husband's on the back cover. The late Myra Finn Hammerstein's alleged dalliances while she was still married to Oscar were vehemently denied in interviews which this writer conducted with her children, William and Alice.

experience with a serious musical. Famous for his parade of either scantily clad or elaborately overdressed showgirls and low comics with interpolated routines, he had never before produced a tightly written musical play where the characters would develop meaningfully out of the plot. The contract he had signed with Kern and Hammerstein[2] obliged them to hand him a suitable script before April first. It was now July and they had done so. But still Ziegfeld dawdled, and the writers feared their producer had either gone broke and could not produce *Show Boat* at all or would commercialize it beyond all recognition.

The delay was fortunate for the work. Between April and August, Hammerstein was able to cut Ferber's book to its essence and add a sense of romance and lightness to the play without violating any of *Show Boat*'s considerable artistic principles. He eliminated whole segments, slicing off at least a decade of Ferber's story and telescoping lengthy scenes that brought the show up to date at the end. He combined characters, and most of all he began to realize that the story's appeal lay largely in its early scenes and period setting.

This pertains even today, for the essence of *Show Boat* is told in its first act. Its last half, with which directors and producers have tinkered on various stages and in several films, has never been solved satisfactorily. Its music is all reprises except for one new song.[3] Even Hammerstein, at the end of his life, admitted that were he to write the work over he would not adhere to the stock operetta ending that unites Magnolia and Ravenal at the final curtain.[4]

Ziegfeld, still somewhat dissatisfied, put the piece in rehearsal at last in October. By the time he read the critiques when it opened in Washington a month later, he realized he had a hit—even if *he* thought it too serious. He hastily moved his reigning success, *Rio-Rita,* out of the Ziegfeld Theatre and began readying the stage for *Show Boat*'s enormous sets and gigantic cast of blacks and whites. Opening December 27, 1927, it held that stage until May 4, 1929.

Show Boat was the most daring musical of the '20s, and history has shown it to be the most successful and the longest-lived. It might well have opened the door to the fully integrated musical long before 1943's *Oklahoma!* had Kern not turned his back and retreated into operettaland and musical comedy. For although *Show Boat* frequently harked back to the unreal and grandiose world of operetta, it was

---

[2] Ziegfeld agreed to pay Kern an advance of $1500 and $4\frac{1}{2}$% of the gross, out of which he was to pay Edna Ferber, with whom he had subcontracted, $1\frac{1}{2}$%. Hammerstein was to receive $1,000 and $2\frac{1}{2}$% when the gross exceeded $30,000 or 2% when that figure was not reached.

[3] So it is in Hammerstein's *Oklahoma!*; the title song is the only major addition.

[4] Ferber's novel has Ravenal disappear after a run of bad luck in Chicago. The story even has Cap'n Andy drown in the Mississippi at about the same time. Obviously the second half of the novel, with only women (Magnolia, Julie, Ellie, Kim, and matriarch Parthenia Ann) surviving, would have to be greatly altered to make a satisfactory libretto for a musical play.

a particularly American period piece. And it broke with convention—at the end of the first act, instead of having a misunderstanding which will be patched up in the second, the young couple are married. But they're married under the shadow of the bride's mother's dire prognostications, which turn out to be true in the second act. The gambler that Magnolia marries deserts her.

Certainly Kern was never to write a finer score, and Hammerstein set his music brilliantly. The show spawned at least half a dozen standards from "Can't Help Lovin' Dat Man," "Make Believe," "You Are Love," "Life On the Wicked Stage," and "Why Do I Love You" to "Ol' Man River," which has become part of the American folklore; the book is arguably Hammerstein's best because hardly a single musical number was inserted that didn't develop the story. Their respect for the seriousness of the theme, which included miscegenation, alcoholism, and desertion, led the two collaborators to abandon their formulaic approach to the musical play.

Show Boat is the first work in this book that merits an in-depth analysis of its score. As mentioned before, the original running order has been varied in each of its many incarnations. New songs were composed and others eliminated during the lifetimes of both Kern and Hammerstein. Things got worse after their deaths, with book changes, lyric changes, even ethnic changes, and, because the show runs long for the current commercial two-and-one-quarter-hour genre of musical, many drastic excisions.

The list below follows the score as sung on opening night at the Ziegfeld Theatre, December 27, 1927. The song titles, although published as listed below, were "cleaned up" for an almost exclusively white audience. During *Show Boat's* run in the '20s, they were programmed as "Can't Help Loving That Man," "Old Man River," and so forth.

ACT I

1. Overture. One senses the seriousness and tragedy implicit in this musical play from the outset. An ominous A-minor chord begins and ends the almost six-minute Overture, which is heavily laden (as is the entire score) with the interval of the fourth as well as the subdominant or IV chord. Kern's bass as well frequently progresses by fourths. The use of this interval and progression, a favorite of Stephen Foster's songs, minstrel shows, banjo plunkers, classic blues, and early soul music, conjures up late nineteenth century America. Unifying the score, the omnipresent fourth allows *Show Boat* to invoke and ride on these musical reminiscences.

Besides its seriousness, the overture presents fragments of most of the main themes that will appear. The only one played in full is "Why Do I Love You?," a late addition to the score inserted as insurance for a hit song.

2. Cotton Blossom. Hammerstein chose a most controversial and denigrating single word, "niggers," as the first word heard at rise of the curtain, but the balance of the stanza below illustrates how deeply he empathized with the blacks on the levee.

> Niggers all work on de Mississippi,
> Niggers all work while de white folks play—
> Loadin' up boats wid de bales of cotton,
> Gittin' no rest till de Judgement Day

has come in for much ethnic criticism and change. In the 1936 movie (supervised by Hammerstein), the line was changed to "Darkies all work. . ."; in the 1946 stage revival it became "Colored folks work. . . ." In *Till the Clouds Roll By*, the truncated film biography of Jerome Kern, the meaning of the line was changed entirely, for "Here we all work . . ." implied that the blacks and whites were all laboring together. The 1991 recording happily restores the original word, for to soften any ethnic slap is to cover it, and, as time goes by and memories fade, the buried outrage is forgotten and eventually condoned. Here is how William Hammerstein, who has several times produced and directed this work of his father's, feels about the line:

> Every time I've staged the show, I've had a talk with the cast; they all agree, but eventually they come to me separately and say, "I can't sing that." "Colored folks" was brought in in 1946 when I was the stage manager. "Here we all work" is Paul Robeson's awkward addition. My father hated it, and I hate it.

Thomas Shepard has similar feelings about the word:

> I think is a shame not to use it. It is the shock value of the word and it is being sung by black people. I can't tell somebody else what they should or shouldn't be offended about. I just wish they weren't. When I worked with the Houston Opera Company on a *Porgy and Bess* recording in 1977, they made the tough choice of not softening the slurs like "keep away from 'em, nigger," which was in the original. It was the way DuBose Heyward assumed the inhabitants of Catfish Row talked to each other.

Throughout the show language is an ethnic divider. All the blacks' speech is heavily dialected and uses simple one or two syllable words, while the whites, introduced as "Mincing Misses and Beaux," are given speeches that would be acceptable in a Noël Coward drawing room. Edna Ferber named the show boat *Cotton Blossom*, but Hammerstein saw the possibilities inherent in the plant name, for it sets up the cast and caste at the outset.

Cotton blossom, Cotton blossom,
Love to see you growin' wild.
On de levee you're too heavy
For dis po' black child!

is sung by the blacks on the levee (note the lower case "b" on the word blossom), while the whites sing

Cotton Blossom, Cotton Blossom,
Captain Andy's floating show!
Thrills and laughter, concert after
Get your girl and go!

From the playwright's point of view this opening scene is masterful. Eight major characters and their relationships are entwined with great economy. Captain Andy, the troupe's leader, and his shrewish wife, Parthy Ann, are the first to appear. But within the first few minutes we are introduced to Julie La Verne, the star of the show boat, and her husband Steve as well as the heavy, Frank. The plot soon involves Ellie—a dancer with ambitions to act on the stage of the show boat—as well as the black cook, Queenie,[5] and her husband, Jo.[6] We learn practically from the outset that Pete, the mean engineer, has a yen for Julie, suspect she has something to hide, and are embroiled in a fight between Steve and Pete which quick-thinking Cap'n Andy pretends is a preview of that night's melodrama.

3. Where's the Mate for Me? This character song first introduces Gaylord Ravenal, the mercurial riverboat gambler. Again the lead-in, although taken word-for-word from Ferber's novel, is telling. The sheriff who warns Ravenal he must leave town in twenty-four hours offers him a "seegar." "What did you call it?" Gay replies. "See-gar," the sheriff repeats. "Optimist" is Ravenal's retort. The exchange sets up the gambler's taste for better things and his snobbishness and fearlessness of the law—all of which will lead to his downfall in Act II.[7] His lyric:

Who cares if my boat goes upstream
Or if the gale bids me go
With the river's flow?

[5] Played by Tess Gardella, a girthy Italian actress in burnt-cork makeup. Miss Gardella had become famous as "Aunt Jemima," her shiny, disguised face selling millions of boxes of buckwheat flour.

[6] Spelled "Joe" in all subsequent productions.

[7] Unfortunately Hammerstein ignores Ravenal's turning the tables on the inelegant sheriff later in Ferber's novel. When the gambler is flush with winnings, he gloatingly sends the sheriff a handmade Havana.

In Show Boat's dramatic opening scene, Steve (Charles Ellis) almost strangles Pete (Bert Chapman) while Windy (Alan Campbell) tries to restrain him. MOCNY.

> I drift along with my fancy,
> Sometimes I thank my lucky stars my heart is free,
> And other times I wonder
> Where's the mate for me

makes the listener realize this is a man who leaves every aspect of his future to chance.

4. Make Believe. Hammerstein has said that the first time he heard Kern play "Make Believe," the words "Couldn't you? Couldn't I? Couldn't we . . ."

Could-n't you?   Could-n't I?   Could-n't we?

popped into his head as being utterly right, but he didn't know what it was the young lovers "couldn't do." Eventually he wrote up to and away from this turn-around point in the refrain. Whatever its genesis, the title and lyric are fresh and the song's situation in the musical is perfect. Throughout the play Gay will be protected and defended by Magnolia. This is written into their opening verses. Gay apologizes with "Your pardon I pray /'Twas too much to say / The words that betray my heart," while Magnolia immediately counters with "We only pretend, / You do not offend, / In playing a lover's part."

It is the first time we have met and heard the two principals sing. It is love at first sight, and Magnolia's placement on the upper deck of the boat with Gaylord on the dock are reminiscent of *Romeo and Juliet*, with all its tragic implications.

Some pedants have criticized Oscar's use of melisma[8] in the verse as lack of technique (see asterisked notes in the example below). Others have stated either Kern—who was notorious for not permitting a single note of his melodies to be altered—or Hammerstein should have yielded to the prevailing rule of one note to one syllable. To his writer these graceful melismatic passages are consistent with the make believe romance the lyric extols.

Our dreams are more ro - man-tic than the world we see.

5. Ol' Man River. Early in the *Show Boat* outline Oscar convinced Jerry that their sprawling show needed a commenting, wise, unifying force, a Greek chorus, if you will. The inexorable force of the Mississippi and the towns on its banks where the show boat cast performed were the true backbone of Edna Ferber's novel. "The river," Magnolia repeats again and again through gritted teeth when daughter Kim is born in the midst of a cataclysmic flood. She intones the words again with ironic tragedy after her father's drowning. It was that force, Oscar believed, that would consolidate the musical. And who better to give words to the rolling river than the lazy, powerful, seething, willful, sometimes tempestuous, and practically idiot-savant Jo? But Kern said he was too busy and that the forcefully unifying song sung by Hammerstein's commentator would have to wait until he completed the considerable music for the first act. Oscar pointed out that if Jerry simply slowed down the melodic line of "Cotton Blossom" he would have the refrain; for the verse he borrowed the

[8] See Glossary.

opening "Niggers all work on de Mississippi," unifying it by having his Everyman state: "Dere's an ol' man called de Mississippi." Kern acquiesced and went farther; turning the melody upside down while stretching the notes out created a stunning motif (see Examples A and B).

Example A

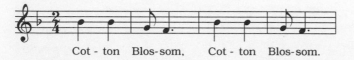

Cot - ton Blos-som, Cot - ton Blos-som.

Example B

Ol' man riv-er, dat ol' man riv-er

Oscar was rightly proud of his lyric's cohesiveness, especially since he had used rhyme so sparingly, being aware of the fact that heavy rhyming equals erudition. "Here is a song sung by a character who is a rugged untutored philosopher," he stated. "It is a song of resignation with a protest implied. Brilliant and frequent rhyming would diminish its importance." In *Lyrics* he gave his readers a practical lesson in lyric writing:

> If one has fundamental things to say in a song, the rhyming becomes a question of deft balancing . . . There should not be too many rhymes . . . Consider the first part of the refrain:

> | | |
> |---|---|
> | Ol' Man River, | He keeps on rollin' along. |
> | Dat Ol' Man River, | He don' plant 'taters, |
> | He mus know sumpin' | He don' plant cotton, |
> | But don' say nuthin' | An' dem dat plants 'em |
> | He just keeps rollin', | Is soon forgotten. |

"Cotton" and "forgotten" are the first two words that rhyme. Other words are repeated for the sake of musical continuity and design. The same idea could be set to this music with many more rhymes. "River," instead of being repeated in the second line, could have had a rhyme—"Shiver," "Quiver," etc. The next two lines could have rhymed with the first two, the "iver" sounds continuing, or they could have had two new words rhyming with each other. I do not believe that in this way I could have commanded the same attention and respect from a listener, nor would a singer be so likely to concentrate on the meaning of the words. There are, of course, compensations for the lack of rhyme. I've already mentioned repetition. There is also the trick of matching up words. "He mus' know sumpin' but don' say nuthin'." "Sumpin" and "nuthin" do not rhyme, but the two words are related. "He don' plant 'taters, He don' plant cotton." These two lines also match and complement each other to make up for the lack of a rhyme.

Although "Ol' Man River" carries no ethic slur as does the opening line of "Cotton Blossom," because of the similarities in the lyric the song was always a matter of contention,[9] and changes were made by individual artists. In the late 1940s Oscar joined the Authors Guild. One of the aims of the organization is the abolishment of stereotypes. For one of their annual meetings he penned a parody of this, his most famous song, which was sung by Guild members representing an Irishman, a Negro, an Italian, and a Jew. The lyrics were reprinted in the organization's newsletter along with Oscar's caution:

It must be remembered that these satirical lyrics were written to be sung before a special audience of *writers* in NYC. We are enclosing them for your enjoyment. The use of them before groups who might not recognize the satire should be very carefully considered. It is entirely possible that they might do more harm than good.

They are reprinted below for the first time.

Quartet: We are as old as the Mississippi
Stereotyped as inferior men.
We're content to be dumb and dippy,

Italian: Angelo,
Irishman: Mike,
Negro: Ebeneezer
Jew: And Ben.

Irishman: I'm a harp, I love to fight,
No one will admit I can read or write.
I get drunk and I throw bricks,
When it comes to intellect
The micks have nix.

REFRAIN
Ol' man author
That ol' man author

He may know somethin'
He may know nothin'
But he keeps writin'
He keeps on writin' along.

He don't plant 'taters
He don't plant cotton
He jes keeps writin'
And writin' rotten
His harmful foolin'
It keeps on droolin' along.

Italian: Rob that bank and tote that gat.
Gangsters are Italians, you all know dat.
I eat garlic and sphighett

Others: No one ever let him eat a lamb chop
yet.

[9] And still the battle rages. In May of 1993, rehearsals of a sumptuous eight million-dollar-revival in Toronto had to be halted because of a dispute that stirred up tensions between the city's 150,000 blacks and a like number of Jewish residents. Notwithstanding Oscar's diatribe against racial hate, a spokesman for the city's black newspaper lamented the Jewishness of the show's creators. The representatives of the black community picketing the theater stated that the musical "reinforces negative images and stereotypes used to undermine the images of black people."

Quartet: We are weary and sick of tryin'
We need a corner to go and die in.
But ol' man author,
He keeps on writing us wrong.
We are as stale as a vote for Hoover,
Put us away with the high wheel bike.
Put us away with the pest remover—
Angelo, Ben, Ebeneezer and Mike.

Jew: I'm a Jew and I like money,
Wealthy Christians think that's funny,
I'm a comic scheming scamp—
Comic as a Nazi concentration camp.

Quartet: We are the men of amusing races
Fated to be eternal jokes.
Dialect men with amusing faces
Never are we like the other folks.

Irishman: I like 'taters,
Negro: An' ah loves cotton,

Jew: And I like blintzes,
Italian: And I don't like nottin'.

Quartet: The worn out bromides
They just keep rolling along.

Negro: I shoot crap, and I steal fowl,
When you hear me laughin'
it'll make you howl
Yak yak yak,
Yak yak yak
Kick him in the fanny
And he laughs right back.

*(As they sing the last lines they throw off their
    hats, wigs, dice, and other paraphernalia as if
    they were freed of chains.)*

Quartet: We keep tryin'
We're in there flyin'
We're in there fightin'
We're in there dyin'
But ol' man author
He keeps on writin' us wrong.

6. Can't Help Lovin' Dat Man. This is another example of the unifying songs that will involve us emotionally while serving to move the plot forward. Julie had taught it to Magnolia, and Julie sings it first. Queenie, overhearing, interrupts with ". . . ah didn't ever hear anybody but colored folks sing dat song," which sets the audience up for the melodrama to follow.

Since it was to to be a "colored folks" song, Kern's manuscript requested the pretentious musical direction "Tempo di Blues." Still, there is nothing snobbish about the song, one of the great examples of the "torch song." The refrain is in pure 32-bar pop song form, with only a touch of blues harmony (see Examples A and B[10]) but the form and lyrics of the *verses* conform rather closely to classic 12-bar blues: three lines, a couplet (8 bars) with the second line repeated (4 bars). The parallel is seen somewhat better in the second verse (reprinted for clarity in blues format):

"De chimbley's smokin, De roof is leakin' in,	
(But he don' seem to care,)	(4 bars)
He kin be happy wid jes' a sip of gin—	(4 bars)
Ah even loves him when his kisses got gin.	(4 bars)

[10] A blues scale frequently incorporates the lowered third, fifth, sixth, and seventh of its parallel major. Example A uses a lowered seventh while Example B employs the lowered sixth and third.

Example A (Bar 1)

Oh, lis-ten   sis - ter

Example B—Refrain (Bar 6)

lov - in'   dat man—   of   mine—

Example C—Refrain (Bars 1 & 2)

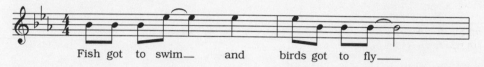

Fish got   to swim—   and   birds got   to   fly—

Hammerstein's honest and heartfelt lyric, coupled with Kern's moving refrain (again featuring the strongly ethnic interval of the fourth—see Example C), is an example of the genre that others were to follow with varying degrees of success for the next thirty years. The elemental technique seems not to waste a word. The splendid release incorporates a sense of movement with the lover going away so that he may return in the final section. And Kern's melody descending in the A section, ascending in the release is exemplary.

Fish got to swim and birds got to fly,
I got to love one man till I die—
Can't help lovin' dat man of mine.

Tell me he's lazy, tell me he's slow,
Tell me I'm crazy, (maybe I know)—
Can't help lovin' dat man of mine.

When he goes away
Dat's a rainy day
But when he comes back that day is fine,
De sun will shine!
He can come home as late as can be
Home widout him ain't no home to me—
Can't help lovin' dat man of mine.

7. Life On the Wicked Stage. Although the song's title doesn't jibe with its opening lyric line, this number adds a necessary bit of levity to a very serious first act. Hammerstein seems less concerned with consistency of character here, for he gives the untutored Ellie, whose final line will be the malaprop "Oh, Frank, don't be so mid-Victrola!," a sophisticated lyric like "Though you're warned against a roué / Ruining your reputation." Oscar seems to have found he has considerable rhyming skill and wants to crow about it by pointing inner rhymes out on paper. (Italics are mine.)

> Life upon the wicked *stage*
> Ain't never what a girl supposes;
> Stage door Johnnies aren't *rag-*
> Ing over you with gems and roses.
> When you let a feller hold your *hand (which*
> Means an extra beer or *sandwich* . . .

8. Till Good Luck Comes My Way. Both this and "Life On the Wicked Stage" are character songs and are performed in front of a drop necessary to ready the large set behind. The entire scene which informs the audience bluntly that Julie is part Negro is termed unnecessary and, says Miles Kreuger in his book *Show Boat: The Story of a Classic American Musical*, "a typical sequence from a 1920s musical" displaying "either a lack of confidence in this own writing or his audience's ability to read between the lines." It shows that "Hammerstein's roots were still firmly planted in musical comedy traditions of the time."

Additionally, "Till Good Luck Comes My Way" (I'll play along / While there's a game on the highway / I'll stray along) is perhaps the weakest melody in the show, and the title seems awkwardly set, emerging as "Till Good Luck Comes My Way."

But the next scene, the miscegenation scene in which Steve, knowing his mixed marriage will bring down the law upon the owners of the show boat, cuts Julie's finger and sucks out some of his wife's blood, is masterful. Confronted with the sheriff's accusation that Cap'n Andy has a white man married to a Negress aboard, Steve asks "You wouldn't call a man a white man that's got Negro blood in him, would you?"

VALLON: No. I wouldn't. Not in Mississippi. One drop of nigger blood makes you a nigger in these parts.
STEVE: Well, I got more than a drop of — nigger blood in me, and that's a fact.

After the crew's engineer corroborates Steve's assertion, the sheriff leaves. But Julie and Steve's days on the show boat are over, too. Soon, Magnolia and Gay will replace them as the leading actors.

The melodramatic moment was further tightened by a musical scene that was deleted after one performance in Washington, D.C. Kern believed strongly in the oppressively minor theme (again the interval of a fourth, this time inverted), used it in the overture, and insisted it be published as part of the piano-vocal score. "Mis'ry's Comin' Aroun'" is certainly the prototype for

Porgy and Bess's "Gone, Gone, Gone," wherein George Gershwin and DuBose Heyward even included a saucer burial. In this case Julie moans:

> When I dies, let me rest
> With a dish on my breast.
> Some give nickel, some give dime,
> All dem folks is fren's o' mine.

It is Julie's catharsis, as though she has been freed from her secret which she has borne for so long.

9. I Would Like To Play a Lover's Part

10. I Might Fall Back On You

11. C'mon Folks. The first of these three songs, all of which are inserted for lightness before the heavily romantic ending of Act I, has a Gilbert and Sullivan lilt. Its lyric tells the audience that Julie and Gaylord are now the successful stars of the show. The chorus boys sing: "Magnolia Hawkes—enchanted name / Your charm for me will never cease / I'll put your picture in a frame / And keep it on my mantelpiece," while the girls counter, singing of Ravenal: "The kissing scene would be sublime / We'd keep rehearsing all the time!"

The next song, which begins, "After I have looked around / The world for a mate / Then, perhaps, I might fall back on you," is necessary to the story because Ellie and Frank will be getting together in the coming act. Its matter-of-factness makes a stunning contrast with the lush romanticism of the upcoming "You Are Love."

"Queenie's Ballyhoo" is a specialty number intending to sell tickets to a black audience. She recounts in sensational terms the plot of a melodrama. This was followed by a scene from *The Parson's Bride*, the evening's show boat melodrama. In the play, after Frank as the villain pretends to throttle Magnolia, a member of the audience pulls out his six-shooter. When Frank retires in terror, Cap'n Andy rings down the curtain and in front of the drop (while the scene is being changed behind) he (Charles Winninger) gets a chance to do his well-known "specialty," wherein he wrestles himself to the floor. Winninger was the highest paid of all the troupe and considered the star of Ziegfeld's production.[11]

12. You Are Love. The scene is the upper deck of the *Cotton Blossom*. Gay, proposing marriage to Magnolia, says he has garnered the approval and enlisted

[11] Winninger's masterful mugging can be seen in the 1936 Universal film adaptation.

the help of her father. They plan to be married tomorrow while her mother is in Fayetteville.

Gay sings first and is joined in the second chorus by Magnolia. The lyric is pure romance. The verse ends:

. . . Then my fortune turned and I found you
Here you are with my arms around you.
You will never know what you've meant to
   me.

Magnolia:
You're the prize that heaven has sent to me.

Ravenal
You are love
Here in my arms
Where you belong,

And here you will stay,
I'll not let you away
I want day after day
With you.
You are spring,
Bud of romance unfurled
You taught me to see
One truth forever true.
You are love
Wonder of all the world
Where you go with me
Heaven will always be.

Of all *Show Boat*'s songs this was Kern's least favorite, although the public and this listener have never ceased to be moved by its melody. One has to admit that the words are indeed treacly. What, for instance is a "bud of romance unfurled" and a "truth forever true"? Singers, too, find it difficult to sing in its high tessitura, replete with many "ee" sounds rather than the more singable "oo" and "ah" syllables.

For the last scene of the act, we are back on the levee. In a dramatic conclusion, Parthy returns before the wedding having found out that in his youth Gaylord killed a man. She insists the wedding be called off, whereupon Cap'n Andy confesses that in his youth he, too, killed a man in a situation similar to Gaylord's. When Parthy faints, in stock musical comedy tradition, Andy bids the wedding continue. The curtain falls to a quick reprise of "Can't Help Lovin' Dat Man."

ACT II

13. At the Fair. One must not forget that this is a Ziegfeld production and that the great showman would arrange to parade his "long-stemmed beauties" somehow. The setting for the opening of Act II is the 1893 World's Fair in Chicago, three years following the end of Act I. Various sideshow acts, a hoochy-coochy dance, and the obligatory Ziegfeld girls are paraded before Ravenal and Magnolia enter. Gay is having a winning streak. "At the Fair" is an operetta-bound chorus number. It is followed immediately by

14. Why Do I Love You? which was a late addition to the score. Although it is generally considered rather charming, this writer finds its square musical

Norma Terris and Howard Marsh on the levee before the wedding scene that closes Act I.
MOCNY.

and lyrical phrases predictable and tiresome. But if the song itself is routine, the musical interlude between its refrains is telling. Ravenal sings, "Darling, I have only / Just an hour to play" to which Magnolia answers, "I am always lonely / When you go away." Then, to the tune of "Can't Help Lovin' Dat Man" he adds, "I'll come home as early as I can / Meanwhile be good and patient with your man." And we are back into "Why Do I Love You?" as he exits. Kern and Hammerstein must have believed heartily in the song, for they plugged it throughout the act.

15. In Dahomey. Again, Ziegfeld created diversion by having a group of supposedly savage singers and dancers come out of the African pavilion. Chanting "dyunga doe! Dyungg hungy ung gunga," they soon frighten off the prissy white chorus, whereupon, as a tepid joke, they reveal their origins are Avenue A in New York City.

This is followed by a short moving scene set in Kim's convent. With its singing nuns it seems like an early precursor of the opening of *The Sound of Music*. The scene originally included a new song, "My Girl," which was sung by Ravenal to his daughter. This was dropped early in previews in favor of a dialogue leading to a song wherein Gaylord tells his daughter that whenever she misses him to "Make Believe" he is nearby.

The next scene set in a dingy apartment is a nonmusical one and is the spot where the believability of the act first comes into question. An Irish landlady is explaining to Ellie and Frank, now married, about the tenants of the rooms she intends to rent to them. After a monologue describing the gambler and his wife who for years have been moving back and forth between these shabby lodgings and the Sherman House, Magnolia enters. Happy to see her old friends, she shows them a picture of daughter Kim, now eight years old and enrolled in a convent school. Soon a letter from Gay arrives, and Magnolia improbably asks Ellie to read it aloud. (How else would the audience know what it contained?) He has enclosed his last money gotten from pawn and asks Magnolia to return to her home on the show boat. Frank and Ellie offer to arrange an audition for Magnolia at the Trocadero, where they will be opening on New Year's Eve, before they exit quietly, leaving Magnolia in tears alone.

Ferber had solved Magnolia's embarking on a career in show business more logically and without the coincidence, but far less movingly than Hammerstein. Realizing at last that Gaylord will always gamble away their future, she decides that she must become the breadwinner and seek some sort of job because she loves—and wants to protect—him so desperately. Remembering her past successes on the show boat and at parties, she goes after an audition alone and practices her songs until she succeeds in landing a job.

16. Bill. Written for the 1918 production of *Oh, Lady! Lady!!*, this rather plaintive melody of Kern's with lyrics by P. G. Wodehouse had been in and out of several shows. Once it even suffered the indignity of being sung to a dollar bill. Its improbable use as a number for a haggard Julie seems incongruous—yet the song never fails to elicit pathos and has become the archetypical torch song. In any case, Helen Morgan, perched atop a piano, created a vogue that carried far beyond the stage of the Ziegfeld Theatre.

Although Oscar magnanimously insisted on a note inserted in the program disclaiming credit for the lyric, he did actually deepen Wodehouse's lyric by eliminating some of its sophistication. Kern, too, improved his original melodic line, removing a double climax. Now it was a far better song and conformed more closely to Julie's character in the play. The changes are noted below:

Helen Morgan, for whom Kern and Wodehouse (with an assist from Hammerstein) created the showstopping ballad "Bill" for the second act. MOCNY.

WODEHOUSE
FIRST REFRAIN
But along came Bill,
Who's quite the opposite
Of all the men
In story books.
In grace and looks
I know that Apollo
Would beat him all hollow . . .

SECOND REFRAIN
He's just my Bill
He has no gifts at all.
A motor car
He cannot steer;
And it seems clear
Whenever he dances
His partner takes chances . . .

HAMMERSTEIN
FIRST REFRAIN
But along came Bill
Who's not the type at all,
You'd meet him on the street
And never notice him.
His form and face,
His manly grace,
Are not the type that you
Would find in a statue . . .

SECOND REFRAIN
He's just my Bill
An ordinary boy,
He hasn't got a thing
That I can brag about.
And yet to be
Upon his knee,
So comfy and roomy
Seems natural to me . . .

17. Can't Help Lovin' Dat Man (reprise). Using the reprise as a generic part of the plot, Magnolia sings the song for her audition. Julie overhears and in a noble gesture decides to leave the Trocadero so Julie can have her job. When the club owner who likes Magnolia's voice says her song is old hat, Frank shows her how to "rag" it and she is hired.

18. Good Bye, My Lady Love

19. Magnolia's Solo ("After the Ball"). "Good Bye, My Lady Love," a well-known cakewalk by Joseph E. Howard, and "After the Ball," a popular teary ballad of the '90s by Charles K. Harris, were used to insert period color into a score that hardly needed it. This penchant of Kern's was seen in many of his shows. The technique, used in *Sweet Adeline* as late as 1937, nearly sank the musical.

In this instance, "After the Ball" is the unlikely number (after "ragging" her audition) Magnolia chooses for her debut. When the audience, already disappointed because Magnolia has been substituted for the pre-announced Julie, grows restive, her father, who happens to be in the audience, coaches her on until she finally wins the ardent approval of the crowd.[12]

The next scene, a short one played before a drop to prepare for the finale on the show boat, merely informs the audience that time has passed and it is now the present—1927. Magnolia has gone on to a splendid singing career, and now even Kim is considered Broadway's finest actress. A gray-haired Jo is discovered whittling. He sings

20. Ol' Man River (reprise) whose bridge and final section are given the following majestic philosophy:

. . . New things come	Ah keeps laughin'
'N old things go	Instead of cryin'
But all things look de same to Joe—	Ah must keep livin
Wars go on	Until I'm dyin'
And some folks die	But Ol' Man River,
De rest forget	He jes keeps rollin' along.
De reason why.	

21. Hey, Feller. Jo's solo is followed by Queenie's up-to-date, jazz-like, almost woman's lib type song. Again we hear, "And if you'll do the askin' / I'll do the yessin." a direct borrowing from Hammerstein's own unused lyric "Come

---

[12] Hammerstein's Magnolia, who becomes a music-hall headliner, is an interesting departure from Ferber's. She makes Magnolia a workaday variety-show artist. Her earnings and modest lifestyle in Chicago and later New York go to support her daughter's true gift for acting. It is Kim who is star material.

On and Pet Me," which attained megahit status as "Sometimes I'm Happy."

In the penultimate scene Cap'n Andy explains Gaylord's presence on stage with a single line, "It seems like fate my bumping into you at Fort Adams yesterday." Audiences seemed to believe so strongly in the characters that any coincidence was acceptable. Later in the scene, left alone, Ravenal reprises "You Are Love."

22. Why Do I Love You? (reprise)

23. Kim's Imitation of Her Mother.
On the road Norma Terris, now adding the dual role of the grown-up Kim to her Magnolia, sang "It's Getting Hotter Up North," a selection from the young star's current Broadway hit, but before the New York opening this was replaced by a reprise of "Why Do I Love You?" Imitations being quite popular in the late '20s, Terris sang the song as it would be performed by Ethel Barrymore, Beatrice Lillie, and Ted Lewis.

24. Finale.
Hammerstein ties all the loose ends in this final scene, which takes place on the upper deck of the refurbished show boat. Frank and Ellie reappear, Magnolia enters and accepts Gaylord's return with joy, and, to round things out, an old lady who attended their wedding makes the understated comment, "I'm glad to see it turned out so well and you're still happy together," while the chorus which began the play closes it with Ol' Man River's philosophy: "He jes' keeps rollin' along."

After the curtain came down on the first performance there was a five-second stunned silence followed by deafening applause. Critics called it "the most distinguished light opera of its generation," "startling and glorious," "a masterpiece."

The show was so acclaimed that Ziegfeld immediately announced a second company which was to feature Paul Robeson[13] as Joe and Libby Holman as Julie. This and the proposed tour were canceled when the great showman decided to keep the show running in New York for 572 performances.

But Oscar's euphoria in his career was mitigated by the failure of *The New Moon*, an operetta he had written with Romberg while waiting for Ziegfeld to stop dancing around *Show Boat*. It played a week in Philadelphia before being withdrawn for repairs.

His emotional life, too, was a shambles. Dorothy had decided to give her marriage another chance and refused to see Oscar. Now she was pregnant with

[13] Robeson was Kern and Hammerstein's original choice, but Ziegfeld had so delayed production that Robeson was committed elsewhere. Jules Bledsoe created the role.

Jacobson's second child. Luckily Oscar had a strong sustaining friendship with Kern and his wife, and so he looked forward to an ocean voyage with them and Myra, ostensibly to supervise the London premiere of *Show Boat*.

Britain's critics did not take as strongly to the tale of the Mississippi as New York had done, but the public could not get enough of Edith Day as Magnolia and Paul Robeson as Joe. Oscar came in for the grayest censure, with James Agate writing in the *London Sunday Times*, "Mr Hammerstein contrived to say nothing whatever, and I doubt whether a whole dynasty of Hammersteins could have said less." St. John Ervine, too, jumped on the bandwagon, saying, "Miss Ferber put the Mississippi into her novel, Oscar Hammerstein 2nd has put a squirt in its place " Yet in spite of the critics the public came eagerly, keeping *Show Boat* running happily at the Drury Lane for over ten months.

That spring in London, Oscar had another look at *The New Moon* when, with the help of Frank Mandel, he decided to begin a complete rewrite. This involved cutting nine of its eighteen songs and rethinking the book, which he based loosely upon the establishment of a French outpost in the Ile des Pins. For the operetta's key song, he penned a lyric containing short, gasping phrases that may have been indicative of his frustrated romantic mood and, quite contrary to their usual method of working, sent it to Romberg for musicalization. The song, "Lover, Come Back to Me," a masterpiece of the succinct—in both lyric and music—has withstood decades of wear from crooners to rock through rap and still shows no sign of fraying:

The sky was blue,
And high above,
The moon was new,
And so was love.
This eager heart of mine was singing:
"Lover, where can you be?"

You came at last,
Love had its day.
That day is past.
You've gone away.
This aching heart of mine is singing:
"Lover, come back to me!"

Rememb'ring every little thing you used to say and do,[14]
I'm so lonely.
Every road I walk along I've walked along with you—
No wonder I am lonely!
The sky is blue,
The night is cold.
The moon is new
But love is old.
And while I'm waiting here
This heart of mine is singing:
"Lover, come back to me!"

The *weldschmerz* in this lyric's progression from the anticipated love in the A^1 to the beloved's appearance in A^2 followed by the release's absence and bitterness in A^3 reminds me of a Goethe or Heine poem. And there is magic

[14] This line was "improved" to "When I remember ev'ry little thing you used to do" for the 1930 film version starring Lawrence Tibbett and Grace Moore. Singers persist in the change, claiming it is easier to sing.

in Hammerstein's joyous use of the "blue sky" and the "new moon" as applied to new love in the first instance which is contrasted with these same conditions, now dolorous, towards the end when "love is old." Of course, Romberg's harmonic intensification of the former phrase (Example A) to the latter one (Example B) is a stomach churner.

Example A

Example B

There are other musical gems to be found in this lush operetta, but hardly any lyric ones. Looking at the dialogue and songs it seems that *Show Boat* never happened and that under Romberg's influence Oscar was back to the artificiality of *The Desert Song.* If "One Kiss," with lines like "with passion all unfurled," and "Wanting You," which included "longing to hold you close to my eager breast" have emerged as operetta standards, it is due to Romberg's generous outpouring of memorable melody. Hammerstein's title "Softly, As in a Morning Sunrise" for decades has been the butt of critics who cite its redundancy—"as distinguished from an evening sunrise."

"Stouthearted Men" (which rhymes the most nonmasculine verb "adore" with "ten thousand more"), fared somewhat better when the lyric was rewritten in the early 1940s as a theme song for the U.S. Navy:

Give me some men who are stouthearted men
    who will fight for the right they adore.
Give me some men who will fight like the men
    who have fought in the navy before!
Oh! Give me some guns for the stouthearted
    sons of the ones who have won ev'ry war.

Then there's not a chance on earth
    for freedom's cause to die,
When stouthearted men
    are on the sea and in the sky!

The New Moon's story bears a strong resemblance to *Naughty Marietta,* Oscar I's enormous hit of almost two decades before. Yet in spite of its costumed story and *déjà vu* plot, it was able to chalk up 509 New York performances and two successful film versions.[15] The Broadway run was even to ride out the coming stock market crash of 1929.

It would seem that with *The Desert Song* on a lucrative tour, *Show Boat* running in New York and London, and now, in September, the successful *New Moon*, Oscar could want for little more, but it must be remembered that his emotional life was a shambles. Dorothy had had her baby and now realized her marriage had collapsed—the child, a girl they named Susan, brought them no closer. Three months earlier she and Oscar had begun seeing each other again. Both asked their spouses for a divorce. Henry Jacobson agreed reluctantly, insisting their young son must live with him. Dorothy, feeling she could not accede to this request to separate brother and sister, felt her situation was hopeless; doubly so when Oscar reported to her that Myra flatly refused to divorce him.

Too afraid of his uncontrollable anger and unable to face Myra during that time, Oscar began spending long hours at the theater or working at collaborators' homes. Besides *The New Moon*, 1928 saw the production of *Good Boy*, a script he wrote with Otto Harbach and Henry Meyers, and *Rainbow*, which had music by Vincent Youmans. Overworked, trapped in his antagonism, and frustrated in his unconsummated love for Dorothy, he grew more and more twitchy, irascible and nervous. Eventually he suffered a nervous breakdown.

After he spent two weeks, most of the time wrapped in icy sheets, in the Leroy Sanitarium, a discreet private hospital in Manhattan, somehow much of the emotional strain was relieved. Now, with more assurance, he was to demand a divorce from Myra, and this time—perhaps because she foresaw Oscar's total collapse if she refused—Myra agreed.

Good Boy, produced by Uncle Arthur, was a dreadful show although it boasted "Boop-oop-a-doop Girl" Helen Kane singing "I Wanna Be Loved by You" and ran for seven months. A sampling of its dialogue (Oscar was not involved with the songs) reminds one of Hammerstein's earliest fiascos:

I wish you didn't look so emaciated.
I was just emaciated into the Elks last Saturday.

Why don't you go see a doctor?
I did see a doctor. He said my chances were fairly good but he wouldn't advise starting any continued stories.

Well, I'll be going now. I've got to go see Samson and Delilah tonight.
Well, bring them along.

[15] The 1930 MGM picture incomprehensively changed the venue from eighteenth-century New Orleans to twentieth-century Russia.

The other show, *Rainbow,* had a glimmer of honesty and was based on Laurence Stallings's story of a young army scout who kills in self-defense,[16] is caught, escapes, and follows the wagon train to the Gold Rush in 1849 California, where he is reunited with his true love. He marries and supports his wife by gambling. The reports from its out-of-town previews predicted that it would be another *Show Boat* (which its plot strongly resembles), but its characters were too unsympathetic to create much audience empathy.

The show's opening night on Broadway was an unmitigated disaster, beginning with the prospector's mule, which defecated profusely on center stage just before the big love scene. Then the scenery for the first act would not move into place, and the audience fidgeted in their seats for half an hour.

The cast was shaken, and, as the second act, which started after eleven o'clock, was rambling on uncertainly, Oscar fled the theater and arranged to meet Dorothy for a drink. That night she told him she had decided she would soon be leaving for Reno to establish residency and sue for divorce.

The news, the promise of their life together, and the critics' almost unanimously kind reviews of *Rainbow* which overlooked the opening night calamity were enough to fill an elated Oscar with what turned out to be a false career hope. The show was loaded with roistering action and teeming life, yet saddled with Oscar's tepid lyrics and Vincent Youmans's score, which contained not a single hit tune, *Rainbow* expired after 29 performances.

Oscar was always pragmatic enough to maintain that if a show closed within a few weeks there must be something drastically wrong with it, but he, Stallings, and Youmans could never understand *Rainbow*'s failure. For years after the show's demise they referred to it their "gorgeous flop."

But now there was no hitch in Oscar's emotional life. Six months after *Rainbow* closed, on May 13, 1929, Oscar and Dorothy were married. They decided to honeymoon in London and Paris, crossing on the *Olympic,* the very liner on which they had first met. The happy voyage seemed to tie with a ribbon and store away in memory the past two successful careerwise but dreadfully emotional years.

[16] Murder in self-defense, Oscar's favorite device in 1927–28, appears in *The New Moon* as well. It happens twice in *Show Boat.*

Hammerstein

1929–1933

UNTIL HE BEGAN TO WRITE SONGS WITH RODGERS, Oscar was more attracted to Kern's music than that of any of his other collaborators. Both men were optimists, and although each could be melodramatic in his creative work, neither could be considered sad or tormented. Intellectually, Oscar also felt closer to Jerry, who could often match him in long-winded rhetoric. Moreover, in the theatrical world where the norm was to indulge in extramarital affairs, Oscar related strongly to the security of Kern's marriage, his love for his wife, Eva, and their daughter Elizabeth.

Kern and his family had been Oscar's rudder when he stayed with them in Palm Beach during the time he was waiting for Dorothy to finalize her divorce. Neither liked nightclubbing or drinking. Both were early to bed and early risers, and both were passionately devoted to their work. Certainly during that time the collaborators talked of the next project, a Gay Nineties romance inspired by Helen Morgan's stories of her adventures as a singing waitress in a beer hall called Adeline's. Miss Morgan was pulled out of *Show Boat* to star in the new work.

Because they both felt guilty that they had cut Arthur out of "the big one" by allowing Ziegfeld to produce *Show Boat*, they decided *Sweet Adeline* would be a family affair. Directed by Reggie, it would be brought in under Arthur's aegis at the Hammerstein Theatre.

Writing *Sweet Adeline* was the fulfillment of a long-term wish for both Kern and Hammerstein. For years Jerry had hoped to write music that reflected the turn of the century. He even eschewed including original music or the then common potpourri of the show's hits in the overture, deciding instead to get the audience "in the mood" by presenting a medley of well-known chestnuts from the period. For his part, Oscar, whose new wife came from a family with five daughters, was eager to get to work on this romantic scenario about one sister's sacrifice for another.

The story, a backstage drama about a bar-maid singer's love affair and her subsequent rise to become a Broadway star, has much in common with *Show Boat*. This time it involved three of Adeline's loves, first for a soldier she loses to her sister, then for a cardboard playboy from a socially prominent family who looks down upon Adeline's humble origins. At last she realizes she has always loved the composer who has written her songs and to whom she owes her start and stardom.

Besides the obvious clichés that pepper the plot, *Sweet Adeline* suffers from a too close paralleling of Helen Morgan's highly publicized and mercurial career. Three of its best songs are of the unquenchable torch variety—all written to capitalize on Morgan's "Bill," which had quickly become a standard. With the singer perched atop a piano brandishing a gauzy handkerchief, "Bill" was performed or mimicked at every '20s party. What Kern and Hammerstein proudly considered a realistic drunk scene, noting, "it was not one of the era's overdone comic bits but a serious attempt to portray an unfortunate weakness in a heroine," was too close a parallel. It was to come tragically true a decade later when Helen Morgan died of alcoholism at forty-one.

But the songs, "Here Am I" (Here is love / Please don't pass us by), "Why Was I Born," and "Don't Ever Leave Me" (which Oscar wrote for Dorothy), were immediate hits and have all become standards. The last shows an uncommon insensitivity—especially for Oscar, whose words imposed on Kern's music are generally most apt—for the emotional words don't seem to belong with the gentle and rather sweet melody. Best of *Adeline*'s songs is "Why Was I Born?," which asks (but never answers) some elemental questions. It is almost as good as Irving Berlin's "How Deep Is the Ocean?," which it would inspire three years later.

Why was I born?
Why am I living?
What do I get?
What am I giving?
Why do I want a thing I dare not hope for?
What can I hope for?
I wish I knew!

Why do I try
To draw you near me?
Why do I cry?
You never hear me.
I'm a poor fool, but what can I do?
Why was I born
To love you?

No reviewer could resist Helen Morgan and her way with these moving ballads. In critic Burns Mantle's opinion, the answer to Miss Morgan's lyric query was moot, but he supposed she was born to sing this kind of song.

With Helen Morgan's fragile intensity and Charles Butterworth's contrasting comedy, *Sweet Adeline* soon became New York's hardest-to-get ticket and Oscar Hammerstein America's most sought-after librettist. The show handily survived Black Thursday, October 24, 1929, the day the stock market collapsed, and went on to play into early 1930. But the deepening of the Depression, which hit all theaters except those featuring revues or mindless escape, forced it to close in March.

It would seem that with the great success of *Show Boat*, producers would fall into line trying to exhibit more serious musicals, but scanning the entire era only *Porgy and Bess* and perhaps *Pal Joey* seem believable and thus revivable today. Why then didn't producers produce more like them? Simply because the public refused to come. Theater has always been an elitist passion. Now the elite were broke. From fifty-one new musicals that opened in 1927 the number dropped precipitously to thirty-one by 1930.

In the early years of the decade the great theatrical empires, Charles Dillingham, Arthur Hammerstein, and even the Shuberts, were all to declare bankruptcy. Only the Shubert empire was ever to rise again.

The "talkies" had taken over.

But the moment the public knew that technology had reached the stage where sound could be recorded on film, they wanted the talkies to become "singies." One could not blame them when for a fraction of the cost of an evening at a Broadway theater, one could see an operetta on the screen, close-ups, singing followed by live vaudeville—or, in those dreary days of long hard evenings, even stay on and see the whole show again. The movies not only emptied the auditoriums of the struggling musicals, the artists and creators on the other side of the footlights were fast disappearing too.

Because the public was clamoring to see their product, all the Hollywood studios which had become rich practically overnight could offer irresistible salaries. New York had hardly enough actors, singers, dancers, choreographers, and especially lyricists and composers to provide grist for the early musicals, which could be assembled and exhibited within a few weeks. Many were lured to settle permanently in that land of perpetual sunshine; only the very successful— or dedicated—like Gershwin, Rodgers, Hart, and Kern were able to shuttle back and forth.

Most wrote down for the celluloid product, knowing their work would be exhibited to the masses in hamlets rather than their accustomed sophisticated metropolises. Worse, others failed to recognize that film was an entirely different medium and treated it as a play. Oscar was soon to join the group and his early experiments were to be guilty of both faults.

Aware that *Sweet Adeline* was a smash hit, Warner Brothers–First National Pictures offered him and Sigmund Romberg a contract to write four musical films. The collaborators were each to be paid $100,000 per picture plus a 25% share of the net profits. Knowing well how Hollywood often cut and bastardized serious work, Oscar insisted the Warner Brothers contract also give him final approval of the films.

Their first effort, *Viennese Nights*, was shot in little over three weeks. Its

Oscar, outside his office in the music department of Warner Brothers. He and Romberg were under contract to write music for four movies, but the first two releases were so unsuccessful that they were paid off not to write for the remaining films. (R & H).

speedy completion was due to the expertise of director Alan Crosland (who had directed *The Jazz Singer*) and a seasoned cast that included Vivienne Segal, late of *Golden Dawn*, Walter Pidgeon, and Jean Hersholt. The libretto—one cannot call it a scenario—a rewrite of *Blossom Time* and *Show Boat*, was trounced in *The New Yorker* for "going to pieces at the end," while the critic for *Judge* went farther, stating, "Why anybody pays money for such tasteless trash is beyond my comprehension."

Their next one, *Children of Dreams*, was largely written while Oscar and Rommy were several thousand miles apart. Oscar had taken Dorothy and baby Susan on a two-month visit to see Dorothy's parents and sisters in Australia. Hammerstein sent story suggestions by wire to Romberg, who composed the huge, almost operatic musical score as they were received.

It must be mentioned that Romberg was notorious for his unique method for turning out enormous scores in a matter of a few weeks. Over the years he had assembled a huge library of all the great operas, symphonies and concertos, operettas, *lieder*, etc., and he would highlight literally thousands of passages, perhaps four or six bars with a notation "good for baritone solo" or "soprano and tenor duet." Then he catalogued and filed these. In writing a show, he would first discuss with his librettist what kind of solo was called for. Then he would turn to his file for motif and inspiration. He was a good enough musician to disguise his "sources," and the origins of his melodies were rarely found out.

During the time Oscar was away in Australia, Romberg, who had been a café pianist, decided to brush up on his repertoire. When he was overheard through the thin walls of Warner's bungalow diligently practicing sections of the Tchaikovsky Piano Concerto, one of the writers nearby is said to have exclaimed, "There goes Rommy. He must be writing a Russian show."

The intensely romantic score Romberg turned out for *Children of Dreams*, unfortunately inspired Oscar to overinflate his simple story, which was set in the orchards of the West where the apple pickers go each year. Now it was only believable in operettaland: Molly and Tommy fall in love, then, to save her father from jail, Molly agrees with the help of a wealthy socialite to embark upon an opera career. There are long operatic scenes, and Romberg even creates an aria from a fictitious opera called *Antonia* that predates the aria Andrew Lloyd Webber wrote and inserted into *The Phantom of the Opera*. Finally, the son of the woman who helped Molly wants to marry her, but she returns to the orchards and to Tommy, her former sweetheart.

Again the critics came down hard on the film, which is underscored practically from beginning to end. The *New York Times* critic wrote that there was "little or no effort at characterization," while *Variety* dubbed it "just blah." Even Oscar himself, writing to daughter Alice, wondered, "Why doesn't anybody like my movie?" Then he answered his own question: "I don't like it myself anymore. It was written in a hurry and it looks it." Realizing that after two cinema flops he was no longer the movies' great white hope, he added, "I have concluded that I better do something good for the stage first. That's the only thing that makes them want you for the movies."

But the remark shows he was certainly not aware of the changes that had been going on around him in his short year in Hollywood, for no sooner had he submitted the idea for his next film, *Heart Interest*, than studio head Jack Warner, fearing another expensive dud, offered to buy out his contract.

"I was given $100,000 *not* to make the remaining two pictures," Oscar crowed as he headed back east to work on the stage. Although *Rose-Marie*, *Sunny*, *Song of the Flame*, *The Desert Song*, *Golden Dawn*, *Show Boat*, *New Moon*, and *Song of the West* (the cinema version of *Rainbow*) were to be filmed before 1935, Oscar and Dorothy left for the East in September 1930, and he would not return to write for Hollywood again until half a decade had passed.

Once settled in New York, Oscar could get on with some of the projects he had begun. First on the agenda was Uncle Arthur's *Ballyhoo*, a musical comedy starring W. C. Fields which Oscar, now the star of the family, had promised to direct.

Arthur had been in deep financial trouble because his last show, *Luana*, an exotic and expensive operetta wherein the heroine kills herself by jumping into a volcano, was (understandably) withdrawn after 21 performances. Worse, the theater and office building he had built in honor of his father and launched with *Golden Dawn* remained largely empty because the noisy El rumbled by every few minutes. Soon the failure of *Ballyhoo*, into which Oscar put all his directorial canny plus some of his own lyrics, was to bankrupt the Arthur Hammerstein empire.

Directing led Oscar into "doctoring." This time it was a musical farce whose themes were the current manias—benevolent gangsterism and psychoanalysis. *The Gang's All Here* was written by Russel Crouse and Morrie Ryskind, with music by Lewis Gensler and lyrics by Owen Murphy and Robert Simon—professionals all and friends of Oscar—but no amount of ministering could help, and the show closed after 23 performances. The critics completed the title line with "what the heck do we care!"

But on March 23, 1931, three weeks after the demise of *The Gang's All Here*, a happy event occurred. Oscar's second son, called James after Dorothy's father, who had died two months earlier, as well as James Nimmo, Oscar's maternal grandfather, was born. Now Oscar had two families, and for the rest of his life he would spend time and emotion shuttling his children back and forth between his home in New York and frequent forays into Hollywood.

Long letters were exchanged with Alice and Bill, who were at sea in their relationship to Dorothy. At home with their mother they were taught to hate their stepmother, and yet when they visited their adored father, he remonstrated them if they did not open their hearts to his beloved.

According to William, Myra felt that "my father had deserted her and treated her very badly. The fact is, he didn't. In fact, after they had been divorced a couple of years, she got married to this *terrible* person. And it only lasted about six months. Of course, once she got married, he no longer had to pay her any alimony—but he did for the rest of his life. He paid her enough so that she always had a decent apartment in the city, and while we were still of school age he paid child support."

But beside child support Oscar had many financial burdens on his plate. The large and heavily taxed house in Great Neck, Long Island, which he and Myra had built, was now empty. Instead he was paying rent on two apartments—a large one for Myra and the children at the Dorset and an equally chic one on Fifth Avenue for his new family.

Perhaps Oscar felt his entire career hung in the balance, for after his two flops he immediately set to work on two new shows. Both were to fail. *Free for All* ran hardly two weeks and *East Wind* less than three at an unlucky theater whose new name must have been anathema to him. The marquee of the Hammerstein Theatre, which Arthur had lost in his bankruptcy, had been replaced. Now the glorious auditorium built to honor his grandfather was called the Manhattan Theatre.[1]

Both shows were wrong for the times. *Free for All*, an intimate and chorusless musical about the opening of a mining camp by a rich man's son who becomes mixed up in radical politics, tried to be contemporary. But even though the score was played by Benny Goodman's band which featured Glenn Miller, the musical, like many in the spartan early '30s, never seemed to come alive.

Oscar and Frank Mandel immediately delved into their trunk and came up with a Sigmund Romberg reject, *East Wind*, which was as opulent and as far away from *Free for All* as possible. Its locale, Saigon, made for interesting costumes and settings, but this story of the love of a French boy for a French girl could just as easily have been set in Bordeaux.

After various critics called his book dull, stupid, and tedious, adding that "there is no inspiration in it and not much sense," Oscar took stock of his recent work. He realized that besides his two film failures, in the year since his return from California he had been deeply involved in four shows that were out-and-out disasters. Not only was his career and reputation in jeopardy, but he had made little money from these time-consuming failures.

Now he "took a deep breath and started again." Remembering how months of polishing *Show Boat* had produced his most glittering success, he promised himself to take a year to work without interruption on his next show. Even though Jerome Kern was commuting between the East Coast and Hollywood, he sought him for composer, rather than Romberg or Youmans, because of the special working arrangement he had with Kern wherein they tended to talk the characters out before they even put pen to paper or hands to keyboard. This

---

[1] The theater, used for television today and formerly known as the Ed Sullivan, was taken over by CBS in 1993.

time he would try for authenticity of setting, realizing that as for the settings of his last two hastily assembled flops, he had never been to a mining camp or to Saigon.

More importantly, he planned to fight his tendency to deal with the superficial while ignoring the fundamental. All through the rest of his life, although he was frequently to forget his vow, he was more often than not to remember to work slowly and to concentrate on the basic emotion.

With all this in mind, Oscar sketched a contemporary story with most of the scenes set in a music publisher's office. It concerned an old music teacher and a young lyricist who write a hit song together. This was obviously inspired by his long collaboration with Harbach and might have made a fine American musical. Oscar told the idea to Kern, who convinced him to change the venue to contemporary Munich and to call their new musical *Karl and Sieglinde*. Kern, a brilliant melodist who could manipulate harmonies in a very sophisticated way, certainly was insecure in the Tin Pan Alley setting his collaborator had elected. His music had never progressed rhythmically into the '30s world of jazz. He was also headstrong and very persuasive.

Since Arthur had declared bankruptcy, the men approached Ziegfeld. Although rumors were circulating that even the great Ziegfeld had fallen on hard times, his lavish revival of *Show Boat* in May disproved that. Unfortunately, the "glorifier [and notorious seducer] of the American girl," who had long burned his candle at both ends and in the middle, died just two months after.

Enter A. C. Blumenthal, who was married to Peggy Fears, one of Ziegfeld's great beauties. He put up the money for the musical but insisted on giving the producing credit to his wife. Before agreeing to shepherd *Karl and Sieglinde* onto the boards, however, he demanded two things: that the production be brought in for no more than $100,000 and that the title be changed.

With Kern's music and the foreign setting, it was hard to keep what soon was titled *Music in the Air*, although called "a musical play," from being much more than a carbon copy of the Kern-Harbach hit operetta of the year before, *The Cat and the Fiddle*—also about songwriters, this time male and female. Once it had been produced, even Oscar was to admit "*Music in the Air* owes a great deal to its predecessor." In both, the hero is seduced by an opera diva but at the final curtain returns to his early true love.

The idea of having the collaborators write a song which would launch the young composer's career was eminently suitable for a musical. Of course, they knew such a key song had better be a great hit. Kern rose to the challenge by writing a fine melody which has since become a classic.

The inspiration for the song came from Kern's fascination with bird calls. He had jotted down the opening motive (see example below) and had written "6 a. m. Bird song from the willow tree outside east window," as the key song to the musical. The bird singing outside his Bronxville home was later identified as a Cape Cod sparrow known to ornithologists as *Melospiza melodia*, and its song is

Eventually published as "I've Told Ev'ry Little Star," the song originally had a charming, wise lyric called "Bright Bird." The "Bright Bird" lyric and philosophy is far more suitable to Kern's twittery melody than the one the collaborators finally adopted. Both are printed below.

Bright bird sitting in a tree
How can I sing like thee?
How can I sing like thee?
Bright bird, winging through the sky
How can I learn to fly?
How can I learn to fly?
Looking wise, he blinked his eyes
To greet a passing dove,
Then he looked at me
From his perch above.
I can teach you how to sing
But if you want to fly
First you must fall in love.

I've told ev'ry little star,
Just how sweet I think you are,
Why haven't I told you?
I've told ripples in a brook,
Made my heart an open book,
Why haven't I told you?
Friends ask me: Am I in love?
I always answer "Yes."
Might as well confess,
If I don't, they guess.
Maybe you may know it too,
Oh, my darling, if you do,
Why haven't you told me?

Music in the Air. L to r: Reginald Werrenrath, Al Shean, Ivy Scott, Katherine Carrington, and Walter Slezak. The magnificent bucolic setting is by Joseph Urban. MOCNY.

The score boasted three other interesting songs, chief among them one of Kern-Hammerstein's most poetic songs, "The Song Is You" (. . . I hear music when I touch your hand / A beautiful melody from some enchanted land . . .), the frisky "In Egern on the Tegern See" (. . . We watch the sunset fade away and melt in the gloam . . .), and the romantic waltz "And Love Was Born" (A warm spring night was stirred by the breeze / And love was born / The moon in flight was caught in the trees / And love was born . . .).

The critics raved and found hardly any fault with it, enabling the show to chalk up a ten-month run on Broadway and to do almost as well in London. Yet this writer finds much of *Music in the Air* tiresome and banal. Only the lighthearted title seems an improvement over the former heavily Germanic one. Since the hit song that promulgates the action is actually derived from a bird call, the show is truly *Music in the Air*. The title also has a nice ring to it, suggesting microphones, radio, and contemporary airwaves.

But nothing can save the talky, tedious scenes. According to Oscar's plot, the diva-siren is trapped in a bad play. Much of the first act is spent exposing to the audience (along with snippets of song) how bad this play-within-a-play is. Unfortunately the whole segment is insufferably dull. Oscar had not yet learned the importance of having his audience laugh with his burlesque rather than at it.

Besides its mannered libretto, one finds songs like "I'm Alone," with lines like "And wishing, dear, / That you were here / To be alone with me"; "When the Spring Is in the Air," a title which contains at least one redundant "the" besides being a mouthful to say in quick tempo; and the especially trite rhymes found in "We Belong Together" (Like birds of a feather).

The pretension in the score was worse. Early in Act I Kern arranged the second movement of Beethoven's Piano Sonata op. 2, no. 3 for chorus and subtitled it "Melodies of May" for a scene entitled *Etudes*. Other scenes were labeled *Pastoral, Impromptu, Nocturne, Intermezzo, Rondo*, etc.

Still, the show was a perfect antidote to the doldrums of the Depression, with not a single minor melody or torch song. On opening night, November 8, theatergoers emerged from the Alvin Theatre whistling, for there was not only music in the air, but a sense, however tentative, of recovery. Friend of the theater and all the arts Franklin Delano Roosevelt had been elected President of the United States that very day.

Music in the Air was the long-awaited hit Oscar wanted so desperately. In sum this success was due to Kern's insistence on setting the operetta to escape to the pre-war and quasi-benevolent Germany of 1932[2] and to Oscar's main-

---

[2] After World War II, with the widespread knowledge that Hitler had already been a powerful force in 1932, the occasional revivals of *Music in the Air* were changed from a German to a Swiss setting.

taining his vow to write slowly and carefully. In pre-Broadway tryouts there was none of the hysterical juggling of scenes or rewriting that made his other shows superficial, frantic, and costly. The show had been brought in handily within Blumenthal's budget, and even with the best seat in the house going for $4.40, it repaid its investment within an unheard-of twelve weeks.

Soon, with both *Music in the Air* and the revival of *Show Boat* selling out nightly, Oscar, his family, and many steamer trunks set sail for England, where he planned to direct the West End production of the former show. He felt it would be seasoning for all the children to live abroad for a while and a chance to escape the gloom of a Broadway throttled by the Depression.

Living in London also afforded Oscar an opportunity to write for the British theater, for his close friend Louis Dreyfus, head of Chappell Music and also a director of the Drury Lane Theatre, was urging him to adapt some operettas for that cavernous house. Most importantly for his own self-image, the Hammerstein name had not been tarnished in Britain. Here his name conjured up memories of *Rose-Marie, The Desert Song,* and *Show Boat.* Here, even he would not be reminded of his four recent failures.

Lerner

1918–1940

IN LATE 1945, WHEN ALAN JAY LERNER'S FATHER, Joseph, was in the Memorial Sloan-Kettering Cancer Center recuperating from a cancer operation that had taken away his power of speech, he became friendly with Francis Cardinal Spellman. The cardinal, then the head of the Archdiocese of New York, often came to call on Damon Runyon,[1] whose room was adjacent to Joe Lerner's. The spiritual leader of New York's Catholics enjoyed exchanging quips with Runyon, well known for his sparkling wit.

After his chat with Runyon, the cardinal would generally step next door for a visit with Lerner, partner in the Lerner Company, a large chain of specialty shops that featured moderate-priced apparel for women. Joe Lerner was also a man of considerable erudition, cleverness, and charm. He kept a pencil and pad with him at all times and had mastered the technique of quickly writing out his answers to the cleric's (or for that matter to any of his many visitors') questions.

Alan, Joe's second-born son, confessed to his father that he understood why the cardinal would be anxious to win as many souls with enduring royalties like Runyon to his cause as possible but could not perceive what his father and the Catholic leader could possibly have in common. In answer to the inquiry, Joe wrote, as Alan recounts, "with a twinkle in his eye: 'We have a lot in common. We're both in the chain store business.' "

That nationwide chain of shops, which Joe thought of so lightly, brought considerable wealth to his family. It was to cloak Joe Lerner's son in its aura, become his passport to education, and make him think of himself as a privileged person. The money, and, later, his fame were to be Alan Jay Lerner's

---

[1] Damon Runyon (1884–1946) was well known for his cartoonlike stories about Broadway characters and a benevolent underworld such as *Guys and Dolls*, which he wrote in 1931.

spur and eventually, because he dispensed both so randomly, his albatross. Having been born into riches, his ambition led him to seek social lionization— which he could never attain—and preeminence in his career—which he could. As a Jew, especially one with *new* money made in the socially unacceptable "rag trade," he would never be able to enter the brahmin society to which he aspired. And so those hopes were always quashed because the only circle that would let him in would be Café Society (which he often confused with the real thing).

As for his other spur—fame—that he certainly did achieve. *My Fair Lady*, which stands out as the undoubted pinnacle of the successful twentieth century musical, attests to his preeminence in the field. Part of the success of this seminal musical is due to Lerner's respect—almost reverence—for Shaw, but the greatness of his other works, *Brigadoon, Paint Your Wagon, Camelot*, and *On a Clear Day You Can See Forever*, attest to Lerner's great love affair with the theater. *Royal Wedding, An American in Paris*, and the imperishable *Gigi* point up his mastery of the film musical as well as his sense of fantasy.

Early in Alan's life, he fantasized that his roots were different from the reality around him. Even though income from the Lerner Company or, as the chain was to become known, The Lerner Shops, was to provide a comfortable cushion for the family, the young man wished for other forebears. In this he was not unlike Oscar Hammerstein, whose grandfather, while a famous figure in the world of opera, was anathema to the family for his costly theatrical fiascoes.

Alan's grandfather Charles Lerner, who was born in 1860 in Ukraine (then a part of Russia) emigrated to the United States in 1880 and settled in Philadelphia. He opened a small store in the teeming ghetto of South Philadelphia, selling new and mostly used carpets. He was hard on his seven children, especially the three boys, Samuel, Joseph, and Michael, who worked long hours in the shop. He is reputed to have been no less tyrannical with his four daughters, and they too resented the heavy responsibilities and lack of freedom he imposed on them.

Samuel, the eldest, angered by the harsh treatment he received from his father, left home at fourteen and eventually became an itinerant carpet salesman for the distinguished West Coast firm of D. N. & E. Carpets. He quickly rose to become their crack salesperson and by the turn of the century was making a then-phenomenal $10,000 a year. Ambitious, he asked to become a partner in the prestigious firm, which is still in existence today. He was rebuffed, let down gently by being told, "No, you never will be a partner. This is a family operation."

Returning to the East with the idea of a family business still implanted in his mind, he got into the manufacturing business, specializing in what was then called women's shirtwaists (now called blouses), a big item in those days. The shirtwaist manufacturing business grew and became very successful. After a decade he changed from the manufacturing to the retailing side of it. Soon he acquired several stores from the Avlon chain that had featured his product. In no time he became the sole owner of the entire Avlon chain, renamed The Lerner Shops and selling shirtwaists and housedresses with simple lines and some style at low, affordable prices. By 1914 there were about forty of them in the New York area. Now he felt that with the aid of his brothers he could expand the chain nationally.

Samuel dominated his brothers and had a great influence on his sisters as well. As the moneymaker in the family, he established a certain authority and an almost paternalistic attitude toward everyone—even his parents, whom he set up in a comfortable New York apartment.

Perhaps he recalled the D. N. & E. Carpet Company when he immediately pressured his two brothers to join him in The Lerner Shops. Joe was practicing painless dentistry on the boardwalk in Atlantic City and was struggling to support his extravagant wife, Edith and their firstborn son, Richard. Edith hated her bourgeois life in the seaside resort and eagerly welcomed a chance to move back to her native New York.

Samuel's youngest brother, Mike, had been his aide and top salesman for the firm for many years. Though he was not very foresighted, he would make a reliable partner. In time equal shares of stock were issued and large offices were outfitted for each of the three brothers.

Frugal and sensible, Samuel Lerner bought a large property in suburban Mount Vernon, where he and his family lived. Although Samuel acquired vacation property in Florida, he and his wife, Eva, were not spendthrift. Their three children were sent to public school in Mount Vernon. Mike and his wife, Helen, childless, were content to live in a house next door to their brother and bask in the reflected glory of the firm's founder and chief workaholic.

Joe was not.

He rented a huge floor-through apartment at 470 Park Avenue, New York's most fashionable and, at that time, predominantly gentile area, which he furnished largely with antiques. He hired a chauffeur and acquired an expensive wardrobe—and then, of course, a maid whose only duty was to look after his suits and haberdashery. Not to be outdone, his wife, Edith, née Edelson in the Bronx, hired her own lady's maid as well as an English nanny to care for four-year-old Richard.

The Joseph Lerners soon began to acquire the culture that had been denied them in their youth. Now they traveled frequently to Europe. They attended all the important Broadway openings, never missing the first night of one of Joe's special passions—the musical theater.

Not unlike Molière's *Bourgeois Gentilhomme*, Joe had teachers of ballroom dancing, piano playing, and foreign languages come to his apartment. Besides his good looks, he had the intelligence and stick-to-itiveness to excel all three of these areas. By midlife he looked like a professional on a ballroom floor, played acceptable Chopin as well as the current show tunes, and was conversant in French and Italian. People he met were magnetized by his ineffable manner and dazzling smile. Joe Lerner had charm that worked equally well on men and women.

Soon Joe began to philander. Edith retaliated by charging more and more expensive clothes and furs to his account, and when this no longer satisfied her anger, left him at home while she traveled. The mature Alan Lerner called his parents' life together "a symphony in three movements: arguing, separating and reuniting."

Thus, it was into one of the outer movements that Alan Jay Lerner was born on August 31, 1918.

By the time Alan had reached the age of six he had already had a year of classical piano instruction and, precocious child that he was, played better than most young beginners and dreamed of becoming a composer. Now he was enrolled in Columbia Grammar School, a long established private school for boys on the West Side of Manhattan. Columbia Grammar had impeccable academic standards and a student body of affluent and largely Jewish liberal intellectuals.

Alan was an adequate student whose charm and good looks let him get away with frequently incomplete assignments. Blond and blue-eyed with creamy skin, the boy glowed when his swarthy uncles joked, "He isn't even related to his parents." So precocious and adventuresome was he that in manhood he was to admit to being "introduced to love and sex by the domestics, around the age of five." Although the youngest Lerner, Robert, was born two years after Alan, it was clear from the beginning that Alan with his early charm was the preferred child.

When he later wrote, "I was my father's favorite and I adored him. However, he made things very difficult for me at home by never attempting to conceal his partiality," Alan must have been referring to his musical gifts, for he played, sang, and accompanied himself often. Because of this his mother, who was musical, was not averse to taking him to matinees of musical plays rather than his older brother Dick, who had no interest in such things.

And what a heady time for musicals it was, with Romberg and Friml as operetta's elder statesmen presenting shows side by side by the up-and-coming Berlin, Kern, Gershwin, Youmans, Rodgers, and Hart. Stars like Ed Wynn, Marilyn Miller, and Fred and Adele Astaire were being wooed by producers like Ziegfeld, George White, and Earl Carroll, whose *Follies*, *Scandals*, and *Sketchbooks* were little more than lavish revues in the "music hall" tradition.

Although Alan certainly enjoyed these excursions with his parents, his music teachers at school and home stressed the Three Bs—definitely excluding Irving Berlin. When he was nine and had entered the fourth grade, all the students in his class were assigned to write a composition after hearing a piece called "From an Indian Lodge" by a well-known American composer. The others described the obvious Indianisms in the music. Not Alan. He produced a dreamy little essay full of longing to escape reality that had very little to do with the description of the piano piece the teacher sought. Still, one can observe that he even saved the composer's name to give his piece a punch-line ending as he was later to do in his songs.

The Little House in the Woods

The little house in the woods was in a very lonely spot. The little paths that led into the woods led into beautiful places. Everything around the house was very beautiful.

The house in which this great composer lived was made of logs. He loved nature very much and therefore he did not cut the trees down around the house but they were hauled from a mountain ten miles away. The house was built on stilts so there was a very good view all around. . . . The walls were made of logs and a lot of bark was put over it. . . . The composer loved to dream and he wrote beautiful songs. He was the greatest American composer. At times he would go into the woods and dream. This great American composer was Edward McDowell.

Throughout the year Alan wrote several short essays, one of which, on Robinson Crusoe, again reveled in the protagonist's isolation. This piece made its way into the grammar school's paper, *The Columbia News*.

Part of Alan's quest for seclusion must have been occasioned by the frequent stormy scenes between his parents at home. Another part was due to the infighting that went on among Alan and his two brothers. To escape this, Alan would generally be found his room with his door locked to his brothers. Joseph Lerner's three sons had as little in common as Joe had with his own brothers. A lonely, introspective youngster, Alan seems to have had no childhood friends, no schoolmates who visited. Since neither parent put any constraint on what their sons bought, Alan and his brothers seemed to take their chief joy from the position, luxe, and ease their wealth brought them.

Occasionally Samuel's son Will would come in from Mount Vernon to spend time with his relatives. Closer in age to Robert than Alan, the country cousin was a very serious boy who, because Samuel had set up trust funds for each of his children, was perhaps even more wealthy than his kin. Will was mightily impressed with the Manhattanites' life style as well as Alan's mastery of the piano. "A night with my cousins at 470 seemed very glamorous to me," Will Lerner recalled. "They would phone in for sandwiches from Reubens on Fifty-eighth Street. And they would often show up for week-ends in Mount Vernon in their chauffeur-driven Packard or Cadillac or whatever. Alan had studied and played classical piano. He switched to modern later, and then he would make the piano come alive."

Alan's ambition was to be involved with musical theater, and to that end he had been writing words and music to songs for some time. His father was not impressed with his music but "he was passionate about the English language," Alan was to write, "and until I graduated from college, I never sent him a letter that he did not return to me with notes in the margin suggesting more interesting ways of saying the same sentence."

Obviously Joe took an interest in improving Alan's literary style, and by the next year at Columbia Grammar one can see a definite improvement in his fledgling essays. There is no doubt that Alan wanted to please his father, about whom his life seemed to revolve. In the essay below, written when Lerner was ten, one glimpses a psyche in which the omnipotent father takes precedence and the mother is totally ignored. Note, too, that by this time Alan has adopted the middle name of "Jay" and would make that a lifelong addition to his signature.

Prometheus, the Fire-Bringer
By Alan Jay Lerner, Grade V

The golden age ended after the war between the Titans and Jupiter. All the Titans were killed except Epimetheus and his brother Prometheus. Prometheus used to sit and watch the poor suffering cold people of the earth, while the proud gods looked down from Mount Olympus and laughed.

One day as Prometheus was watching the people he had an idea. If he could steal fire from the sun, he could teach the people how to keep warm, cook and do many other things. He broke a reed and as the sun came up he caught a spark in the hollow reed. He then rushed to his home land. After he had given the people fire, everything went well for a while. But one day as Jupiter was sitting on his throne he looked at the earth. Lo and behold, the people were using fire!

"Who has done this?" he roared, his eyes filled with anger.

"Prometheus, the Titan," said someone.

"He must be brought before me at once," said Jupiter. Prometheus was brought before Jupiter, who chained him to the highest peak of the Caucasus mountains where a vulture tore at him.

Some years later Jupiter's son, Hercules, was roaming through the mountains when he spied Prometheus and freed him. Ever after, Greek fathers taught their children to honor Prometheus.

In 1931, when Alan had turned thirteen, the age at which most Jewish boys are Bar Mitzvahed, there was no ceremony for him. His father had avoided any connection with church or synagogue since his youth, even denying to his children the existence of God. Two years earlier, Alan had inquired where we go when we die. "Nowhere," Joe replied immediately, "and if anyone tells you differently, he's lying to you. You go to sleep and never wake up." For many nights afterward Alan had lain awake, pondering his father's dire prediction. The fear was to stay with him; all through his adult life he was to need no more than a few hours' sleep. When he was working on a show, he would sometimes stay awake for days.

The next summer, 1932, the odyssey of summer holidays in Europe that had begun with The Grand Tour for the three Lerner boys was curtailed because Joe Lerner believed his sons needed to improve what he considered was "their abominable English." All three were enrolled in the Bedales School in East Hampshire, England. Bedales was famous for its program of Arts and Crafts and like most British public (equivalent to American private) schools did not indulge its students. One of Alan's classmates, Justin Brooke, remembers how awkward Alan looked in the short trousers English schools required boys of Alan's age wear. Smaller than his classmates, Alan was especially embarrassed by this uniform because Bedales was coeducational—his first exposure to a coeducational school—and he was, by now, becoming very aware of girls.

Accommodation at the school was spartan, with no heating in the dormitories. The boys all slept with windows open. The food was wholesome but plain. The cuisine, which boasted a mid-morning snack of bread soaked in bacon drippings, could not have gone down well with the Lerner clan, coming from their background. Perhaps the most embarrassing aspect of that memorable summer at Bedales to Alan, who in later life kept his body and clothes scrupulously clean almost to the point of paranoia, was the primitive form of sanitation available at the school. Alan's squad was forced to line up after roll call each morning. Then each boy was allowed only five minutes to perform his bodily functions. All this was not done in a proper lavatory but in an earthcloset. Coming from a seventeen-room Park Avenue apartment which boasted five marble bathrooms, Alan was kind to Bedales when he acknowledged that "if nothing else, it instilled in me a certain regularity of habit."

He and his brothers pleaded with Joe to be spared future terms at the British boarding school and promised to study their English if only he would not send them away again. Alan kept his word and that year had a good, if not an outstanding, record at Columbia Grammar, which he attended through the spring of 1933, leaving before his fourteenth birthday. In his last year there, he became much more socially aware. Although not generally the kind of student to cram in subjects that held little interest for him, now he began to study hard and even managed to get himself elected to the Freshman Honor Society.

His demeanor and outlook were to change even more the following year, all because of his Uncle Samuel and Sam's only son, William. Sam decided that Will, a very good student, had grades strong enough to be accepted into Choate, one of the most prestigious preparatory schools in the country. Choate, in Wallingford, Connecticut, a stepping stone to entry into top-notch Ivy League universities, was known as a very selective school whose enrollment had been culled from America's most prominent families. But the school's excellent academic record was what appealed more mightily to Samuel, who expected Will one day to take over the management of all the Lerner Shops, by now a chain with stores on Main streets throughout America. As soon as Will was accepted, Sam's partner Harold Lane (originally Lavine) entered his two boys as well.

Joe, who was not to be outdone by his older brother, immediately applied to enroll Alan and Bobby. (Richard had graduated from high school by that time.) It was the height of the Depression, and private schools could hardly afford to say no to almost any applicant who had the money to "contribute" to a endowment fund. And so in the fall of 1933 these five young Jewish boys, all connected through the same business enterprise, journeyed to a small Waspish New England town where they were to study to prepare themselves for entry into Yale or Harvard.

They were joined there by Dan Sicher, an acquaintance Alan and Bobby had met two years earlier when Edith had taken the boys for a summer at the Lido in Venice. Dan, whose grandfather was a Bloomingdale, was unhappy at Choate. He sensed a strong sense of anti-Semitism at the school. "There were only seven Jews in the school out of an enrollment of about five hundred," he recalled later. "If you were a good athlete you did fine. I didn't have a problem, except that I felt it, and I am not any more sensitive to being Jewish than anyone else. It never bothered Alan."

Dan's statement about Lerner being unaffected is totally wrong, for it was at this time that Alan developed an aloof quality that might have passed for snobbishness. But to increase his sense of belonging, he entered into extracurricular activities with a relish he had never shown before at Columbia Grammar. Well-coordinated, with a feisty manner, although small-boned, he immediately went out for football. Later he was to add basketball, baseball, crew, golf, and boxing to his recreational sports. He and his cousin Will used to practice sparring when Alan would visit them in the country.

"He was small and reasonably well coordinated," Will recalled. "I remember he took boxing more seriously than I did. Once I was startled that he actually was out to win, and I said, 'We're just sparring here.' But it looked as though he was trying to beat me . . . then we just stopped."

But except for boxing he was soon to become half-hearted about sports and never rose above second stringer in his prep school years. What interested him most in the three years he was to remain at Choate was working on *The Brief*, the school yearbook, and eventually, in his senior year, becoming its advertising manager. In this he was stepping close to the footsteps of John F. Kennedy, who served as the yearbook's business manager the year before. But the only area where he truly shone was at the piano in the common room, evenings before and after meals.

"He and two or three others would sit down at the piano, each topping the other in a display of virtuosity that was dazzling," his cousin Will recalled. "We would call out songs, and it seems as though Alan had total recall of every single song he'd ever heard. We would stand there enthralled as these guys would slip off the piano bench only to be replaced by the next one who would say, 'Oh, I know that one.' It was almost magical."

It was during his first year at Choate that Alan began, as some of his family members have put it, "to invent himself." He shared a room with John Jones and told his roommate as well as all his other gentile classmates that he was only half Jewish. Sometime he would go so far as to say his mother was born a Catholic. Later in life he would spread the same myth. Doris Shapiro, who worked for Alan later in his life, said, "Alan hated to be reminded that he was Jewish."

But his imagination took him to invent other stories as well. "One day he came back from a trip to New York and said that he had visited the Gershwins," Jones was to report. Soon, ignoring his Uncle Samuel, he was to tell everyone that his father had founded the Lerner Company. Later he stretched his six-week term at the Bedales School into "having been educated in England."

Jones felt that at least some of these fantasies were prompted by his parents' separation, which led eventually to their divorce. In later years Alan elicited a great deal of sympathy as having been a child of divorce—although the divorce was not to come until 1944, long after he would have finished Harvard.

A six-time divorcé by the time he wrote his biography, *The Street Where I Live,* he still makes light of the manner in which the break between his parents was to evolve into the chasm of divorce. But the story is so engagingly told that its time frame is irrelevant. It must be added at this point that although Alan Lerner was to reinvent or exaggerate his private life and turmoil, he never overstated his theatrical accomplishments. He was scrupulously honest about his art. In the Choate years two aspects of his future obsession were already apparent: his love of and prodigious memory for show music. In 1974 he recalled the days in the '30s when he was at school:

I was induced into a trance by the pipes playing "Dancing In The Dark," "Just One Of Those Things," and "Embraceable You." To me, an exotic aphrodisiac was a pair of great legs on a girl in the front line. And the gurus who led me on to the next plane of happiness had nice Occidental names like Gershwin, Rodgers, Porter, Hart, Berlin, Dietz and Schwartz. I knew every song and every lyric they wrote, including the verse and second chorus, as well as I did the Lord's Prayer. And I knew them within a week after each show opened.

Returning to the saga of his parents' divorce, Alan begins by explaining that his father attended the boxing match at Madison Square Garden every Friday, and adds: "When I said that my father went every Friday night, I should have said almost every Friday night, for on many occasions his taste for combat drew him to other more quilted arenas."

He then goes on to tell of the Saturday morning when his mother, in an unusual departure, asked her husband who had won last night's bout. It was a heads or tails gamble, and Joe Lerner, whose seat at The Garden had remained vacant the previous evening, murmured the name of the fighter who was the favorite to his wakening spouse. Alas, when he opened *The New York Times* to the sports page over his morning coffee, he realized he had picked the wrong man.

"He methodically finished his breakfast," his son recalled, "and went downtown to the office. Once there he called the house. There was one maid specifically assigned to looking after his clothes. He told her to pack everything and the chauffeur would call for his luggage shortly. By the time my mother fully awakened, my father and all that was his were gone."

Joe Lerner moved into an apartment at the Waldorf Towers, where Alan was to visit him whenever he could get away for a weekend from school. One weekend he returned to the campus and introduced his schoolmates to Vernon Duke's sophisticated "April in Paris." He told his roommate that his father had taken him to the theater to see the revue *Walk a Little Faster*. The urbane fare Joe chose to attend with his son had already made him more theatrically sophisticated than anyone in the school.

By now Alan was gung-ho for the theater although Joe had loftier ambitions for him, wanting him to go into politics, even hoping he might become an ambassador. "Do anything, but don't go into business," he warned. But the anything never included writing for the theater. Fortunately for posterity, Alan was not only hooked but determined. So deeply involved was he that in his senior year his roommate, F. Sigel Workman, with whom Alan had a running game of identifying show tunes, asked to be moved to another dormitory because "Alan kept waking me up in the middle of the night to sing esoteric tunes which he asked me to name," Workman recalled. The same sleeplessness that was to plague him throughout his life was already a part of his psyche.

Several of his classmates noticed Alan's nervousness and his severe nail biting—sometimes until his fingers bled—but they chalked it up to his being high-strung. He had also begun to smoke, and although all of Choate's students had signed a pledge that they would not smoke at school, Alan and some of his friends often could be found inhaling cigarette smoke on the golf course adjacent to the campus. He broke many rules just because his charm and position let him get away with much. Some considered him haughty; many felt he placed himself above what he considered petty constraints. But no one could resist his ingratiating personality. And all were aware of his talent as a composer.

In the autumn of his senior year he and Carlton Tobey wrote a football song, "Cheer For Choate Today." It certainly does not compare to Cole Porter's "Bingo Eli Yale," but still it has a certain lilt to the melody, which is mostly Lerner's, and a growing intensity at its end ("We've the might to win, We've the fight to win"). The lyric is nothing special and even contains words like "fray" and "victory," the latter with accent on the last syllable—faults the mature Lerner would never condone.

By April Alan's grades had sunk to barely passing. After spring recess he returned late to school in a car, and while he was confined to the campus for

this transgression, he and two friends were caught smoking in a small roadhouse which was not far away from the school. In his autobiography he wrote that he was expelled, again an exaggeration. The headmaster, Dr. St. John, took the very unusual solution of stating that if no other boy in the school smoked for the next month, then Alan could stay at school, go to classes, take his college entrance exams, and finish the year. Joe Lerner was very grateful to the school for allowing Alan to graduate.

In spite of Alan's penchant and gifts for music and writing, the father had always dreamed his son would study for the foreign service beginning with a year at the Sorbonne in Paris. Now, after the smoking incident, he suddenly changed his mind, canceling hopes for any diplomatic career, and "sentenced" Alan to what he described as "four years of hard labor at an American university, which was a little like punishing a prisoner by kicking him out of jail." Thus it was that in September 1936, Alan entered Harvard as a candidate for a Bachelor's degree in, of all things, sociology.

Once he was settled in Boston, Alan looked up Ruth Boyd, whom he had met some years before in Venice when he was on one of those perpetual Grand Tours during which he had been dragged over the map of Europe by his mother. As teenagers, they danced well together and kept up a kind of glib repartee. Ruth came from a socially prominent newspaper-publishing family, and, as previously noted, Alan considered that a plus. At seventeen he was an inveterate snob, eagerly wanting to sweep his garment district Jewish roots under a social carpet. Relationships have been built on less.

Alan, who was basically shy and overly conscious of his shortness, began asking Ruth, with whom he felt at ease, to the Harvard dances. It followed that she would be "his girl" and that they would eventually wed. "They used to dance at the Stork Club, and they were like two little marionettes," said one of his college chums, Ben Welles. Welles would go out with many girls, but Alan was always with "Ruthie."

Stanley Miller and Ben Welles, both upperclassmen, besides remaining life-long friends of Lerner's would become his collaborators on two of the Hasty Pudding shows. Harvard's Hasty Pudding Club, established in 1847, presented an annual all-male show as famous and professional then as those of Yale's Whiffenpoofs and Princeton's Triangle Club. Alan's song (he wrote words and music) "Chance To Dream," from So Proudly We Hail in his junior year, shows remarkable music and mundane lyrics (see example below).

The following year, for Fair Enough, the Pudding show that celebrated the New York World's Fair of 1939, he wrote "Home Made Heaven" and "From Me to You." Both songs have catchy and rather daring melodic and harmonic lines, but, as the excerpts from their refrains below will testify, far-less-than-professional lyrics.

Refrain

When I hear a song of trees and flow'rs and A-pril show'rs, And think you're whis-p'ring to-night is ours! Oh, that's my chance to dream.

Street lights and noises of horns behind us
And where no one can find us
We'll have a HOME MADE HEAVEN
Sunlight that's pouring through ev'ry room, dear
Chasing away the gloom, dear
We'll have a HOME MADE HEAVEN . . .

FROM ME TO YOU, I'll let a secret out,
FROM ME TO YOU, I've learned what love's
about.
I've just felt my head reeling the funniest way
And heaven has no ceiling since love bid good day . . .

During those undergraduate years Alan spent little time studying for his degree in sociology. He took elective courses in French and Italian but spent the majority of his energy off campus. Boston was, after New Haven, the favored tryout stop for shows headed for Broadway. Alan saw every play and musical en route to Broadway. Then he spent his summers taking courses in piano and theory at Manhattan's Juilliard School of Music. Quite apart from his theatrical pursuits, he took flying lessons so that if war came he could join the Air Corps.

At the university he continued the boxing he had begun at Choate, aiming to make the varsity team. It seemed a perfect sport for the 5' 6" feisty young man, especially since it was the only sport that interested his father. One day in the boxing ring, he says, "My mind wandered, my guard dropped and a left hook to the side of the head removed all sense from my expression."

His vision had been seriously impaired. Upon examination it was discovered he had lost the retina of his left eye and had bruised the right one; all vision in the left eye was lost, and he was in danger of losing the sight in the other as well.

He was in the hospital for eight months, forbidden to move or bend and cautioned not to sneeze. One of the family servants stayed with him to ease things for him. Towards the end of the spring term he was allowed to go back to school, but all exercise was forbidden for five years.

Throughout his life Alan was conscious of this visual deficiency but took care to hide it from the public. Yet people who worked with him were admonished "to sit the other side of him or face to face," as his secretary, Stef Sheahan, recalled. "Sometimes when we were walking on a narrow sidewalk he would bump into a parking meter and then he'd apologize thinking it was a person he had bumped into. And he was a disaster at dinner or lunch because he'd just knock things over the whole time that were out of his vision. He'd then get right up and be terribly apologetic, and probably tip something over at the same time. One evening he actually tipped two bottles of red wine right over Princess Michael of Kent."

But once he was graduated[2] in 1940, his flying lessons were to be of no avail when he saw his brothers and classmates march away to war the following year. Rejected because of his impaired vision, but accustomed to overriding the rules, he made a futile appeal to the Surgeon General's office. He was later to talk of how anguished and ashamed he was that he could not join the Air Corps.

Returning to New York but not to his home on Park Avenue and now determined to initiate a career in the musical theater, he moved into the Lambs Club on West 44th Street. If he could not actually write for Broadway, he could at least be near enough to see all the latest shows. Moreover, from that vantage point he could pretend to be a member of the theatrical fraternity by living among those who actually were.

[2] Although Lerner finished his course of study with passing grades, his diploma from Harvard never arrived—nor did he ever inquire why. A university rule denied a diploma to any student whose campus bill was unpaid. Alan owed $80 for tutoring he deemed unsatisfactory. Sixteen years later, a week after the opening of My Fair Lady, his diploma arrived, quite uninvited.

Hammerstein

1933–1939

I N LONDON, *MUSIC IN THE AIR*, although tepidly received by critics, went on to become a popular favorite, while the two new shows Oscar worked on, *Ball at the Savoy*, which he had adapted from a German success, supplying only lyrics, and *Three Sisters*, an original with music by Kern, were failures.

Ball at the Savoy was a silly piece, not unlike *Die Fledermaus*, about the imagined infidelities of a French marquis and marquise during a masked ball. Even with the lavish production it was given, it is difficult to understand what Oscar saw in the story that would make him accept its adaptation.

James Agate impaled the musical on his pen in the *London Sunday Times*, writing "the scenery was pre-anybody with any kind of taste. . . . the music, except for the patter numbers, was exactly what we have been hearing for years in scores of other musical comedies . . . the plot bored to tears even our grandmothers . . . the dialogue had not a line of wit from beginning to end."

When *Three Sisters* opened in the spring of 1934, Agate was even more venomous. This time he aimed his barbs directly at Hammerstein. Noting that Britain's premier theater had been occupied exclusively with Hammerstein's works since 1926,[1] he asked, "How long is Drury Lane to be the asylum for American inanity?" Lauding only Gladys Calthrop's scenery and Kern's score, he added, "Probably there is no other country in the world in which visual and aural embroidery of such delicacy is tacked on to calico so coarse." Then, taking a final swipe at Oscar's long tenure in what had become a strongly chauvinistic Britain, he concluded: "Mr. Hammerstein is no Shakespeare."

---

[1] 1926's long running *Rose-Marie* had been followed by *The Desert Song*, *Show Boat*, a revival of *Rose-Marie*, and *The New Moon*.

Agate was telling Oscar nothing new, for the librettist had felt insecure about this musical and had begged Otto Harbach to come in with him in a letter he wrote before coming to London.

> . . . I can't go ahead with the production of *The Three Sisters* with Jerry. I sincerely believe that with comparatively little work it can be made into a great show. Would you be interested in taking it over and splitting 50–50 on author's rights? . . . I have told Jerry that I'd like you to adopt the child and bring it up with the breeding that only you can and he is very enthusiastic about the idea. I'll see that you get the script and hope you like it, and even if you don't I'll always be your greatest admirer, most grateful pupil and dear friend.

But Harbach turned the project down, and *Three Sisters*, a British story of siblings, one of whom loves an Irish constable, another an aristocrat, and the third a busker, was patently phony and annoyingly pretentious besides. Kern turned out a few buoyant melodies, and Oscar set them professionally if not inspiredly: "Hand in Hand," tenderly nostalgic (We'll grow old and walk together / And smiles will sweeten our tears, / While we gently talk together / Of bygone beautiful years); "A Funny Old House," a sort of Irish recitative; and "I Won't Dance."[2] Best among them all was a duet, "Lonely Feet," whose title words seem mismatched at first, but once one gets used to them, the song seems to possess a naïve bumpkin-like charm.[3] The opening and closing lyrics are given below:

Eustace:
Lonely feet,
While others are gliding by.
Lonely feet,
Waiting to dance.
Lonely arms,
All eager to hold a waist
Well embraced,
Lonely arms
Wait for a dance . . .
Oh, to feel
The thrill of a joy that's new
Feeling two
Other feet
Stepping on mine.

Tiny:
Lonely feet,
While others are gliding by,
Lonely feet,
Waiting to dance.
Lonely waist,
Intended for arms to hold,
Lonely waist,
Unembraced,
Waits for a dance . . .
Oh, to feel
The thrill of a joy that's new
Feeling two
Other feet
Stepping on mine.

[2] The song, revised by Kern and given a new lyric by Dorothy Fields, was sung by Fred Astaire and Ginger Rogers in the film version of *Roberta*.

[3] After *Three Sisters* concluded its brief London run, it was decided to use the song in the film version of *Sweet Adeline*, which was made the following year.

When *Three Sisters* folded in June 1934 after only 45 performances, Oscar realized he did not have his finger on the British pulse any more than Broadway's. His dreams of offering his children and Dorothy the experience of living abroad had not paid off and would have to be put on hold. Financial security, even when it was unrealistic, had always been inordinately important in his life scheme, and once again it took over.

Since Broadway had shuttered so many of its theaters, Oscar asked Kern, "Where next?" Kern's terse reply was, "Hollywood. For good." Since Kern and Romberg, the only two composers with whom he had been able to turn out recent hits, would be living in that land of perpetual sunshine, Oscar quickly negotiated a contract with Metro-Goldwyn-Mayer, rented a house at the top of La Brea Terrace, and moved everybody to Hollywood.

Metro was fast becoming the preeminent studio for production of musicals. They had Lawrence Tibbett, Grace Moore, Jeanette MacDonald, Nelson Eddy, Myrna Loy, Sophie Tucker, Evelyn Laye, Charles Butterworth, Rosalind Russell, and Maurice Chevalier under contract and soon were to add Mickey Rooney and Judy Garland. Besides the chance to work with this choice galaxy of stars, Oscar's agreement brought him $2,500 a week to write a film with Romberg. It was not a "per picture" contract, but an open-ended one, implying that he could take his time to turn out a quality product.

For Oscar's first assignment, he and Kern wrote a blues song built around the title of a film, *Reckless*, that featured Jean Harlow and Alan Jones. For his next film, an operetta called *The Night Is Young*, Oscar only contributed the lyrics to eight songs. With his collaborator Romberg writing his swirling melodies, the movie well deserved *Life's* quip: ". . . a preposteroperetta with a couple of good tunes and a plot that is as stale as they come."

Of the "good tunes" cited by *Life*, one was to emerge and rise to the top of all popularity charts and become an enduring standard. Its title, "When I Grow Too Old To Dream," suggested itself to Oscar immediately when he began working on Romberg's melody (see example below).

But once the lyric was written, he did not know what it meant and felt the public, too, would find it obscure. "Is anyone ever too old to dream?" he asked himself, but could find no answer. For days he tried to come up with another opening line, but nothing satisfied him as well. At last, when he had completed the remaining seven lines of this short refrain, he concluded that the rest of the song with its nostalgia of separated lovers remembering past kisses held well together.

When I grow too old to dream,
I'll have you to remember,
When I grow too old to dream,
Your love will live in my heart.

So kiss me, my sweet,
And so, let us part.
And when I grow too old to dream
That kiss will live in my heart.

Oscar soon rationalized that Romberg's soothing music compelled the listener to accept the title phrase as a lulling package, assuaging all their questions. But this writer finds perhaps another reason. Feeling "too old to dream" is only possible when one has lost heart in the future or is afraid to daydream of sexual or career fantasies. Certainly, spending most of his creative life working with far older collaborators, facing his own fortieth birthday, the traditional time for the male menopause, and selling out to the Hollywood moguls would make anyone feel too old to think up a new dream.

Before *The Night Is Young* was released, Oscar had decided to give the royalties from "When I Grow Too Old To Dream" to Dorothy. The song would soon be crooned on every radio in America, and now Oscar, feeling his oats, informed MGM that he and Jerry Kern, his collaborator this time, were not about to write their next picture in the orthodox way. "I know that before dialogue is written," he wrote to production chief, Sam Katz, "there is usually a treatment concocted by a producer and two or three writers in constant conferences—an elaborate scenario which contains all sorts of details, devices and indeed everything but dialogue and a soul."

With that they submitted a libretto as it would have been written for Broadway and five songs for a film to be called *Champagne and Orchids*. Two of the songs are worth noting. "Dance Like a Fool," while another of the how-cruel-show-business-is genre and pretentiously labeled by Kern as "très sauvage," has some cogent things to say and says them well: "Men who admire / Your one or two tricks / Till they desire / A girl with new tricks. / That's all they want, / If that's all you want, / Fool, go on and dance!" The other, "When I've Got the Moon," contains a rather philosophical lyric exposing the country's malcontentedness. The naturalness of speech, the paucity of rhyme and the simplicity of thought were harbingers of things to come in the Rodgers-Hammerstein era. The verse and part of the refrain of "When I've Got the Moon" are printed below:

What's my future?
Where am I going
Don't ask me . . .

When I've got the moon
I'm wishing for the sun.

When I'm sitting in the sun
I'm wishing for the moon.
When I've got no job
I'm as blue as I can be
When I've got some work to do
I am longing to be free . . .

But in spite of the go ahead signal that the script and score of *Champagne and Orchids* were totally acceptable and the casting of Jeanette MacDonald, Nelson Eddy, Wallace Beery, Clifton Webb, and Constance Collier, the picture was postponed and eventually never made. Worse, Oscar's Metro contract was not renewed.

Undaunted by this failure and his previous one at Warner Brothers, he turned to Paramount, and before he entrained with his family back to the East Coast to write yet another operetta with Romberg, he secured a contract that not only guaranteed him $150,000 but *allowed* the studio to invest heavily in his future Broadway ventures. With shows closing every day, 1935 was a time when angels were rarer than virgins on Broadway. Having Paramount in his pocket eager to support his product on both coasts placed Oscar in the driver's seat.

Now all he needed was the product.

May Wine, which opened on December 5, was certainly not a *premier cru*. Oscar's contribution had been merely to supply lyrics to Romberg's eleven musical numbers, and though Rommy tried to be contemporary in this tale of Vienna featuring a Freud look-alike, the result was mostly *alt Wien*. However, in one song they created a masterpiece of free sexuality that seems quite alien to any of their previous work. The perennial cabaret favorite "Just Once Around the Clock" is more appropriate to the sleep-around '60s than the repressive '30s in its advocacy of a one-night stand, here only twelve hours.

This is Hammerstein (and Romberg for that matter) in a Cole Porter mood and helps counterbalance the slurs of saccharinity that have so often been directed at them. Oscar actually outdoes Porter, who always needed to soften the blow by adding romantic reasons for the affair. From the verse with its "nothing to regret," the lyric is cold as ice. There is not even a hint of romance in this sexual encounter. Romberg's melody with short gasped phrases must have inspired Hammerstein to write this untypical but unforgettable line, "Just as we met with the coming of the moonlight, we will part with the rising of the sun."

VERSE
Nothing to regret
Tomorrow.
Nothing to regret.
You'll find me
Making no demand or claim upon you.
Kiss me and forget,
Kiss me and forget.
Kiss me while you may,
Tomorrow
Is another day
So while you're
Glowing with the sweet warm flame upon you,

Let it burn away,
Let its life be short
And gay.

REFRAIN I
Just once around the clock
And then goodbye, dear,
Love likes to fly by night
So let it fly, dear.
Love likes to fly by night
So let it fly, dear.
And when it's over
Bid me goodbye, dear.

Just as we met with the coming of the moon-
    light,
We'll part with the rising of the sun.

REFRAIN 2
Just once around the clock
As happy strangers

Who seek the joys of life
Without its dangers.
Just once around the clock
Let beauty lead you,
And when it's over,
Darling, God speed you.
Just as we met with the coming of the moon-
    light,
We'll part with the rising of the sun.

Only *Time Magazine* mentioned the song, but grudgingly added, "Lyricist Hammerstein sets no record for originality." Yet *May Wine*, with a hefty invest- ment from Paramount Pictures, was able to stay afloat for six months in a season which brought forth a meager dozen musicals, only one of which, *On Your Toes*, realized a substantial profit. By the season's close, the critic's darling *Porgy and Bess* had chalked up a money-losing run of just 124 performances, Billy Rose's extravaganza *Jumbo* operated at a loss, and even Cole Porter's much heralded *Jubilee* had run out of customers. Paramount was obviously Oscar's savior, and since there was no money to be made on Broadway, he packed up his family and once more headed back west.

Dorothy bought their Hollywood house with the money she had amassed from owning the copyright of the tremendously popular "When I Grow Too Old To Dream." Having a unique talent for decoration, she was able to deco- rate this large, comfortable, French-provincial home and yet make it cozy and comfortable. So adept was she at decorating that when others of the Hollywood community saw her work they began to ask her advice and for her help. Soon she had rented houses and decorated them for rental and eventually opened a shop on Wilshire Boulevard.

The Hammersteins were to remain in Hollywood for three years, during which time Oscar was to be involved in three films for Paramount and one film each for Universal, Columbia, Metro, and RKO. Few of the group had even a modicum of originality, and, aside from his adaptation of his stage success, *Show Boat*, none of them made money.

At first Paramount stood doggedly behind him. Before Oscar could even begin to work on the first, *Give Us This Night*, a film for which he supplied lyrics to Erich Wolfgang Korngold's operatic score, the studio granted him a leave of absence to go to Universal and write the scenario and additional lyrics required for the film version of *Show Boat*. After a long search Alan Jones was found and made an ideal Ravenal, while Paul Robeson, the original choice for the role, was cast as Joe. There never had been any doubt that Irene Dunne, who had played the role on tour before she came to Hollywood, would be cast as Magnolia.

Although he stuck closely to the plot as it had been done on stage, in writing the screenplay Oscar opened the action into cinematic terms. A typical change was a new love scene for Magnolia and Gay in which they go to their respective rooms and remove their makeup and costumes. Magnolia hangs her stockings on the line outside her window while Gay, hooking a stocking with his cane and bringing it in through his window, sings "I Have the Room Above Her," a new duet with the camera shifting back and forth that would hardly have worked on stage.[4]

The film, released the end of April 1936, was a critical and popular success and helped to temper some of the brickbats that Oscar had taken just a few weeks before when *Give Us This Night* opened to such devastating reviews.

Of all Oscar's further film credits, which included co-writing the screenplay for *Swing High, Swing Low,* writing screenplay and lyrics for *High, Wide and Handsome,* co-writing lyrics for *The Lady Objects,* lyricizing Johann Strauss II's music for *The Great Waltz,* and co-adapting the scenario of *The Story of Vernon and Irene Castle,* only *High, Wide and Handsome* with a lovely score by Kern and *The Great Waltz* deserve special mention.

For *High, Wide and Handsome,* Paramount's Adolph Zukor had spared no expense in attempting to match Metro's class and way with musicals. Oscar wrote a fine screenplay about the birth and growing pangs of the oil industry. The story concerns Peter, a farmer in the mid-1800s who befriends Sally, an entertainer in a medicine show. They fall in love and happily are drenched in oil on their wedding day. Oil wells soon spring up all over the area and Peter, organizing the group of farmers, decides to fight the railroads who are charging ruinous freight rates to bring the oil to the refineries. Peter inaugurates a pipeline soon destroyed by the adversaries. But despite all the railroadmen's efforts to stop them, the farmers prevail.

The story had an importance and realistic approach rarely seen in singing movies of its day and because of that did not do very well at the box office. Perhaps the fault lay in its very conception, which made it fall between the stools of a macho western and a story with a predetermined philosophy of the individual defeating big business. This, coupled with operetta songs, especially when sung by the operatic soprano of Irene Dunne, did not seem to go down too well with the American public. Watching man's man Randolph Scott covered with oil from his newly discovered gusher is thrilling, but willowy Irene Dunne rolling in "black gold" is more than miscasting—it became sacrilege. Some of the fault is attributable to director Rouben Mamoulian, who placed "The Folks Who Live on the Hill," written, as Hammerstein's unused verse attests, to be sung by a man, into the light realm of Dunne's voice.

Of the six period songs Oscar and Jerry wrote, two, "Can I Forget You" and

---

[4] Hal Prince, the stagecraft wizard, had a special set built so he could include this charming song in the *Show Boat* revival that hit Broadway in 1994.

"The Folks Who Live on the Hill," have achieved standard status. The latter seems to presage the mood Hammerstein was to capture completely some years later in *Oklahoma!*

This is normally known as "charm." But here, the charm is uplifting without becoming preachy. The message is to have a house of one's own, not in a dreary valley; kids to raise to maturity; and a love affair that continues into old age. In some twenty lines the lyric works perfectly for the mid-nineteenth century of the film while it encapsulates the lifetime dreams of every American couple emerging from the Depression. Closer to home, it mirrors Oscar's own sentiments concerning his future with Dorothy, a dream he was to fulfill when he purchased Highland Farms some years later. The lyric also boasts two of my favorite inner rhymes with ". . . adding a thing or two, a wing or two," and "our veranda will command a . . .":

VERSE
Many men with lofty aims
Strive for lofty goals
Others play at smaller games,
Being simpler souls.
I am of the latter brand;
All I want to do
Is to find a spot of land
And live there with you. . .

REFRAIN
Someday
We'll build a home on a hilltop high,
You and I,
Shiny and new,
A cottage that two can fill.
And we'll be pleased to be called
"The folks who live on the hill."

Someday
We may be adding a thing or two,
A wing or two,
We will make changes as any fam'ly will.
But we will always be called
"The folks who live on the hill."
Our veranda will command a view of meadows green,
The sort of view that seems to want to be seen.
And when the kids grow up and leave us
We'll sit and look at the same old view,
Just we two.
Darby and Joan
Who used to be Jack and Jill,
The folks who like to be called
What they have always been called,
"The folks who live on the hill."

Once the rushes of *High, Wide and Handsome* were viewed by Hollywood mogul Adolph Zukor, it was decided that the picture was strong enough to stand a high-priced two-a-day reserved-seat showing in metropolitan centers. But business soon slackened, and the film was withdrawn. Now Zukor, who had been so enthusiastic on the night of the movie's premiere that he pumped Oscar's hand, raving, "The greatest picture we ever made!," walked right past Oscar, ignoring him. Hammerstein's contract was not renewed, and the studio invested no further money in any of his future musicals. Oscar moved to Columbia for one picture.

His only contribution there was "I'll Take Romance," the lovely title song written for a Grace Moore movie. This is one of the few individual songs, not part of a score, that Hammerstein ever wrote. In its lyric Oscar used a concept of loving—"While my heart is young and eager and gay / I'll give my heart away"—with which he would always be identified. He was to expand it in "Sixteen Going On Seventeen" from *The Sound of Music* when he wrote: "Love in your heart wasn't put there to stay / Love isn't love 'til you give it away." The song from which the original concept came, with Ben Oakland's buoyant melody, has since become a standard.

In September 1937 Oscar moved back to Metro for *The Great Waltz*, a big-budget musical that was not released until a year later. Although *The New York Times* complained that "no studio could make so big a picture out of so small a script," and the film is generally dismissed as "sentimental claptrap," this writer considers it one of the most lavish musical films ever made. Beautifully cast with Fernand Gravet as Strauss and Miliza Korjus as a vamping opera singer, it also boasted Luise Rainer, who had already won two Academy Awards and nearly won a third for her heartfelt interpretation of Johann Strauss's sacrificing wife.

Oscar's poetic settings of "Only You," "Tales from the Vienna Woods," and "I'm in Love with Vienna" are less arty than any he wrote to the music of Friml or Romberg and have all deservedly become standard concert fare. Best of all are the seemingly effortless words to "One Day When We Were Young," which are printed below.

One day when we were young,
One beautiful morning in May,
You told me you loved me,
When we were young one day.
You told me you loved me
And held me close to your heart.

We laughed then, we cried then,
And swore we'd never part.
When songs of spring are sung
I'll think of that morning in May
You told me you loved me,
When we were young—one day.

Although he had been able to produce occasional brilliant lyrics, one cannot call Oscar's contributions to cinema history anything more than workaday. The greater part of his heart was always on Broadway, and throughout those California years he never ceased meeting, phoning, and writing with Harbach, Kern, and Romberg, working on shows or squirreling away ideas for future ones.

Many months after Oscar had finished his last screenplay,[5] for which he had originally borrowed the imaginative title *Castles in the Air* (the film was eventually released under the prosaic title of *The Story of Vernon and Irene Castle*), he and Dorothy decided to head back to New York. Dorothy was the first to mention that her husband was going nowhere in the film colony. "You know,

[5] No new lyrics were written because it was decided to use only old songs, but one new song, "Only When You're in My Arms," by Con Conrad, Bert Kalmar, and Harry Ruby, was interpolated into the score.

Ockie," she said, "it's better to wear out than to rust out—and we're rusting out."

And with that they made their plans to move back east. Dorothy magnanimously closed her by-now thriving decorating business, planning to reopen in New York, and Oscar took up where he had left off with Harbach and Kern on a Civil War musical that was to become a disturbing millstone around their necks for the next five years.

The story of *Gentlemen Unafraid,* which concerned West Point cadets from the South who are faced with the problem of fighting against their native states or rebelling against the government, was interesting, having its roots in the nostalgic Americana Kern, Harbach, and Oscar reveled in. But perhaps more fascinating were the production problems: the mountains of communications, rewrites, and recriminations that went on among longtime collaborators in mounting this last triple collaboration that suffered the indignity of never being given a Broadway performance.

Harbach writes: "Oscar brought a script to me from a short story by [Edward] Boykin, and Jerry Kern said he would do the score so we went to work and it was a big show. Then the War scare came along in 1939–40. It got to be very unpopular to mention a soldier, although this play itself was about the time the Civil War about to break out."

Unfortunately, when the show was ready to be tried out in St. Louis, illness prevented Otto's collaborators from being there. Kern had had a cerebral hemorrhage shortly before, and Oscar was recuperating from a severe bout with bronchial pneumonia. When expected bids from Broadway were not forthcoming, Harbach blamed "Dick Rodgers. . . for throwing cold water on the thing. He was a great friend of Dreyfus who we thought was going to put money in the show and Rodgers didn't like it at all. It was a new technique for here again we had numbers that were there because they had to be."

After the fiasco in St. Louis, Harbach went to work to rebuild the show, now called *Hayfoot, Strawfoot.* At the same time, after Oscar received the seventeenth version of the play he wrote to Kern: "Otto has as always improved his previous versions, but I am so certain that I am not interested in the music or the story that I finally wrote to him urging him to continue his efforts to have the play produced because of his faith in it but very clearly removing him of any further obligations to me as a collaborator. I've said in all fairness to him and me 'count me out.' "

When Kern, too, bowed out, Otto—who was then in his middle sixties—carried on alone for twenty years, patching and polishing the sinking musical almost until his last years. And although their friendship continued until Oscar's death, so ended the lyrico-dramatic collaboration of the two who had almost singlehandedly shaped the musical of their day.

But if they were no longer to work together, Oscar never forgot his debt to Harbach. By 1949 with *Oklahoma!*, *Carousel*, and *South Pacific* to his credit, when he published the collection he simply called *Lyrics*, he had absorbed the Harbachian principles of relevance—writing slowly, polishing, and, above all, integrating the songs within the libretto. In the introduction he wrote:

> I was born into the theatrical world with two gold spoons in my mouth. One was my uncle, Arthur Hammerstein, who took me into his producing organization after I left law school and gave me wise guidance. It was he, too, who supplied the second gold spoon. Otto Harbach. Otto Harbach, at my uncle's persuasion accepted me as a collaborator. It is true that this was not entirely a gesture of friendship on Otto's part. I know that he thought well of my talents at the time and saw promise in me. From the very start our relationship was that of two collaborators on an equal footing, although he was twenty years older than I and had written many successes while I had been going to school and college. His generosity in dividing credits and royalties equally with me was the least of his favors. Much more important were the things he taught me about writing for the theater. Otto is the best play analyst I have ever met. He is also a patient man and a born teacher. Like most young writers I had a great eagerness to get words down on paper. He taught me to think a long time before actually writing. He taught me never to stop work on anything if you can think of one small improvement to make. To speak of his non professional qualities as a civilized human being is completely irrelevant to these notes. Please, nevertheless, let me record that he is the kindest, most tolerant and wisest man I have ever met.

But although Oscar, observing the principles of his mentor, spent a full year polishing his two next theatrical fiascos, *Very Warm for May* and *Sunny River*, and although they both had integrated musical scores, Oscar omitted three caveats. Perhaps he had never learned them in his years with Harbach. Certainly he was to utilize them from the beginning of his collaboration with Rodgers: relevance, freshness, and, above all, a good love story.

Hammerstein

1939–1943

EXCEPT THAT *MUSIC IN THE AIR*, which was Kern's original idea, had been such a great success, at first glance it is hard to see why Oscar allowed himself to be persuaded to adapt the nostalgic story of talentless, stagestruck Elisabeth Marbury, who had produced Kern's intimate Princess Theatre shows, as his next project. Compound that with the fact that the irascible Max Gordon, who got Kern's and Hammerstein's dander up in previous shows, would be producing this one, and the duo's involvement in *Very Warm for May* becomes doubly incomprehensible.

But one must consider Oscar's eagerness to get back to the *original musical theater* and his overwhelming awareness of Kern's instincts for choosing successful properties. Hadn't Jerry brought him in on *Show Boat* and *Music in The Air*, his major successes of the last decade?

Emphasizing the words *original musical* is advised, for since his return from Hollywood, after the *Gentlemen Unafraid* debacle, Oscar had involved himself in writing, directing, and producing *Knights of Song*, a cobbled-together fiction based on the lives of Gilbert and Sullivan. Then, for the first time in his life, he took over the reins as producer of two straight plays: *Where Do We Go From Here*, a fraternity house comedy, and *Glorious Morning*, a valiant but preachy diatribe extolling religious freedom. None of these last three played more than a dozen performances. Adding *Gentlemen Unafraid*, which never even made it to Broadway, one can understand why Oscar, after four successive washouts and feeling his Hollywood years had dried out his sense of what was workable on stage, would latch onto Kern, who believed strongly in his talent and for whom he felt enormous admiration. Very near his nadir, Oscar would have accepted anything Kern suggested. Now he was especially intrigued by the show business background of *Very Warm for May*.

Together they had just toyed with and abandoned a gigantic musical based

Hammerstein watches admiringly as Jerome Kern plays songs from the score of Very Warm for May. *The show was scorned by the critics.* (R & H).

on the adventures of Marco Polo. Now in December 1938, after a visit to New York to immerse himself in theater, Kern wrote Oscar that the dearth of good musicals on Broadway[1] would turn almost anything they wrote into a hit.

> The situation is now past the pleasant, joshing point, and you and I both have got to bang through with something powerful for the stage. We have both been much too long off the boards. While we naturally do not want to roll up on Broadway with just a show, I find the prospect very encouraging. . . the target for a bull's eye has pretty good visibility.

---

[1] Although quite a bit more sassy than the typical Kern-Hammerstein musical, *Hellzapoppin, Leave It to Me,* and *The Boys from Syracuse* all opened in the autumn of 1938.

But the struggles of a young singer who runs off to join an avant-garde the-ater troupe to extricate her father from gangsters to whom he is heavily in-debted was already old-hat. What appealed to Oscar was the possibility for show-within-a-show that had been so successful in their former collaborations, *Show Boat*, *Sunny*, and *Music in the Air.*

Most of the blame for *Very Warm for May's* short (59 performances) stay on Broadway has been laid to puritanical Max Gordon's decision to delete all refer-ences to gangsters, to fire director Vincente Minnelli, and to tone down the effete pretensions of a stock character, the director of the summer theater May (yes, indeed, the heroine is named May) runs away to join.

But exasperating and dictatorial as the diminutive producer might have been after the Wilmington previews on October 20, 1939, he did not, as theater pundits have been claiming for decades, singlehandedly sink *Very Warm for May.* Oscar's book was at best discursive, at worst trivial. He later admitted to interviewer Arnold Michaelis that "the architecture was wrong. The story we had chosen and the building of that story was faulty and everything tumbled down around us."

His lyrics, except for the bittersweet "All in Fun," were either pedestrian or pretentious. Even Kern's music, replete with beguines and ballads, sounded like he was trying to be Cole Porter. Worse, except for "All the Things You Are," this much-heralded score contained not a single hit.[2]

For days Oscar worked to improve the last couplet of Kern's surprising mel-ody, which eventually became a hit and is perhaps Jerry's most artful and en-during composition. But Oscar's lyrics are less inspired, although he admitted that he tried without success to eliminate the ultimate hackneyed "divine" and "mine." While he was at it he might well have rethought the trite release and the entire A[3] section to eliminate the awkward "happy arms." The last half of the burthen[3] follows:

You are the angel glow	Someday my happy arms will hold you,
That lights a star.	And someday I'll know that moment divine
The dearest things I know	When all the things you are
Are what you are.	Are mine.

But the music and lyrics for "All in Fun" show that both men could rise to breathtaking heights. Using (then) contemporary terms like El Morocco, the

---

[2] The aforementioned *Leave It to Me* introduced Cole Porter's "My Heart Belongs to Daddy," "Tomorrow," "Most Gentlemen Don't Like Love," and "Get Out of Town" while Rodgers and Hart's *Boys from Syracuse* featured "Falling in Love with Love," "This Can't Be Love," and "Sing for Your Supper." All of these have since become standards, and most of them were hits.

[3] Kern used this parochial, antiquated, and rather pretentious word to mean chorus or refrain.

Stork Club, or "Cholly" Knickerbocker, Hammerstein sets up a black-and-white madcap movie romance that might star Carole Lombard and William Powell. Everything is a laugh to these "good-time Charlies" from the social register. Aware of the rumors that they might really be in love, the lyricist walks the tightrope of not corroborating or denying the gossip—that is, until the last devastating syllable.

VERSE
We are seen around New York,
El Morocco and the Stork,
And the other stay-up late cafés,
I am on the town with you these days,
That's the way it stands.
Just a fellow and a girl,
We have had a little whirl,
And our feet have left the ground a bit—
We've played around a bit,
That's the way it stands.
For we are strictly good-time-Charlies
Who like to drink and dance around
And maybe kick romance around,
And that's the way it stands.

REFRAIN
All in fun,
This thing is all in fun.
When all is said and done
How far can it go?
Some cocktails, some orchids,
A show or two,
A line in a column
That links me with you.
Just for laughs
You're with me night and day
And so the dopes all say
That I'm that way 'bout you.
Here's the laugh,
And when I tell you,
This'll kill you . . .
What they say is true.

After Gordon's considerable revisions, the New York critics were only tepid to the Kern score, but truly devastating to Oscar's contributions. Even the moving "All in Fun" was trivialized in the revised libretto by being misplaced in a summer stock rehearsal. Brooks Atkinson led the naysayers, proclaiming "VERY WARM FOR MAY NOT SO HOT IN NOVEMBER" and calling Oscar's book "a singularly haphazard invention that throws the whole show out of focus and makes the appreciation of Mr. Kern's music almost a challenge." Others termed the show "labored," "routine," "exasperating," even "stupid." After the press did its hatchet job, only twenty customers showed up for the musical's second performance.

Kern was supportive, but nothing could help Oscar shrug off his depression, brought on by the worst reviews he had received in his lifetime, until Romberg, who had been urging Oscar for months to collaborate on another operetta, wrote his old friend. "I just want to tell you that I still love you," Rommy gushed in December, shortly before *Very Warm for May* shuttered permanently, "and I'll tell you one thing more. Out of one hundred outstanding authors, writing dramatic plays, I don't think you'll find three who have the knowledge, the feeling and the sentiment to know how to write a book to which music is supposed to be interwoven with the action. You are one of the few."

Romberg's letter brought about the desired collaboration, but that had to be deferred until Oscar completed a pageant he had contracted to write for the New York World's Fair. Still, by late 1940 Oscar had sketched an original story replete with southern dialect, which he called *New Orleans* and which was set in that city at the beginning of the nineteenth century.

It concerned a café singer and a socially prominent debutante who are in love with the same man. After the debutante tricks the hero into marriage, the café singer goes to Europe and (it could only happen in operettaland) returns as a world-famous opera singer. She is determined to wrest her man away from the socialite until she realizes how much the latter loves him. All this amorous struggle is futile, however, for the hero is killed fighting with Andrew Jackson in the War of 1812. As the play ends the two women commiserate by mourning his loss.

With the conflict in Europe now ever nearer, and after the collapse of *Gentlemen Unafraid*, Oscar should have realized how ill advised war and death were— even if removed by a century—as a subject for operetta. More currently, he might have remembered his own depression when Paris fell to the Nazis on June 14, 1940. He was unable to concentrate on *New Orleans* and realized he had to "get this off his chest."

In a rare departure from his usual collaboration with Kern, and an even rarer one from his own habit of creating songs for specific situations in shows, he wrote "The Last Time I Saw Paris" as a poem.

"First I started to think of Paris rather superficially; what it had been before the Germans had come in. . . . I thought of Mistinguette . . . and Chevalier in his straw hat. And then I started to think of the taxi cab horns and all the lighthearted gaiety and then the beauty of Paris and the romance."

Fearing the Paris he loved might never come back, he began to personify the city. "I began to write words about the last time I had seen her she was beautiful and all that I loved about her." Then he phoned this lament, a tender personification of the city he had fallen in love with when he spent time there in his youth, to Kern for musicalization.

Kern lost no time, and the nostalgic words coupled with an insouciant melody soon became a hit. Eventually the song was added to the MGM film *Lady Be Good* in 1941 and went on to win an Academy Award. But Oscar and Jerry, although pleased with the accolade, felt the song, an interpolation, should not have been eligible for the award. They were upset that another nominee, the Harold Arlen-Johnny Mercer song "Blues in the Night," a masterpiece actually written for the screen, had not won. So disturbed were they that they brought the matter before the Academy board and were instrumental in changing the rules which, since 1942, have made eligible only those songs *specifically* written for a particular movie.

Finally, after Oscar had made sixteen trips to the West Coast to work with Romberg, by the spring of 1941 *New Orleans* was ready for production. It was first seen at the open air theater in St. Louis where *Gentlemen Unafraid* had bombed three years prior. To make matters worse, the irascible Max Gordon, now broke, was the only producer interested in optioning the show, but he had to delay Broadway production for half a year so he could raise the money.

During that time Oscar worked to achieve naturalness in what looked like his last chance on Broadway, but Romberg's music was firmly implanted in the past. The struggle was almost palpable, with Oscar's lyrics reaching for the fresh, uncontorted speech which was to become his hallmark ever after; but his words could not pull their tunes into the 1940s. Of the two lyrics below, one might be a sketch for the King's philosophy a decade thence in *The King and I*, while the other is a young, uncomplicated love song.

THE BUTTERFLIES AND THE BEES

Observe the bee	If you want me
And see the way that he	To do the same for you
Persuades a lovely flower	You just imagine you're a bee
To give him honey free.	And do what he
	Would do.

MY GIRL AND I

VERSE

Please pardon me if I seem incoherent	A patch of earth and sky
My talk doesn't sound very sane.	Put by for us alone.
How can a man be concise and coherent	My girl and I,
When a man's got a girl on his brain?	A very vain, conceited pair are we—
	And we know why,
	For I have her

REFRAIN

My girl and I	And she has me!
Are in a world of our own,	

On December 4, 1941, *New Orleans*, now retitled *Sunny River*, finally gave its first of what would be only 36 Broadway performances. Critics echoed all the adjectives with which they had impaled *Very Warm for May*, adding "stuffy," "ponderous," and "hopelessly old-hat."

It was the worst time to open a war musical like *Sunny River*, for the only musicals making money were escape vehicles. Two mornings after the opening the newspaper headlines screamed the shocking news that Japan had attacked Pearl Harbor and destroyed a large part of the U.S. fleet.

Sunny River limped on into the new year, but Oscar's letter to Max Gordon shortly after the show closed shows the depth of his depression:

January 8, 1942

Dear Max:

   Thank you for your letters. I feel sure that you did everything that was humanly possible to give the show its chance to find a public, and it didn't. I don't believe there is one—certainly not in New York. Operetta is a dead pigeon and if it ever is revived it won't be by me.

   I have no plans at the moment and I don't feel like making any . . .

   We are in the midst of a terrific cold spell. It's so little above zero that it might as well be below. The snow in the country is beautiful, though, and if you were here I'd dig up a pair of galoshes for you and we'd tear off a few brisk miles.

Sincerely,
Oscar

The letter was written from Highland Farm, a 72–acre working farm in Bucks County, Pennsylvania, that Oscar and Dorothy had bought three years before. Oscar loved the country: walking, jogging, writing, or just sitting and contemplating. Moreover, it was obvious to him that America had to get into the war, and he was uncertain as to what would happen to the family, show business and whatever money he had managed to put by.

"Dad decided right then and there that we should get a farm so the family would have a place of refuge so that if we lost everything we could still grow our own food," his son William remembered.

We found a Pennsylvania farmhouse that had been built in 1818 and had a large dairy farm and a big chicken house and sixty acres. And dad bought it for $23,000.

   Well, things didn't turn out that badly and so the farm was bought as a homestead and it turned out to be a country club because Dorothy turned it into a highly decorated mansion and it was very beautiful; and people came for the weekends all the time and we never had to do the retrenching because, oddly enough after the war started, dad and Dick began working together and having their amazing success. So everything went in the opposite direction than what we had surmised it might go when the war began.

Dorothy was able to manage handily, especially now since she had reopened her decorating business. In one of her first moves, she had redecorated the master suite on the second floor to make a comfortable studio-cum-fireplace where Oscar could work uninterruptedly. It was there he went to seek solitude and lick his wounds after Sunny River, and there the idea of adapting Carmen into Carmen Jones came to him.

   He recalled that day in Hollywood back in 1934 when he had first discussed a cinematic version of the opera. Somehow the project had slipped through his fingers then, but now, although he might be down he would not be out if he could pull off this tour de force. At worst, this intellectual exercise would could fill his days until someone came forward with an offer. Certainly he couldn't approach Kern or Romberg so soon after his recent botches. Even Richard Rodgers, who had offered him a carrot six months ago when he suggested Hammerstein join him and Lorenz Hart in musicalizing Edna Ferber's *Saratoga Trunk*, was forced to take it out of reach when Ferber and Warner Brothers began "acting up."

But listening over and over to a La Scala recording of Bizet's *Carmen* brought a modicum of solace. Oscar realized as he proceeded with the work that for first time in his life his collaborators, Meilhac and Halévy, the original librettists who had adapted Prosper Mérimée's novel, were dead and no one (as Kern so often did) could naysay him. He plunged into the opera—amazingly—without deadline or thought of Broadway production.

Dropping the long sung recitatives written by Ernest Guiraud after Bizet's death which were inserted between musical numbers so *Carmen* could be played at the larger opera house[4] made the work suddenly seem like a play with music. Next, Oscar contemporized rather than reconstructed the plot. He created urgency by turning Seville's cigarette factory into a South Carolina parachute factory during the current war. Don José became Joe, a GI corporal who is seduced and talked into deserting by Carmen. With so many Americans fighting and dying in Europe, the plight of a soldier consumed by love and trapped by fate into going AWOL became real and poignant. Using the same parallels from the opera but moving them to contemporary Negro culture, Micaela becomes Cindy Lou while Escamilio the toreador translates to Husky Miller the prizefighter.

With characters thus intensified, Oscar began to translate the arias. He said he was not aiming for that "phony sob-in-the-throat trick of grand opera singers in emotional passages [but was] portraying two human beings in terrible trouble, two confused souls moving towards their destruction with every word they sing." One gets an idea of how well he succeeded throughout by looking at the words he supplied to the intense "Flower Song." This confessional sung by Joe exposes the sensitive core of a draftee whose passions are strong enough to lead him to murder.

Putting aside the question of the continuing validity of *Carmen Jones* and pinpointing this particular lyric, it seems far more than a translation from the Meilhac-Halévy libretto. As Alan Lerner did in *My Fair Lady*, the liberties taken with the source pay off in fleshing-out character. Here it takes us into the heart of an antihero's obsession. Joe is openly sexual where Bizet's Don José is repressed. In a place where the original *Carmen* libretto talks about "not

---

[4] Nineteenth-century Paris boasted two opera houses (it now has three, with the addition of the new Opéra Bastille.) But at that time the repertoire at the Opéra Comique was expected to have intermittent spoken dialogue, while at the grander Opéra all the works were through-sung.

believing in oneself," or "blasphemy," Oscar talks about his "craving" and desire. And all this is set in natural speech. Hammerstein is always aware of the passion in the French composer's score, and, master of technique that he is, helps the singer get it across to the audience by setting his climactic pitches on words like "I" and "are," open vowel sounds.

MEILHAC-HALEVEY	LITERAL TRANSLATION	HAMMERSTEIN
La fleur que tu m'avais jetée	The flower that you threw me	Dis flower dat you threw my way
Dans ma prison m'était restée	remained in my prison	Has been my fren' by night an' day.
Flétrie et sèche, cette fleur	faded and dry, this flower	I saw it fade an' lose its bloom,
Gardait toujours sa douce odeur	Still kept its sweet smell	But still it kept a sweet perfume.
Et pendant des heures entières	And for hours at a time	In my cell through ev'ry darkened hour,
Sur mes yeux, fermant mes paupières	on my eyes, closing my lids	On my lonely eyes lay dis flower
De cette odeur je m'enivrais	I became drunk with this fragrance.	An' so I'd sleep de whole night through
Et dans la nuit je te voyais!	And in the night I saw you	An' dream of you, an' dream of you.
Je me prenais à te maudire	I began to curse you,	Den I'd wake up, wid no one near me,
A te détester, à me dire:	To detest you, to tell myself:	An' talk fo' de jail walls to hear me—
Pourquoi faut-il	Why should it be	"She ain' de bes'
que le destin	that destiny	Dere all de same!
L'ait mise là	placed her there	Like all de res'
sur mon chemin.	in my path.	She jes' a dame."
Puis je m'accusais de blasphème,	Then I accused myself of blasphemy	Den I tol' myse'f I wuz ravin'
Et je ne sentais en moi même	And I no longer believed in myself.	Dere wuz jes' one t'ing I wuz cravin'—
Je ne sentais	I felt	It wuzn' food,
qu'un seul désir	only one desire,	It wnzn' dough,
Un seul désir	One desire alone,	I guess you know
un seul espoir	only one hope,	Dat it wuz you.
Te revoir ô Carmen,	To see you again, oh, Carmen,	I only saw you once . . .
Oui, te revoir!	yes, to see you again!	Once wouldn' do!
Car tu n'avais eu qu'à paraître	For you only needed to appear	I don' know anythin' about you
Qu'à jeter un regard sur moi,	And throw a glance at me	I don' know much about a shinin' star.
Pour t'emparer de tout mon être,	To enter my being,	Jus' know de worl' is dark widout you—
Ô ma Carmen!	Oh, my Carmen!	Dat's all I know . . .
Et j'étais une chose à toi!	And I belonged to you!	I only wan' you as you are . . .
Carmen, je t'aime.	Carmen, I love you.	Dat's how I love you.

All the other arias were translated with equal fluency and similar use of the vernacular. The sinuous "Habanera" became "Dat's Love" (Love's a baby dat grows up wild / An' he don' do what you want him to); the "Toreador Song" was contemporized as "Stan' up an' fight until you hear de bell / Stan' up an' fight like hell!").

Carmen Jones may have been created when Oscar's peers jeered, "Hammerstein? Why he can't write his hat!," but its adaptation, libretto, and lyrics show him to be at the height of his artistic powers. Gone are the inverted speech forms he so often formerly affected. Gone, too, are the forced rhymes. Perhaps because he was writing of ordinary black people he was able to recapture some of the naturalness that made his songs for blacks in *Show Boat* so artless.

The exciting Spanish dance number in Carmen Jones, *which Hammerstein titled "Beat Out Dat Rhythm on a Drum."* MOCNY.

With nothing to work on and nothing to do after he completed *Carmen Jones*, only a few days went by until his life was to change dramatically. One might call it karma, for there were many forces working on all sides that would bring about his next show, *Oklahoma!*, the inaugural collaboration of one of the greatest musical phenomena of the twentieth century.

The catalyst in all this was Theresa Helburn, co-director with Lawrence Langner of the Theatre Guild and a true theater visionary. The Guild had been the first to take a chance on the saucy songs and sketches by the fledgling team of Rodgers and Hart back in 1925. For many years Helburn had harbored a vision of a new type of play with music, as she wrote in her autobiography, *A Wayward Quest*,

> not operetta in the old sense, but a form in which the dramatic action, music and possibly ballet could be welded together into a compounded whole, each helping to tell the story in its own way. . . . I kept toying with various possibilities. One was to have *The Devil's Disciple* done with music.

I wrote in great excitement to Shaw about it and got a firm rejection: "My dear Tessie, after my experience with *The Chocolate Soldier,* nothing will ever induce me to allow any other play of mine to be degraded into an operetta or set to any music except its own."[5]

Not disheartened by Shaw's brush-off, Helburn persuaded the Guild in 1931 to produce Lynn Riggs's *Green Grow the Lilacs,* believing in this play from the beginning. Even then sensing its musical implications, she had suggested there be a few folks songs like "Skip to My Lou" and "Git Along Little Dogies" interspersed in the action. When the play was revived in 1940 for a short run at the Westport Playhouse in a staging by Gene Kelly, the choreographer kept the cowboy songs and added some dances, notably a square dance. Now it was coming closer to Theresa's dream.

But in the early forties, the Guild, after producing serious plays that a war-weary public was not interested in, was staggering on the verge of financial ruin, with admittedly only about thirty thousand dollars in the bank. Helburn decided "this was the moment for the fulfillment of my dream, the production of a totally new kind of play with music. I was also in the position of finding investors who would put up the estimated ninety to one hundred thousand dollars."[6]

Helburn's colleagues at the Guild called her attempted resurrection of the flop *Green Grow the Lilacs* "Helburn's Folly," saying that one didn't offer possible investors a play that hadn't done well in the past. "Musicals," everyone added, "don't have murders in the second act." But Helburn realized not altogether altruistically that since her organization still held the option on the play, if by some chance it should become a hit, that would spell salvation for the Guild. She marched straight to the top—asking Richard Rodgers, whose recent *Pal Joey* had been a *succès d'estime* and whose *By Jupiter* was now the reigning hit on Broadway, to come up to Westport to have a look.

"It was an amazing coincidence," Oscar admitted, explaining his side of the happenstance that was eventually to become *Oklahoma!* "I was in California not to write a picture but to witness a production of *Music in the Air* in Los Angeles and make a few changes in it to accommodate a big theater."

While there, he talked to Jerome Kern about *Green Grow the Lilacs.* "I told him I thought it would make a good musical play, and I even read some of it to him. His response was enthusiastic in a way. He liked the lyric quality of

---

[5] Shaw was to change his mind two decades later when he allowed Gabriel Pascal to offer his *Pygmalion* first to Rodgers and Hammerstein, then eventually to Lerner and Loewe.

[6] In contrast to today's multimillion-dollar musicals, *Oklahoma!* was brought to Broadway for a total cost of $87,000.

Lynn Riggs' play[7] but he thought that the third act was pretty hopeless—and as a matter of fact, it was. That was the reason why the play failed."

Back in New York a week later, Richard Rodgers called, and over luncheon told Oscar the Theatre Guild had asked him to do a play, *Green Grow the Lilacs,* and he thought that Oscar "ought to read it because he would like to do it with me if—they had suggested it too—I liked it. Well, I said I didn't have to read it; I have read it three or four times and I would love to do it."

Oscar's reaction to the invitation was not entirely unexpected. Everyone in theatrical circles knew Rodgers and Hart's course was not running smoothly. There had been that "feeler" back a year ago when Rodgers invited him to join Hart and himself in a version of *Saratoga Trunk.* More recently, when Dick was in Philadelphia at a tryout of a play he had invested in, he paid a social call on Oscar at nearby Highland Farm. In the course of the afternoon the men had a heart-to-heart about Dick's troubles with his longtime collaborator and his fears that Hart would not be able to finish the lyrics on a play to which they were committed, Ludwig Bemelmans's *Hotel Splendide.* Again he asked Oscar to join them.

Oscar's response to his friend's dilemma was to suggest that Dick keep working with Hart as long as he was able to function. "But," he said, "if the time ever comes when Larry cannot function, call me. I'll be there."

The realization that the collaboration was crumbling had come to Dick even before *By Jupiter.* It was perhaps foolish for the two of them to take on the heavy task of writing its libretto *and* score in the face of Hart's serious drinking problem. The year before, 1941, was the first year Rodgers and Hart didn't have a new show on Broadway, because Hart would often disappear for days at a time.

One must realize that Hart, who was barely five feet tall in elevator shoes, was a rather unattractive homosexual whose lifestyle was mercurial. Rodgers, on the other hand, who lived an extremely ordered existence, was always punctual and would hit the ceiling when an unkempt, hung-over Hart returned. He tried withholding his money and confiscating Hart's liquor bottles. He even arranged for Larry to see a psychiatrist, but nothing worked.

As the opening date of *By Jupiter* neared, Rodgers's worst fears were realized. Weeks had gone by without hearing from Larry. Now he insisted that Hart check into Doctors Hospital in New York, ostensibly to oversee his drinking problem but in reality so that Rodgers could keep an eye on him.

Doctors Hospital was the society spa of its day, where celebrities checked in

---

[7] So did George Gershwin. He was so enchanted by Lynn Riggs's poetic style that in 1936 he commissioned Riggs to write a libretto, *The Light of Leamy,* for his opera after *Porgy and Bess.* Tragically, the composer died the following year, while the project was in its infancy.

and out at will, ordering luxurious meals sent in from Le Pavilion or "21." Although it had a superb staff of doctors, it was more famous for its guests. Rodgers rented the room next door to Hart's. He had the piano Cole Porter had donated to the hospital, which was kept in the doctors' lounge, put into that room and was prepared to meet his deadlines. With Rodgers hovering over Hart, spurring him to write more and more verses, they turned out a light-hearted sparkling score, even though Hart occasionally was able to sneak liquor into his room. Although they ignored it they both knew that the libretto was not much above the level of a college smoker.

They had written twenty-five Broadway musicals and three special ones for London as well as nine movies. They had even done a straight play script together. Now Rodgers knew he had come to the end of the road that he and Hart traveled together for twenty-four years, although ever after Rodgers claimed the anxiety of *By Jupiter*'s creation was too much for him to risk further collaborations.

Listening to this score one notices there is more to it. Rodgers's music tries constantly to break out of the 32-bar refrain mold as though it had reached a dead end in what he could do with the musical comedy form. He made no secret of his overwhelming interest in Gershwin's folk opera *Porgy and Bess* of a few years before and that he wanted to write that naturalistic kind of folk play.

After Oscar and Dick's fateful luncheon meeting, the newspapers prematurely announced that Rodgers, Hammerstein, and Hart would be working on Lynn Riggs's *Green Grow the Lilacs*, but although Dick had proposed to Larry that the trio join for the project, Larry responded flatly that he didn't see any "fun" in the Western. He needed a rest, he declared, and was going off to Mexico. They were in Max Dreyfus's office at Chappell, Rodgers recounts in his autobiography, *Musical Stages*, when Rodgers laid down an ultimatum:

> "This show means a lot to me," I told him. "If you walk out on me now, I'm going to do it with someone else."
> "Anyone in mind?"
> "Yes. Oscar Hammerstein."
> Even the realization that I wasn't bluffing, that I actually had someone else waiting to take over, couldn't shake him. All that Larry said was, "Well, you couldn't pick a better man." Then for the first time he looked me in the eyes. "You know, Dick," he said, "I've really never understood why you've put up with me all these years. It's been crazy. The best thing for you to do is forget about me."

Hart could not resist a parting shot as he walked out the door. "I really don't think *Green Grow the Lilacs* can be turned into a good musical," he said over his shoulder. "I think you're making a mistake."

Making the offer has to have been Rodgers's finest moment, for in his heart of hearts he must have known Hart was too far gone, actually too ill, too alcoholic, too self-destructive to tackle such a big project.

William Hammerstein feels that had Hart "not been drinking himself to death Rodgers would have remained on course with him and they would have gone on writing the same kinds of shows." This writer disagrees, feeling that Rodgers himself must have sensed the untapped creative springs that make his oeuvre with Hammerstein so different from his work with Hart.

So it was dissolved, the twenty-four-year partnership that seems to have brought Hart his only happiness. Many have said the enormous achievement that became *Oklahoma!* broke his heart, for he died not long after its opening. How lucky for Rodgers and posterity that Hammerstein was waiting in the wings.

Although Rodgers himself was a competent lyricist[8] as well as a rapid-fire composer, he respected Hammerstein's literary prowess, his sense of poetry and craftsmanship. From the beginning of their collaboration it was decided that he would set his music to Oscar's lyrics. This was contrary to the way he had worked with Hart, but a far more preferable way of achieving an integrated score. "I was brought up that way, to make my music fit the text," he was to say. He had always wished for a collaboration in which he would be given words to musicalize but knew that unless he had a completed song ready to show Hart, Larry might not show up at an appointment. As it was, he was generally several hours late. Rodgers feared that were he to ask Hart to bring in a sketch or a completed lyric, he might never show up at all.

Oscar's promptness was a great relief for Dick, a harbinger of a smooth relationship that was to make all their creative work and even the shows they produced run like clockwork. "When he would say I'll meet you at two-thirty, he was *there* at two-thirty," Rodgers beamed. "That had never happened to me before."

From a theatrical team they were suddenly transformed into an organization. Now working clearheadedly, the new partners became deeply involved in so much preliminary discussion of plot, placement, and motivation that Rodgers felt if he took his cue from Oscar's words, his music could more easily say what the play was striving for.

And from this initial partnership Oscar sensed he could take his time with lyrics and libretto. He polished "Oh, What a Beautiful Mornin' " for three weeks until it gleamed. For a full week he agonized as to whether or not to use

[8] When Hart was on a bender, Rodgers had often supplied additional lyrics to their songs. After Hammerstein's death, Rodgers wrote words *and* music for his *No Strings*.

Creating Oklahoma! *Rouben Mamoulian, Hammerstein, Theresa Helburn, and Richard Rodgers in the planning stage.* MOCNY.

the "Oh," but since the lyric he had written was clearly in triple meter, eliminating the syllable meant opening the song with a rest, making it begin rather torchily, like "The Man I Love," rather than naïvely. And after all, he knew that the very opening is the most important moment in any musical—even more so in this case, as it was to be their first show. Borrowing his clue for the words from the poetic stage direction at curtain's rise, he found Lynn Riggs had written:

> It is a radiant summer morning several years ago, the kind of morning which, enveloping the shapes of earth—men, cattle in the meadow, blades of the young corn, streams—makes them seem to exist now for the first time, their images giving off a visible golden emanation that is partly true and partly a trick of the imagination, focusing to keep alive a loveliness that may pass away.

"When Oscar handed me the words of our first song together," Rodgers remembered "I was a little sick with joy because it was so lovely and so right."

After studying the lyric Rodgers was to say, "I had a feeling I must say something very important musically. It isn't often a composer gets a chance at lyrics as perfect as 'Oh, what a beautiful morning' or 'the corn is as high as an elephant's[9] eye.' "

And he did. A new Rodgers is apparent in the opening line of the refrain of this very song. On the whole, Rodgers's music written to Hammerstein lyrics is far different from what he turned out with Hart. Tom Shepard feels that "when Rodgers wrote with Hammerstein a lot of the rhythm of the melody disappeared. There was a lot more straight on the beat stuff whether it's 'chicks and ducks and geese' or 'oh, what a beautiful.' Then when you compare that with a song like 'Mountain Greenery,' you notice a rhythm, not exactly a jazz rhythm, but a feeling for syncopation in Rodgers's melody. When he worked Hammerstein, you got into songs that were more 'on the beat.' It didn't make it worse music, in fact, it was gorgeous stuff, but it *is* different."

"Oh, What a Beautiful Mornin' " ushered in a new period, a new sensitivity that would have been impossible in his reverse way of working with Hart, whose lyrics, while often overwhelmingly sensitive as a whole, were rarely *sensitive to the pitches of the song.* For example, Hammerstein's lyric-poetry aside, Rodgers's use of the D-flat, an accidental, on the syllable "morn-" not only turns a typical Country & Western song into an interesting show tune, but gives a feeling of the day not having been quite started yet. The flatted seventh of the key wants urgently to be resolved.

When this is contrasted, the next time the phrase rolls around in the twelfth bar with a D-natural on "feel-" the resolution suggests the sun has fully risen in the sky, and the day (and even the play) can begin (see Example A below).

Example A

---

[9] Originally Oscar wrote "as high as a cow pony's eye," but walking through the cornfield of his neighbor at Highland Farm in August 1942, he realized that was too low. "Elephant's eye" suggests some of the braggadocio of cowpuncher Curly and it sings infinitely better.

With the opening out of the way, Hammerstein went to work on the book. His main hurdle was to eliminate Riggs's unresolved third act wherein Curly, having killed the villain, Jud,[10] is in custody and allowed to spend his wedding night with Laurey before being taken off to jail.

"I think the author wrote a beautiful lyric play for the first two acts, and in the third act he went way off into some kind of Freudian darkness which disappointed the audience." Oscar explained. "We had to invent a new finish and what was the third act of this play is all covered in about five minutes of our second act." Here Hammerstein brings out his favorite device: he has Curly kill Jud with his own knife and in self-defense, just as he wiggled out of a tight spot in *Rose-Marie, Show Boat*, and in several other plays. But somehow here, the device works.

The other criticism frequently leveled at *Oklahoma!*'s libretto is Hammerstein's building up the creaky dramatic device of subplot. But again in this case, Ado Annie's dilemma—the story Oscar created of her involvement with both Will Parker and Ali Hakim, which goodnaturedly parallels the relationship of Laurey, Curly, and Jud—seems inspired.

"Ado Annie was a very shy girl in the play. I made her the opposite. I made her a girl who didn't see why she should ever say 'no,' who was very primitive about it and not at all sophisticated."[11]

With the libretto roughed out, Oscar went back to the songs. After "Oh, What a Beautiful Mornin,' " which he decided to introduce in the libretto as an unconventional *a capella*—a fact that raised many eyebrows—he tackled "The Surrey with the Fringe on Top." Again, he took his cue from the play. Printed below are, first, several speeches Riggs gives Curly to intrigue Laurey into going to the dance with him; next, the verse and first refrain Hammerstein wrote for the song; and finally (Example A), a quote from the country barnyard, hen-pecking melody, a perfect onomatopoeia for making these fowl scurry away from the wheels of Curly's magnificent rig.

[10] Riggs had named his character "Jeeter," but "Jud" was chosen to avoid the connotation of Jeeter Lester, the unpleasant grandpaw in the long-running *Tobacco Road*.

[11] William Hammerstein finds Ado Annie in Lynn Riggs's play "a kind of moronic dim-wit, and she becomes the butt of jokes. My father took her and made her a real buddy of Laurey's, and I thought that was a marvelous use of character and comedy. I was rehearsing a girl in the part who had never seen a performance of *Oklahoma!* and on the first day of rehearsals she began singing 'I'm Just a Girl Who Cain't Say "No" ' in a wanton way, like a girl who just wanted to climb into bed with everybody. I had to convince her that the humor is derived from Ado Annie's innocence, it had to be the comedy of charm. Just as in the finale when she comes out of the barn with Will Parker and she's got some hay stuck to the back of her dress, it's charming. She's discovered something she didn't know before."

A bran' new surrey with fringe on the top four inches long—and *yeller!* and two white horses a-rarin and faunchin' to go! You'd shore ride like a queen settin up in that carriage! Feel like you had a gold crown on your head, 'th diamonds in it big as goose eggs . . . And this yere rig has got four fine side-curtains, case of a rain. And isinglass winders to look out of! And a red and green lamp set on the dashboard, winkin' like a lightnin' bug! Don't you wish they was sich a rig though? Nen you could go to the party and do a hoe-down until mornin' 'f you was a mind to. Nen drive home the sun a peekin' at you over the ridge, purty and fine.

VERSE

When I take you out tonight with me.
Honey here's the way it's goin' to be,
You will sit behind a team of snow white horses
In the slickest rig you ever see!

REFRAIN

Chicks and ducks and geese better scurry
When I take you out in the surrey,
When I take you out in the surrey with the fringe on top!
Watch that fringe and see how it flutters

When I drive them high steppin' strutters—
Nosey-pokes 'll peek thru' their shutters and their eyes will pop!
The wheels are yeller, the upholstery's brown,
The dashboard's genuine leather,
With isinglass curtains y'can roll right down,
In case there's a change in the weather.
Two bright side lights, winkin' and blinkin'
Ain't no finer rig, I'm a thinkin'!
You c'n keep yer rig if you're thinkin' 'at I'd keer to swop
Fer that shiny little surrey with the fringe on the top.

Example A

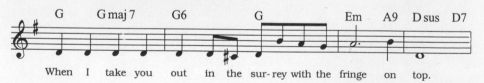

Once Oscar and Dick had "spotted" where the remaining songs would go, things became a bit easier. In a shorter space of time, "Kansas City," "Out of My Dreams," and "Many a New Day" were written. Then came the book songs, the cynically amusing "Poor Jud," and a true art song, "Lonely Room."

"Lonely Room" fleshes out and humanizes the character of Jud. With passionate lines like

. . . when there's a moon in my winder
And it slants down a beam 'crost my bed,
Then the shadder of a tree starts a-dancin' on the wall
And a dream starts a-dancin' in my head.

And all the things that I wish fer
Turn out like I want them to be,
And I'm better 'n that smart Aleck cowhand
Who thinks he is better 'n me!
And the girl that I want

Ain't afraid of my arms,
And her own soft arms keep me warm.

And her long yeller hair
Falls acrost my face
Jist like the rain in a storm! . . .

we are given a mini soliloquy that in its resolve certainly is a twin brother to Billy Bigelow's credo in *Carousel*. But here Hammerstein amplifies the humanity of villainous Jud and Rodgers sets his words to the most erotic music in the entire play in a way Puccini did for *Tosca*'s Baron Scarpia. Just as Scarpia did in the opera, with this number, Jud becomes believable, human, and much more threatening.

Believability coupled with a hoe-down animation is also evident in "The Farmer and the Cowman" (One man likes to push a plow / The other likes to chase a cow / But that's no reason why they cain't be friends), which opens the second act spiritedly but never became one of the score's popular numbers. The song and its message—Jud being the farmer, Curly the cowman—encapsulates the whole struggle within the play. Beyond the borders of the stage, it is the impossible dream which outlines Hammerstein's optimistic philosophy.

When it came to the major love ballad, "People Will Say We're in Love," Rodgers wrote the tune first, and here both men seem to be aiming for the hit which they got. Although the lyric goes overboard in sophistication, the song fits Laurey's flirtatious character, teasingly argumentative with Curly, to whom she is attracted. And although this is a lyric of denial, the music tells us we are hearing a love song. At the end of the musical the song is reprised with the lyric message it has been giving us all along. "Let people say we're in love!" shouts Curly, once the couple becomes engaged.

But whether it is a song of affirmation or denial, what made this the hit of perhaps the most revolutionary show in all American musical theater is its universality. Every song that hits is always applicable far beyond the bounds of an orchestra and balcony. That people who are in love sigh alike, have the same glow in their eyes, hold hands, and have been known to dance all night makes this lyric appeal to any listener. That it amplifies character when fitted into the show is a bonus.

Don't throw bouquets at me,
Don't please my folks too much,
Don't laugh at my jokes too much—
People will say we're in love
Don't sigh and gaze at me
(Your sighs are so like mine)

Your eyes mustn't glow like mine—
People will say we're in love!
Don't start collecting things
(Give me my rose and my glove)
Sweetheart, they're suspecting things,
People will say we're in love!

With the book and major songs in place, the search for other creative people began, but as was to happen with the difficulty of raising backing, top-drawer, proven theater people saw very little promise in the venture. Eventually

Rouben Mamoulian, who had directed *Porgy and Bess* seven years earlier, was chosen after Josh Logan, Brentaigne Windust, and Elia Kazan said no. Other refusals were in the offing: both Shirley Temple and Mary Martin rejected the role of Laurey, while Charlotte Greenwood wired "no thanks" to the role of Aunt Eller. Choreographer Robert Alton had rejected the project as well and no other seemed forthcoming when Lawrence Langner, director of the Guild, received a note from Agnes de Mille, proposing herself as choreographer and inviting him to a performance of the Ballet Russe de Monte Carlo to see her dance her own ballet, *Rodeo*.

"Lawrence passed the letter on to me," Helburn recalls, "and I persuaded Dick and Oscar to go with me. After the performance I wired Agnes: 'We think your work is enchanting. Come talk with us Monday.' "

De Mille, by her own account, had been fired from every musical she had worked on, and although Oscar was all for hiring her, "because I had been bored with musical comedy dancing for a very long time, and thought she had something very unusual," Rodgers was wary and made his fears known to her.

"Rodgers thought my work awfully nice," the diminutive choreographer told this writer, "but he added, 'I don't know whether you could possibly handle the wear and tear of a Broadway musical because that moves swiftly, and you've got to be right on the clock.' "

Eventually she was hired, and soon more conflict erupted as de Mille and Mamoulian began daily skirmishes. On her career journey from choreographer to director of musicals, de Mille was not above mixing into the creative processes of the show.

When Oscar gave her the original scenario for the ballet, de Mille remembered,

He said, "There is a big circus ballet to end Act One. Aunt Eller is riding around in the surrey with diamond wheels, Laurey is the aerialist, Curly is the ringmaster with a whip and so forth." He had the whole thing worked out. And I said to him, "Mr. Hammerstein, *what* has this to do with *Oklahoma?*" And he said, "You've got to have an up ending. You can't send everybody out into the lobby at intermission, gloomy and worried." I said, "You leave Laurey scared to death and apprehensive, then you go into this fantasy—which is supposed to be her dream," and I added, "that's not a dream, that's not what young girls dream of." Then I said, "Where's the sex?" He said, "What sex?" And I said, "What indeed! There isn't any in this show. Well, there is now." Then I said, "Where are the post-card girls?" And he looked at me, and then he picked up the phone and said, "Dick, get over here right away."

So I threw out that ballet, and I wrote a ballet containing great apprehension, and threat—and also violence—and even fun. And I gave it to Rodgers in written form. And he put it in his pocket—didn't pay any attention. And I said, "I would like you to compose some music for this, because there's no music for a fight to

Agnes de Mille invented the "postcard girls" to put some sex into Laurey's dream ballet, which ends the first act. MOCNY.

the death in this show." And he said, "You have my songs." I said, "Don't you want to do a score?" I had just done a ballet with Aaron Copland. He wrote a score. What a score! But Rodgers had no intention of doing it.[12]

And when we got to the first run-through of the fight at the end of the ballet, I said, "There is no music, because no music has been composed and we will go through it silently. But it's very violent, and since we go on counts, I'm afraid if you mistake your counts you'll hurt each other. So the piano is just going to play

---

[12] In general practice Broadway composers do *not* write the music to the dances. The convention began when choreographers set dances late in the previews while composers were frantically busy creating new songs or replacements or, if they were capable of it (like Kurt Weill), orchestrating the score. It must be added that few Broadway composers had the technique to write a quarter-hour ballet. Richard Rodgers, however, was entirely capable of it: witness his "Slaughter on Tenth Avenue."

chords at the space of intervals which will be landmarks for you." And that's what they did. And Dick came up the aisle and said, "I love it! We'll have this!" And that's what we have until this day. Except that Russell Bennett did some little hurry music, lahdle-lahdle, lahdle-lahdle, that's it.

De Mille's ballet, which closes the first act, was indeed one of the glories of the show. For the first time dance, instead of stopping the story, carried the narrative forward. Now Laurey dreams of a fight in which Curly is killed. As the act ends Jud is awakening her, and she will go off to the dance in terror.

Oklahoma! was an unconventional show, and Laurey's dream ballet did send theatergoers out to intermission "gloomy and worried" that the show might have a depressing denouement since the honesty they sensed involved them in a way no tired-businessman-show on Broadway had before. But the freshness of (then) non-stars Alfred Drake and Joan Roberts, the songs that all grew logically out of the story, the fresh perspective in Lemuel Ayres's atmospheric sets, and de Mille's inventive dances brought everybody back eager for the second half.

Before the show was put on its feet, while it was still called *Away We Go* and that second half had consisted of a comedy song, "All or Nothin'," for Ado Annie, a chorus number, "The Farmer and the Cowman," and reprises of most of the other numbers, Theresa Helburn suggested an ending with a song about the earth.

In a taxi on their way to do a backers' audition, Oscar asked Helburn,

"What do you mean, Terry?" and she said, "I don't know. Just a song about the land." I forget what I said—I thought it was one of the silliest and vaguest suggestions I had ever heard. The strange thing is that two days later I wrote a lyric which I never intended to write. I described a brand new state that was going to provide barley, carrots and potatoes and pasture for cattle, flowers on the prairie. I spoke of wind sweeping down the plain and how sweet the wheat smelled when the wind came behind the rain. I introduced a couple who expressed happiness that they belonged to the land and that the land they belonged to is grand.

With this song completed, Oscar wanted to change the title of the show to *Oklahoma.* He was overruled when it was deemed the title would remind theatergoers of the Oklahoma dustbowl, vagrancy, and the Depression. But later, after the first previews, when his duet was orchestrated and amplified by Robert Russell Bennett with the addition of a whooping chorus, and when director Mamoulian lined up the company on the stage apron, practically into the laps of the audience, *Oklahoma!* became the only possible title. The addition of the final exclamation point suggested by Lawrence Langner was insurance to theatergoers that they were in for a rousing, lusty evening.

The whole cast moved down to the footlights for the rousing "Oklahoma" finale. After the ovation accorded to the song there was no doubt that the show's title must be changed from Away We Go. MOCNY.

With no well-known stars in the show and with Hammerstein's dismal recent record of flops, the team found raising the money a humiliating struggle. "Dick and I would go from penthouse to penthouse giving auditions. Terry Helburn would narrate the story. Dick would play and I would sing, 'Pore Jud Is Dead.' " They were not very successful, but at last producer Max Gordon and Harry Cohn of Columbia Pictures made considerable investments. Helburn shamed playwright S. N. Behrman into investing $20,000 because the Guild had produced so many of his works. Although he laughingly called Terry Helburn's insistence "blackmail" at the time, his $20,000 stake eventually netted him six and half million dollars.

With the word from its tryout in New Haven being Mike Todd's disapproving description, "No Gags, No Girls, No Chance," there were empty seats in the St. James Theatre on its opening night, March 31, 1943.[13] But from the

[13] The oft-quoted gag, probably apocryphal, has been attributed to Todd, his secretary, and Walter Winchell.

second performance until it closed over five years later Todd's well-known phrase was parodied as brokers complained, "No Gags, No Girls, No Tickets!" The touring company that was formed that autumn gave its last performances nine and a half years later. For eighteen years, until 1961, *Oklahoma!* held the record as the longest-running musical in Broadway history.

Many have given reasons for *Oklahoma!*'s success and why, with over six hundred productions worldwide *every* year, it remains Rodgers and Hammerstein's most popular show. James Hammerstein, who has directed several revivals, prefers to mount it over any other of his father's works because it affords him "the sweetest seduction of all. It isn't by applause or laughter, it's that when you look over the audience's heads in Act Two and you see their attention on the stage, it's as if they're sitting in a pink cloud of enjoyment."

Ethan Mordden reasons that the show's perennial popularity has to do with its social references, which never date: "a consideration of how a relatively primitive but orderly community deals with brutish lawlessness in its preparation to enter statehood."

But one must look even deeper to realize that *Oklahoma!*, premiering during the dark days of World War II, was full of spitfire, honesty, and promise. The resolve with which these young pioneers accepted their coming hardships in the early twentieth century was not unlike what American youth were facing as they went off in uniform or to jobs in parachute factories. As it continued to run, American servicemen in particular seemed to have a special love for the show. Many soldiers shipping overseas from New York actually made a point of seeing *Oklahoma!* sitting next to "the girl they were leaving behind." Its homey quality represented to them part of what they felt memorable about the country they were fighting for.

After the war, with optimism still its byword, the show ushered in two decades of confidence and joy—the Golden Era of the American Musical—which lasted until the assassination of John F. Kennedy.

But most of all it was youth, youth behind the footlights and the youthfulness, not in age but in spirit, of its creators. Although seasoned professionals, Oscar at forty-seven and Dick, forty, were creating the musical they would have liked to have written while in their twenties—had they only known how. Iconoclastic and honest, brooking no "shtick" or compromise, this was their first, the announcement of what they stood for, their manifesto. They would write deeper, more human, more moving, and sometimes more patently contrived musicals but never would recapture that bright-eyed, vigorous, exuberant feeling. And who would want them to.

Lerner

1940–1947

ALAN LIVED ONLY SIX MONTHS AT THE LAMBS, but while there he crowded his days and his largely sleepless nights with show-business experiences. During the day he earned a living by writing advertising copy for the firm of Lord and Thomas, but he was never really serious about his job. He, Ben Welles, and Stanley Miller would usually lunch together after which one or the other of the trio would make the rounds of the Brill Building, Broadway's then-substitute for Tin Pan Alley, trying to peddle their songs to the many publishers who had offices there. Nothing was ever sold.

Nights were spent in more exciting pursuits. For Alan, dinner at the Lambs was like hobnobbing with showbiz professionals, although it must be admitted that except for nights when they appeared to attend meetings, only neophytes or respected failures were to be found in the Lambs public rooms. But Lorenz Hart, one of the giants—although less than five feet tall—of the art of lyric writing, now well on his way to confirmed alcoholism, was frequently at the bar. Alan respected Hart's work more than that of any other lyricist and had tremendous empathy for him.

"Because of his size," Alan wrote many years later, "the opposite sex was denied him and so he was forced to find relief in the only other sex left. But all this I only heard about from others and never saw a sign of myself. He was kind, endearing, sad, infuriating and funny, but, at the time that I knew him, in a devastating state of emotional disarray."

For his part, Hart enjoyed the adulation of the young man. Although very prolific during this period,[1] he was interested in Lerner's inquisitive mind and

---

[1] Within the space of fourteen months, from October 1939 to December 1940, Hart had turned out splendid lyrics to three important shows. (All had music by Rodgers.) *Too Many Girls*, which pro-

never seemed to be too busy to look over Alan's latest song. Both insomniacs and both tremendously intrigued by musical lore, nonstop they talked the nights away.

Lerner's later work was to bear far more kinship to Hart's, with its dazzling rhyming and his use of the vernacular, than to that of any of the many lyric giants that seemed to abound in the mid-twentieth century. One notes a touch of Howard Dietz's iconoclasm, a bit of Ira Gershwin's erudition, and some of Yip Harburg's zaniness in Lerner's mature work, but it is Hart's gentle warmth, his oblique way of looking at life, and his somewhat mystical penchant that were to influence Alan most. (Hart's lyric for "Where or When," with its allusions to reincarnation, was Lerner's special favorite. Its philosophy, that lovers take up with the same lovers in afterlife, was to be fully developed in Lerner's *On a Clear Day You Can See Forever.*)

But eventually Alan was to go Hart giant steps farther. As a playwright in the Hammerstein tradition, he was to imbue his lyrics with *character*, always being careful that the language he chose was natural to the personality he had created. Perhaps more than any lyricist except Hammerstein, he was careful to be consistent with tone. King Arthur sounds Arthurian and Higgins, Shavian. Lolita sings and speaks in the whiney vernacular of '70s preteen, and Coco Chanel as limned by Lerner is no more or less than a determined French spinster with a passion.

Even Lerner's most extractable songs, those aimed for hit status, seem to spring naturally from the soul who sings them in the show, whereas Hart's do not. The uneducated dance hall hostess in Rodgers and Hart's "Ten Cents a Dance" would never say "though I've a chorus of elderly beaux, stockings are porous with holes at the toes" (much less would she recognize that despite its spelling, "beaux" rhymes with "toes"). In every Hart lyric we hear the voice of the lyricist often drowning out the voice of the singer.

If Lerner's lyrics don't flesh out the persona, then they achieve the other goal of a show song. They move the plot ahead. This is something that Hart's lyrics are rarely interested in doing. But it must be admitted that in Hart's heyday, the '20s and '30s, entertainment was the byword. The least important aspect of a musical was plot mobility.

Alan was content in those months at the Lambs, but conformity in the guise of love and marriage was to become more important to him than his dipping

---

duced the memorable "I Didn't Know What Time It Was," and *Higher And Higher,* the following April, from which came "It Never Entered My Mind," were followed by *Pal Joey,* which contained a plethora of gems, including "Bewitched, Bothered and Bewildered," "Take Him," "Zip," and "I Could Write a Book." It was only after this period in 1942 that the inspiration and the will to continue his career dissipated.

his toes into theatrical waters. Alan and Ruth Boyd were married in a Roman Catholic ceremony when Alan was twenty-one. Certainly it seemed to be a love match, since both parties had no need of money.

Lerner moved from the Lambs directly after the marriage in 1940 and rented a suite at the Royalton Hotel, where he and his new wife resided for a year and a half. Further up the block, across from the Algonquin, the Royalton was home for many denizens of showbiz.

Lerner was then working as a radio writer. He said this experience taught him never to be afraid of deadlines, no matter how heavy the work load. Writing five daytime shows a week, comedy for Victor Borge, material for "The Raleigh Room" and "The Philco Hall of Fame," as well as continuity for "The Chamber Music Society of Lower Basin Street," taught him much about the craft of writing dialogue, at which he was to excel. At the time he was also writing a play which was never to be produced and the only way he could make the schedule work, he says in his typically wry manner, "was not to sleep on Monday night."

In addition to this heavy schedule Alan took on a labor of love: writing lyrics and continuity for the annual Lambs Gambol. He laid aside his considerable prowess in writing melodies and performing on the piano, and gave himself entirely over to turning out cogent lyrics. Creating this witty showbiz roast filled up the few remaining hours that Alan might have spent at home. But Ruth accepted the inevitable: that her husband would pass what little free time he had at the club, hoping to get his career into orbit. "Everybody in show business leaves his wife alone, like people in politics," Ruth confessed to biographer Gene Lees, while at the same time admitting her great love for Alan.

It was on one of these nights in late August of 1942 that he was to meet "a short, well-built, tightly strung man with a large head and hands and immensely dark circles under his eyes." Frederick Loewe, known as Fritz to the Lambs membership, was seventeen years older than Lerner.

The men recalled the meeting differently. In his autobiography Lerner says Fritz was on his way to the men's room, took a wrong turn, and then spied Alan. Before completing his call of nature Fritz stopped at Lerner's table and, after briefly admiring Lerner's lyrics for the Lambs Gambol, asked the younger man if he would like to collaborate. "I immediately said 'yes,' " Lerner writes.

Loewe has a different version. "I told him I had been commissioned to do a musical based on the play *The Patsy*," he told a reporter for the *New York Daily News*, "and . . . it was scheduled to go into rehearsal two weeks from the following Monday [seventeen days thence]. Alan answered that he was busy. But on Sunday I phoned him that I had a $500 advance. Alan's reply was 'I'll be right over!' And he was."

Lerner gave yet another version of that first encounter, to reporter Tony

Thomas some years later, saying that *he* was on his way to the men's room when Fritz stopped him and made the offer to collaborate, stating, "I don't have a lyricist. Do you want to write with me?" Lerner says he answered "yes" immediately because he felt that "anybody who was a member of the Lambs must be a genius."

Whatever the true story of the meeting it is almost immaterial. Lerner and Loewe, however stormy their personal relationship was to grow, were to work together for the next eighteen years, and their collaboration would become one of the glories of the American musical theater.

Frederick Loewe was born into an acting family in Vienna on June 10, 1901. His father, Edmund Loewe, a tenor whose most famous role was Prince Danilo in *The Merry Widow,* noticed his son's musical precocity and sent him for piano lessons at an early age. By the time the boy was nine, young Frederick was writing songs, one of which was interpolated into a sketch his father used in a European tour. But his mother, Rosa, born into a strict German-Jewish family, objected to her son spending so much time in backstage dressing rooms and insisted her son should be enrolled in a military school.

The boy was unhappy and tormented by his German classmates because of his small stature and his Jewishness, so his father's wishes eventually prevailed and Frederick entered Stern's Conservatory in Berlin the next year. At Stern's he prepared for a career of concert pianist. He made rapid progress with his serious concert music but never ceased putting down his Viennese *schlages,* romantic melodies, and frothy waltzes on paper.

At fifteen he wrote a novelty song called "Katrina," whose lyric talks about a girl with the best legs in Berlin. It caught on and swept Germany and Austria, eventually selling two million copies of sheet music. In the ensuing years the royalties were gambled away by Frederick's father,[2] but from that time on Frederick was liberated enough to be a young Casanova in Vienna.

Slender, handsome, with sandy-colored blond ringlets, his artistic personality made him popular with girls. (He boasted that the had his first sexual experience by the time he was three and his first affair with a governess at nine.) "I thought I was abnormally precocious," he told a reporter for *Time* in 1960, "until I read Kinsey. Once, when I ran out of credit at a brothel, I paid off my debt by entertaining on the madam's piano."

In 1923 Edmund was contracted to appear in a David Belasco production and able to bring Rosa and Frederick with him to New York. No sooner had they arrived and begun the rehearsal period when Edmund died.

---

[2] Although Fritz had no love for his father, he bridled when Alan, referring to his partner's well-known penchant for gambling, said it was a trait inherited from Edmund.

After Frederick became famous for his collaborations with Lerner he often told interviewers elaborate stories of the various far-flung jobs—a boxer, a horseback riding teacher, a cowboy, a moving-picture pianist in Montana—the twenty-two-year-old undertook to support himself and his mother. He went back even farther, telling journalists that in Berlin he studied with Ferrucio Busoni and Eugene d'Albert. In these inventions he is not unlike a great many "artistic" émigrés who blew up their pasts. Lerner, whose own background contained much self-invention, of course reveled in Fritz's outlandish stories. Each man's overstatements contributed to a tacit understanding that brought great rapport. Alan's favorite yarn concerned the bout with Tony Canzoneri.

Fritz would recount to his fellow boxing enthusiast how he would spar and dance on the retreat, always keeping out of Canzoneri's reach which he erroneously assumed was a bit shorter than his own.

"I was quite a dandy, very agile, quick on my feet and very deft with a slap-like jab. I'd won eight fights in a row in the 120-pound class, and then I ran into an up and coming fighter named Tony Canzoneri. He took me out with one punch in the first round."

It seems unlikely that a concert pianist would jeopardize his hands in boxing matches, and there was no newspaper reportage of this major bout. Canzoneri was by no means "up and coming" but already a champion, and any fight of his would have been written up.

Loewe returned to New York and got jobs playing the piano in various gin mills that were scattered throughout the theater district. He had only been in New York a short time when he met and married a sloe-eyed Viennese beauty, Ernestine ("Tina") Zwerline in 1931. They lived in one room on Lexington Avenue, Fritz practicing and writing songs on a rented upright piano. To help make ends meet, Tina went to work as a model for the fashionable hat designer John Frederics.

In 1932, a decade after he had arrived in America, Loewe was having an occasional song interpolated into a show. One was even included in *Walk a Little Faster*, the revue starring Beatrice Lillie that young Alan Lerner's father had taken him to see on a weekend away from Choate. It was not Loewe's song that Lerner came back to school eternally whistling to the consternation of his classmates, but Vernon Duke's "April in Paris." Duke, born Vladimir Dukelsky, had a tandem career in classical music under his real name. His output and training bears a close parallel to that of Frederick Loewe.

By 1934 Fritz had taken a steady job as show pianist. Kitty Carlisle remembered his playing the piano in the pit of her first Broadway show, *Champagne Sec*. His Viennese know-how would certainly have been put to good use, for the show was an adaptation of Johann Strauss's *Die Fledermaus*.

"He was a lowly pianist in the pit," Kitty Hart remembered recently, but he seemed to like me, so every evening he would walk up three flights of stairs to my dressing room and he would stand behind me while I got made up . . . and he used to say to me in those days, "someday I am going to write the best musical on Broadway" and I would think "you and who else!" There were thirty musicals in those days, and thirty pit pianists. It never occurred to me to pay any attention.

Opening night of My Fair Lady I was standing against the back wall of the theater—the wall Moss always stood against and paced up and down. He and Alan were pacing up and down while the show was in progress, and in the middle of the first act Fritz came over to me and he said, "Well, I wrote the best musical on Broadway."

After several more years of an interpolation here, a theater orchestra or gin mill job there, he got the opportunity to write his first full score. Great Lady starred Norma Terris, Helen Ford, and Irene Bordoni,[3] three of Broadway's greatest ladies. Unfortunately they were all near the end of their careers. More unluckily, the impossibly complicated story of this quasi-historical musical dragged these stars into a plot that shuttled between the French Revolution and the present day (1939). All of this quite overwhelmed the adequate music Loewe had written. Great Lady, however, deserves a footnote in history: before it folded after twenty-four performances, it introduced the dynamic dancing of Jerome Robbins to Broadway.

Understandably discouraged by his lack of success in composing, Loewe wrote nothing except an occasional song for the Lambs Gambol. Now he returned to his first career and at forty-one decided to give a recital at Carnegie Hall. The critics were not enthusiastic about his pianism, and once again Loewe floundered between performing and composing until producer Henry Duffy asked him to write the music for The Patsy. This farce by Barry Connor had already been adapted into a musical with a few songs by Loewe and his former lyricist, Earle Crooker, but it was far from being a complete score.

The fateful meeting of Lerner and Loewe took place in August 1942, six months before the premiere of Oklahoma! which was to change forever the way musicals would be written. Even at that time anyone with show business savvy could tell you that slipping the score into an already written libretto was a sure way to guarantee a flop. But in spite of their naïveté, the new team tackled the project with unbelievable élan. It seemed an entrée for twenty-four-year-old Lerner and a last desperate chance for forty-one-year-old Loewe.

Lerner was unduly hard on the first L & L collaboration when he said, "Had you been away for the weekend you would have missed it." Alan's radio-

---

[3] Norma Terris's most famous role was Magnolia in Show Boat; Irene Bordoni is perhaps best known for introducing "Let's Do It" in Cole Porter's Paris; and Helen Ford created the title roles in Rodgers and Hart's Peggy-Ann and Chee-Chee.

deadline experience came in handy, for the fourteen-song score for *The Patsy*, now retitled *Life of the Party*, was written and rehearsed in twelve days. It had a respectable run of nine weeks in Detroit. Although the show never made it to Broadway, the experience showed both Alan and Fritz that they could work smoothly together.

The following year conductor and producer Mark Warnow engaged them to write a musical to star Jimmy Savo, a burlesque comic. This time the romp, called *What's Up?*, *did* make it to Broadway, and most of the critics took note— at least of the songs of the new team. Alan, assisted by Arthur Pierson, wrote the libretto, which showed only too clearly his inability to underscore a particular performer's special aptitude. Not until he wrote *My Fair Lady* to the sound of Rex Harrison's voice would he again attempt a vehicle for a star.

The thread of *What's Up?*'s déjà vu plot concerned the Rawa of Tanglinia, an Eastern potentate whose plane crashes on the lawn of an exclusive school for girls. After some seemingly naughty but actually harmless mix-ups during which everybody in the school is quarantined by an outbreak of measles, everything is set right. Coming six months after *Oklahoma!* had revolutionized all musical theater with a fresh, innovative book and score, Lerner and Loewe's pedestrian burlesque plot, and so-so songs made no dent on the public. Savo was like a fish out of the sea, for Alan's script mistakenly called for him to trade his trademark baggy trousers for a morning suit and fez. Despite the show's inventive choreography by George Balanchine and imaginative settings by Boris Aronson, the critics agreed that *What's Up?* was thumbs down. The miracle was that Mark Warnow kept it open for eight weeks.

But the team was off and running. Alan and Fritz rented a suite on the twelfth floor of the Algonquin and began work on their next show, an integrated story dreamed up by Alan. Although Ruth had recently given birth to Alan's first child, Susan, Lerner and Loewe rarely interrupted their round-the-clock schedule to spend time with their families. They telephoned only intermittently.

By now Loewe's wife, Tina, was as deeply involved in her career as Fritz and Alan were in theirs. Tina had left the John Frederics firm and worked her way up to become the executive in charge of the hat division at Hattie Carnegie.

So immersed were L & L in their creative effort that nothing outside their twelfth-floor suite of rooms seemed to matter. "Fritz would only begin after I'd given him the title and talked about what the song was going to say," Lerner said explaining the method they had evolved.

We had a sort of shorthand. And then he'd start to compose and I'd be in the same room. He'd start to improvise on that title and I'd say, "Oops, that's it!" and he'd say "What did I play? Oh, that's terrible" or "Oh, that's not bad." All the time I'd be thinking of things. I wouldn't write the words right then, but I was so involved in the very creation. Not that I was the composer, but a mood was created by the hours spent in the room. I was so involved that sometimes I would almost cry.

Once the score was completed, Lerner, the usual spokesman for the pair, seems to have had no trouble interesting John C. Wilson, who had recently had a hit producing the Arlen-Harburg *Bloomer Girl,* in his idea for a new musical built around the story of a love affair with a decade's interruption. Wilson, a canny theater man[4], not only agreed to produce *The Day Before Spring* but was eager to direct it as well.

The story revolves around a newly published book, *The Day Before Spring,* written by the handsome hack Alex Maitland. It recounts with thin disguise an affair as it might have been that he and his former college girlfriend, Katherine, had ten years before. As the musical opens, Katherine, who has since married Peter, a reliable but plodding businessman, learns that Alex is going to lecture at their college reunion and scurries to Harrison U. with Peter and a subplot couple. She lets us know that in reality their affair never was consummated because Alex's car broke down on the way to the lodge where the tryst was planned. Peter came along and helped to fix the auto with such dispatch that Katherine began the relationship that led to her present comfortable but rather routine marriage.

At the get-together Alex's flame for Katherine[5] is rekindled when she enters looking more beautiful than she did in her college years. Bored with her marriage, she is seduced by the romantic passages she had read in Alex's book. They plan to carry their thwarted affair to its conclusion by eloping that very night. As anyone might imagine, Alex's car breaks down in the very same spot it had a decade earlier. Peter comes searching for his wife, and, after Katherine realizes that Alex will always philander (this last dawns on her during the obligatory second act ballet), she goes back with safe, lovable Peter.

While Lerner's protagonists act more like high school sophomores than thirtysomething college alumni, the dialogue has enough dash to be intriguing. As with any artist's early work, one can notice themes that will become lifelong obsessions and will recur in all of Lerner's shows.

The time warp would be a persistent Lerner theme. It would find expression

[4] John C. Wilson (1899–1961), producer and director, had a long association with Noël Coward, presenting three of his musicals. After *The Day Before Spring,* his major successes were *Kiss Me, Kate* and *Gentlemen Prefer Blondes.*

[5] Lerner's stage direction for Katherine's entrance is a gem: "(She is wearing the most beautiful, daring and expensive evening gown that John C. Wilson can afford)."

in the same characters going through different generations as in both *Love Life* and *1600 Pennsylvania Avenue*—incidentally, two of his worst scripts. But this concept of reincarnation was to be fully developed in the paranormal reincarnations in *On a Clear Day You Can See Forever.*

"Lerner was always attracted to the wistful plot," said record producer Tom Shepard. "Somebody always wanted something more or other than what they had, whether it was Katherine in *The Day Before Spring,* King Arthur or Eliza Doolittle (of course he didn't write that story, but he was attracted by it). That quality of yearning to be more or other made him the incredible romantic that he was. And it had everything to do with the way he lived his life."

"He had a longing for immortality—in any way, shape or form he could get it," Stone Widney, his associate for thirty years, told me, "whether it was through his work or the changes he wrought." Widney was referring to Lerner's involvement in things political and his ardent campaigning for more humanistic, liberal candidates, but he might as well have been alluding to Lerner's eight marriages and the four offspring he produced.

Just as important as the next life was his relationship with the next wife. Lerner would always relocate with his current inamorata before he had quite moved out on the former one. Obviously, the wedded state was not at all sacred to him. "He asked me to have lunch with him the next day, and I did," says Katherine early in *The Day Before Spring.* Her next line, "and that was the beginning of Peter and the end of Alex," presages a lifelong Lerner pattern. Even in this early libretto he would make fun of marriages that lasted beyond a few years.

His own was in a shambles. He would call Ruth at the Royalton every few weeks. According to conductor Maurice Abravanel, both Alan and Fritz were neglectful of their wives, and their orchestra leader was not afraid to tell them so. "I couldn't help feeling sorry for the two women, and I told them, 'that's not the way to treat your wives.' "

It almost seems as if Alan and Fritz, holed up in the Algonquin, were each so wrapped up in the enormous joy of at last finding a collaborator with whom he could work smoothly, a producer with a track record of hits, and the romanticism of their characters that anything of their lives before they began this project was an intrusion.

The characters, the music had captivated them. The almost autobiographical soliloquy below with which Alex seduces Katherine is typical of Lerner's romanticism. It was to become the lifelong motif of his work and might as easily have come from *Brigadoon, Gigi, Camelot,* or *Lolita.*

I'll tell you a story and you'll understand. It's a story I read years ago and never forgot. It seems a man after many years returned to his hometown where he'd once been young and happy. He could hardly wait to see it again. But when he got there, his heart broke with disappointment. Nothing was the same as he remembered it. The houses were dingy and depressing; the trees along the streets were cold and leafless; and the people were downcast and unfriendly. So in despair he walked down to a lake where he'd once spent so much time. There, suddenly he met a girl, a beautiful girl, and just as suddenly he fell hopelessly in love with her and she with him.

(The music begins softly)

Well, they turned and walked up toward the village hand-in-hand. And when he looked all the houses were warm and inviting; the trees were green with leaves and all the people were friendly. Kath, that's what happened to us today.

The two ideals, eternal youth and romance, fit together in *The Day Before Spring*'s seduction scene.

The sands of time will never harm   Let others age and lose the spark,
The wond'rous life you're dreaming of.   And leave their happy song unsung,
So face the world without alarm   Our life will be a joyous lark—
For you'll be young if you're in love.   We'll be in love and ever young.

But even though the message of this, the first *real* musical of Lerner and Loewe, is juvenile, and despite the hacklike construction of plot cum subplot replete with a Gilbert and Sullivan commenting chorus, the dialogue and stagecraft are generally spritely. The *New Yorker* critic noted this, observing that "the most original touch in the show comes when the husband is disgustedly reading aloud from his rival's book while the lovers act the same passages out behind him on stage, freezing in their tracks when his voice stops and even obligingly repeating themselves when he loses his place. Tiresome as this may sound on paper, it works out very well in the theatre, and I'm surprised that nobody seems to have thought of it before."[6]

Yet what allowed *The Day Before Spring* to make it through 167 performances was certainly its inventive score. Loewe's music is thoroughly American while unabashedly romantic. He hasn't found his own voice yet, borrowing heavily from Cole Porter and Richard Rodgers, but he almost creates a standard in the title song. For his part Lerner has given him many sassy lines such as "We could have such fun / Clean or maybe un" and titles like "A Jug of Wine, a Loaf of Bread, and Thou, Baby" to lighten the proceedings.

The spirit of Lorenz Hart, who died two years earlier on the same date as *The Day Before Spring*'s Broadway opening night, hovers over the whole proceedings. "God's Green World" is only one step away from Hart's "Mountain Greenery," and a rueful song for a love-smitten coed, "My Love Is a Married

[6] Lerner's idea did not surface again (so far as I know) until *City of Angels* in 1989.

Irene Manning, playing a fantasy-prone wife, helps husband Bill Johnson remove his coat preparatory to a showdown with her romantic ideal in The Day Before Spring, *Lerner & Loewe's first, albeit mild success.* MOCNY.

Man," could just as easily have been penned by Larry. I quote some of its best lines below:

My lonely bitter heart has needed him to make
    it sweeter
He came, he saw, he conquered and then sic
    transit Peter.
And now I shouldn't cry, I should be brave in-
    stead,
But bravery is cold in bed.

My love is a married man,
I'm a marital also ran.
Though I love him so,
Does he love me? No!
I'll never enter his life
Because he's true to his wife.

My love is a married man . . .
My dreams abundant
Are redundant
And they fall very short.
The ship I hoped for
Sat and moped for,
Docked in some one else's port.

He's gold I can never pan,
He's the ember I'll never fan,
And I know it well,
But, oh, what the—tell
Me what to do with a married man.

Opening on Thanksgiving Day, 1945, the show was to become the first acceptable collaboration of the new team. Several critics went overboard and called Lerner & Loewe worthy successors to Rodgers & Hammerstein, their more venerable colleagues having produced *Oklahoma!* and *Carousel* by that time.

Even before the Broadway first night, Alan interested his agent, Lillie Messinger, in the show. Lillie, a diminutive lady with a dynamic spirit, was to remain a lifelong friend of Lerner's. She always kept one foot in Hollywood and the other on Broadway. Now, while the musical was in early rehearsals, she sold the script to Louis B. Mayer for the then staggering sum of $250,000. Although the film was never made, the huge advance was another vote of confidence for the collaborators. For Alan, it represented the success he sought so avidly; Fritz, who would later become a sybaritic gourmet, was to go on record by saying "for twenty-one years Rikers [a chain of restaurants with rock-bottom prices] kept me alive. I knew just how many free rolls went with the 45 cent meal. Now for the first time I began to eat properly—and regularly."

While Alan and Fritz were in Hollywood, ostensibly to sign their deal, they auditioned a young starlet, Marion Bell, who was under contract to MGM at the time. "I was quite taken with Alan and wanted to know him better," Marion Bell said recently. For Alan's part, he couldn't forget her charm and the beauty of her voice, and he promised to audition her for his next show. As it turned out, that Hollywood excursion was to provide him with *Brigadoon*'s leading lady as well as his next wife.

On first glance *The Day Before Spring* and *Brigadoon* seem to have nothing in common except the decade lapse in the former and the century reawakening of a little Scottish village in the latter. But looking beyond the illusory romance or a similar comic subplot about an aggressive female and passive male, the thrust of each musical is that faith can rewrite history.

Alan said the idea for *Brigadoon* came to him from a casual remark of Fritz's—"Faith can move mountains"—and that he first began sketching a play about belief actually moving a mountain. "From there," he added, "I went to all sorts of miracles occurring and eventually faith moved a town."

The mystical story concerns Tommy and Jeff, two Americans on a hunting expedition in Scotland who become lost in a forest. Tommy, who is engaged

to be married, confesses that even this last stag expedition is really a postponement of a marriage about which he has doubts. Suddenly they hear muted
voices chanting an almost hymnlike melody and see a distant fog-shrouded village that is not on their map.

When they arrive at the quaint, old-style place, they discover the inhabit-
ants, dressed in native costumes, have never heard of telephones and refuse to
take American money. They also discover a wedding is about to take place.
Tommy falls in love with the bride's older sister, while Jeff is seduced by a
village girl, Meg, a ringer for Hammerstein's Ado Annie. As the marriage ceremony takes place Tommy notices the groom has signed the Bible with the date
of 1746. "What is going on?" he asks Fiona.

Fiona takes Tommy to Mr. Lundie (a figure not unlike the starkeeper in
Carousel), who explains that back in 1746 their minister was so disturbed by
encroaching witches that he prayed to God that Brigadoon might vanish into
the highlands. But not forever. It would return for one day every hundred
years. It has done so twice and so only two days have passed for Brigadoon.
"But," Mr. Lundie warns, "should any of its occupants leave, the spell will be
broken and the town will disappear forever." When Tommy, now caring deeply
for Fiona, asks if someone from the outside could stay, he is told that he'd have
to forswear the world forever.

Shortly afterwards at the wedding celebration, when the bride's rejected
suitor, Harry Beaton, kisses her passionately, a fight breaks out between him
and the groom. As the first act ends, Harry threatens to leave and so destroy
the miracle.

The second act opens with a balletic chase after Harry. When he is found
and accidentally killed by Jeff, the town is saved, but Tommy senses the violence directed at outsiders from within the town, and his resolve to stay forever
in Brigadoon with Fiona is shaken. He and Jeff leave.

Back in New York, Tommy realizes his nagging fiancée will make his life
impossible and once again postpones his upcoming marriage to return to Scotland. Before long he and Jeff have found the spot where they first heard the
voices. Seeing no town, they are convinced it was all a dream. Soon they hear
the chanting again, and Mr. Lundie comes forward and says, "Ye mus' really
love her! Ye woke me up!" As Tommy walks quizzically towards him, Mr. Lundie adds, "Ye shouldna be too surprised, lad. I told ye when ye love someone
deeply anythin' is possible. Even miracles."

Although the story, especially in Tommy's second-act motivation to leave
and return, doesn't always make sense, Alan had tremendous faith in his book
and the songs he and Fritz had written. He was later to say that every theatrical
producer had turned it down. The Theatre Guild, with uncanny lack of perception, wanted to move the story to the United States. Lerner wouldn't hear
of it.

Oliver Smith's other-worldly setting for the wedding in Brigadoon. *The shards of a chapel make a perfect backdrop for Agnes de Mille's dramatic sword dance that ends Act I.* MOCNY.

At last Cheryl Crawford, who had produced the successful revival of *Porgy and Bess* as well as *One Touch of Venus* (another show that plays havoc with the centuries), heard the score and agreed to produce it. *Venus*, although created by the distinguished triumvirate of S. J. Perelman, Ogden Nash, and Kurt Weill, had been a *succès d'estime*. But Cheryl Crawford, who had never had a hit beyond her revival, was not a "money" producer and so had difficulty raising capital. Even though *Brigadoon*, with a unit set, was budgeted at only $175,000, Lerner and Loewe had to give almost sixty auditions—complete run-throughs—before that sum was raised.

To set the show on its feet Crawford approached Agnes de Mille, fresh from her triumphs in *Oklahoma!, Bloomer Girl,* and *Carousel*. "I had heard the music and I thought it was lovely," de Mille remembered, "and also the lyrics." She singled out the wedding song "Come to Me, Bend to Me" as having swayed her, adding "bend to me—now that's poetry."

But the choreographer did not like many of the lyrics in the show. "Alan did a thing called 'Almost Like Being in Love,' and I said, 'don't use that song, throw that one out.' Well, of course, it was the hit of the show."[7]

Although the overall director of *Brigadoon* was Robert Lewis, de Mille had great input in fashioning the proceedings. For one of the most memorable and

[7] Not always the best judge of hits, de Mille had told Rodgers four years earlier when she was choreographing *Oklahoma!* to discard "People Will Say We're in Love," saying bluntly, "It's not up to your standard!"

exciting first-act endings in all of musical theater, Agnes persuaded Alan to cut the dialogue in the spot before the rejected suitor threatens to leave the village, and fashioned a dangerous, violent, and ultimately exciting sword dance which she says she "stole" from Scottish history.

"They had the wedding and it was interrupted and they had reprises of every song in the first act. And I said, 'this is just garbage,' and crossed the whole thing out, and said, 'we'll put the sword dance here. We'll do it in pantomime, do it in dancing.' And Bobby Lewis, the director backed me up absolutely."

If Agnes and Robert Lewis supplied *Brigadoon*'s élan, there is no doubt Marion Bell furnished the romance. "The day I arrived in New York, Alan called me to audition." Within a week there was no doubt that Marion, curvaceous, raven-haired, with flawless skin, would be his Fiona and the next Mrs. Lerner. But his choice was far from nepotistic, for she won the critics' hearts as well as a Donaldson Award as the best female debut artist.

Marion was born in St. Louis, the daughter of a railroad freight agent, but the family had settled in Los Angeles by the time she was fifteen. A gifted singer with a strong soprano, she went to Rome after her high school graduation to study with Mario Marifioti, who had taught many who later made it into the Metropolitan Opera—notably Grace Moore. Returning to the United States as war threatened, she studied with Nina Koshetz, who brought her to the attention of the director of the San Francisco Opera, where she made her debut. Spotted by a talent scout for MGM and signed to a film contract as their latest addition to a burgeoning roster of young female opera stars, she made a stunning appearance in the movies singing a duet from *La Traviata* opposite James Melton in *The Ziegfeld Follies*. That was where Alan first saw her.

In 1944, in Hollywood, he had auditioned her for a role in *The Day Before Spring* but told her she was too young for the part. He must have been mightily attracted to her for when she invited him to her home, Marion recalled, for a rationed dinner, he said, "I can't. You see, I'm married."

When she heard that Lerner and Loewe were casting *Brigadoon*, she left the stock company with which she was touring and came to New York, checked in at the Waldorf-Astoria, and had her agent call to arrange an audition. Before he heard her sing, Alan invited her over to the Lyons Agency to hear her read and to set the date for her vocal audition. "He had been working toward a deadline," she told Gene Lees, "and was unshaven. He told me he wanted to see me home, so would I wait until he got shaved? And we went home in a taxi."

After the audition, Alan, Fritz, and Cheryl Crawford felt Marion had the acting ability and vocal quality they wanted in Fiona—not an easy role to make believable, for the hero has to fall desperately in love with her, enough to want to give up the outside world—all within the space of a single day.

Marion Bell remembered the rehearsals at the cavernous Ziegfeld Theatre and recalled that once the choreography was set, there were remarkably few changes in the dialogue. Whenever they were called for Alan showed them to Marion first. "He was shy, and afraid my reaction might not be good," she remembered, adding that he wrote Fiona's confession of her love, "Dinna ye know, Tommy, that you're all I'm livin' for?," as an inside avowal of her love for the author. "He loved to hear me say the line from the stage every night."

The Boston tryouts had their few mishaps (one night the scenery was loaded inversely and when the forest drop came down the audience saw the roots of the trees sticking up in the air), but the reviews were splendid. Yet no one was prepared for what New York would think of such a fantasy.

As they got closer to opening night, tension between Fritz and Alan began to build and they would quarrel openly in front of cast members. Perhaps Loewe disapproved of Alan's living so openly with Marion after he had left Ruth and his daughter and told him so. Certainly they were competitive. "When Alan bought me a full-length ranch mink coat," Bell recalled, "Fritz bought my understudy, Virginia Oswald, a mink coat too—just like mine." The act seemed to confirm the backstage speculation that Loewe was sweet on Oswald. In the ensuing weeks, the quarrels grew more violent. "One night, shortly before the opening," Marion said "we were confronted by Fritz in the lobby. 'Don't touch me or I'll kill you,' Alan screamed at Fritz."

Part of the tension was caused by the unpredictability of that particular Broadway season, one in which critics and audiences came down hard on the depressing books of the Duke Ellington-John Latouche Beggar's Holiday and the Elmer Rice-Kurt Weill Street Scene (both critical successes because of their stunning musical scores). Yet, with the war only two years behind, the public was longing for escape, hope, and illusion. That they got in healthy measure when the unheralded Finian's Rainbow broke through like a thunderbolt in January. Finian's lively music was the work of Burton Lane, with whom Lerner would eventually write two shows, On a Clear Day and Carmelina, and two movies, Royal Wedding and the aborted Huckleberry Finn. Finian's book was pure fancy with a dash of Yip Harburg's perpetual polemics thrown in. It dealt with a leprechaun and a bigoted southern senator who is transformed into a black gospeler by the heroine's wish. When Brigadoon opened two months later on March 13, 1947, the critics now had no trouble swallowing Lerner's illusion of a mirage-like village that awakens every hundred years.

"To the growing list of major achievements of the musical stage add one more—Brigadoon. For once, the modest label 'musical play' has a precise meaning. For it is impossible to say where the music and dancing leave off and the story begins in the beautifully orchestrated Scotch idyll," proclaimed Brooks Atkinson in The New York Times, while Robert Coleman in the Daily Mirror looked even deeper. "It took courage to produce *Brigadoon*, an unconventional show of marked originality. But in this instance courage will be richly rewarded. Compromising less than *Finian's Rainbow* with Broadway and Tin Pan Alley . . . as artistic as a modern dance recital, it still manages to pack a tartan full of popular appeal." And although seven of the nine members of the press had cheered *Finian's Rainbow*, *Brigadoon* immediately joined the pantheon of shows receiving unanimous raves. Although Agnes de Mille and Michael Kidd, *Finian's* choreographer, tied for that award, *Brigadoon* was judged best musical by the Drama Critics and handily swept the Tony Awards.

In Lerner's total output *Brigadoon* ranks only slightly below *My Fair Lady* in number of far-flung revivals. This is due to the vitality of its score. Loewe mitigates his slight overuse of typical Scottish 6/8 rhythm with delicious ballads in the usual 4/4. And Lerner's lyrics show a great advance in technique over those he wrote for *The Day Before Spring*. Notice the lightness of the ballyhoo to come to the fair, "Down On MacConnachy Square," which has funny lines treading lightly between the Broadway and Scots tradition.

Now, all of ye come to Sandy here
Come over to Sandy's booth.
I'm sellin' the sweetest candy here
That ever shook loose a tooth.
I eat it myself and there's no doubt
'Tis creamy an' good an' thick.
So laddies, I hope ye'll buy me out,
'Tis makin' me kind o' sick. . . .

I'm sellin a bit o' milk an' cream
Come sip it an' ye will vow
That this is the finest milk an' cream
That ever came out a cow.
Though finest it is, the price is small
With milk an' the cream alack!
There's nothin' to do but sell it all—
The cow winna take it back.

But perhaps the most interesting aspect of *Brigadoon* is its blatant romanticism. This is nowhere more evident than in the subtle way Lerner interlaced the emotion with his commercialism. He chose to work in reprises late in the second act. But it is more than that, he builds them into what is called the "eleven o'clock number"—the emotional high point of the show.

Reprises were the stock and trade of the commercial musical theater of that time because each replaying drove the song further into the audience's subconscious and gave the show a possible hit. Richard Rodgers loved them. Stephen Sondheim hates them. Lerner knew he needed hits after three flop shows. He created the diversion of subtly putting the titles into the mouth of Tommy's New York fiancée, Jane, and having them picked up by the heroine, Fiona. Jane says: "After all darling, I did think the minute you'd get to town you'd call me . . . or come to me . . . or in fact, why didn't you . . ." As she drones on Fiona appears and sings

Come to me, bend to me,
Kiss me good day!

Darlin', my darlin',
'Tis all I can say.

Jus' come to me, bend to me   Gie me your lips
Kiss me good day!   An' don' take them away.

Soon Jane begins to talk about where they will live after they're married and Lerner's dialogue leads her into, "Why don't we take Mr. Jackson's house? It's far away and right on the top of a high, beautiful hill." Fiona immediately reappears and sings:

. . . Through the heather on the hill.   If ye're not here I won't go roamin'
But when the mist is in the gloamin'   Through the heather on the hill.
An all the clouds are holdin' still,

Now, at last, Tommy calls off the wedding and Jane says, "I refuse to stand here and argue with you in this bar! Let's go home." Another cue, this time to bring in Charlie, the bridegroom, and the townsfolk singing, "Go home, go home with bonnie Jean."

Before Jane walks out in a huff she says, "If you think anyone else is going to put up with your nonsense, you're stark raving mad. So think that over, Mr. Albright, when you're all alone!" Now Fiona is very near Tommy and reprises a moving line she had uttered when he was leaving *Brigadoon*, "I think real loneliness is not bein' in love in vain, but not bein' in love at all." This cues a reprise of their love song, "From This Day On." At the end of the scene he calls Jeff and they resolve to go back to Scotland. The scene is masterful.

Brigadoon ran for 581 performances during a great many of which Marion Bell, Mrs. Lerner since the fall of 1947, starred. Performing the arduous role was to affect her health and keep her out of the cast for months, during which Priscilla Gillette filled in. When she returned to the cast and wanted to go to Philadelphia to audition for Eugene Ormandy to do a concert with the Philadelphia Orchestra, Alan objected. "Maybe he was trying to protect me," Marion offered, for her health was fragile. More likely he was trying to keep her close at home as an anchor of security.

Shortly after opening night, once all the rave reviews were in, Fritz flew off to Hawaii. The newspapers announced that Loewe was off on a well-deserved holiday and that the brilliant new team would soon begin working on their next musical. Only their intimates knew that Loewe left no forwarding address and had sworn before he left that he never would work with Lerner again.

Hammerstein
1943–1945

ON THE MORNING AFTER THE PHENOMENAL OPENING OF *Oklahoma!*, reading of how they had revolutionized the musical theater, seeing avid customers queueing around the block for tickets, Oscar Hammerstein and Richard Rodgers both realized they were no longer simply professional lyric-librettist and composer. They were ampersanded and suddenly the most celebrated team on Broadway.

Eager to protect that status, when Dick heard a few weeks later that Oscar's lawyer might be scouting a collaboration with Kern for a new musical, the composer arranged a luncheon during which he brought up the matter.

"I told him of my conviction that it would be a serious mistake," Rodgers said, "for either of us to do anything professional without the other. Oscar was in complete agreement."

But each man had some unfinished business before he could give himself wholeheartedly to the collaboration that was to grow into the biggest creative musical conglomerate of the twentieth century. In an attempt to "help postpone a dear friend's drinking himself to death," as Rodgers referred to it, he involved himself with Hart in a revival of *A Connecticut Yankee*[1] while Oscar pushed forward with the production of *Carmen Jones*.

Almost simultaneously, they went about insuring their empire by seeing to the widespread dissemination of their own works. First they established Williamson Music (named after both their fathers) to publish their works without sharing any of the profits with a publisher. Williamson's initial publications were the piano-vocal score of *Oklahoma!* and its fast-selling hit songs.

Then, when Jack Kapp of Decca approached the partners and proposed that

---

[1] Less than a week after the opening of this well-received revival, in November 1943, Lorenz Hart died.

his firm issue records on which the actual cast, conductor, and theater orchestra would perform, they realized the enormous potential of the idea. This was the first original cast album—which was truly that, an album with four sleeves containing 78-rpm acetate records. The idea caught on, and *Oklahoma!* quickly sold over a million albums.

Aware that each was somehow able to outdo himself when they worked together, they looked for other properties but found nothing they could agree upon. Eager as they were to become known in theatrical circles as a team, but reconciled to postponing their own new musical (when they found it) until the furor over *Oklahoma!* died down somewhat, they decided to produce someone else's show. This was, as many of the business suggestions were to be, Rodgers's idea, but does not seem outlandish when one considers that Rodgers had co-produced *By Jupiter* and was sole producer of *A Connecticut Yankee.* Oscar, of course, had gotten wide experience in producing during all the years he had been associated with Uncle Arthur's shows. The partners moved eagerly into the new venture, renting an office on Fifth Avenue and later a suite on Madison Avenue.

Their first attempt, a creative adaptation of a sheaf of short stories titled *Mama's Bank Account,* was adapted under their aegis by John Van Druten, who transformed it into a folk-play, not unlike *Oklahoma!* R & H cast it with Mady Christians and Marlon Brando (in his first Broadway role). Retitled *I Remember Mama,* eventually it was to become a hit play, a film, a TV series, and even a musical. Because of their long experience in the theater and uncanny sense of taste, R & H, as the organization soon came to be nicknamed, quickly developed into a creative force in musicals and straight theater as well.

Of the five plays they presented between 1944 and 1950 only two struck out. But even these failures brought them prestige, having been dramatic works of Graham Greene and John Steinbeck. *Happy Birthday, John Loves Mary,* and *The Happy Time* were all successful, to be outdone only by the R & H production of a super-hit: *Annie Get Your Gun.* They were good, they knew it, and they capitalized on their preeminence.

By the mid-fifties, as Frederick Nolan says in his book, *The Sound of Their Music,* "collaboration with R & H meant R & H got 51 percent of the credit, 51 percent of the billing, not to mention the action . . . [Soon] their firm was grossing over fifteen million dollars a year by which time it had also bought The Theatre Guild's investment in the early Rodgers and Hammerstein triumphs. Dick and Oscar owned a hundred percent of everything they wrote, and a good sized piece of everything else."

In late 1943 they were both in the enviable position of having *Oklahoma!* selling out nightly. *Carmen Jones* was being readied for a New York opening under the aegis of Billy Rose, all of which gave Oscar the unaccustomed luxury of being able to choose his next project without rushing.

Both he and Dick constantly discussed what would appeal to them for their next undertaking, knowing too well they would be in for the inevitable comparisons. Samuel Goldwyn, after seeing *Oklahoma!*, had reportedly sidled up to Dick, pumped his hand enthusiastically, and uttered his famous remark: "Terrific! Do you know what you should do next?" When Rodgers shook his head, Goldwyn sputtered, "Shoot yourself!"

But Oscar was beyond the joking area. Like many show-business personalities, he listened to everyone but took no one's advice. In his quest for a project his first impulse, hoping to get as far away from the hugeness of *Oklahoma!* and *Carmen Jones* as possible, was to suggest a small, chorusless, urban domestic comedy: *Life With Father.* But that was soon vetoed by Dick, who wanted to try something more meaningful.

They were in the habit of meeting with Theresa Helburn at Sardi's for "gloat luncheons" every Thursday, and her co-director Lawrence Langner joined them at one of these midday discussions early in 1944. The indefatigable Terry put her finger to her lips so that no member of the theatrical fraternity sitting at nearby tables might overhear and whispered the name of Molnár's 1909 play, *Liliom,* which The Theatre Guild had produced quite successfully back in 1921.

Both Dick and Oscar had seen the recent revival starring Ingrid Bergman and Burgess Meredith and had been unimpressed with this depressing and mystical story of an inarticulate carousel barker who kills himself after an aborted attempted robbery. Allowed to return to earth sixteen years later, he visits his wife Julie and daughter Louise, but does little to redeem himself before he is called back to spend eternity in Hell.

The collaborators' major reasons for rejecting the play were primarily its "down" ending and its Hungarian locale, for it looked like Hungary would shortly be going communist. After further meetings Terry swept away their objections to the finale by pointing out how Oscar had transformed the dismal ending—indeed, the entire last act of *Green Grow the Lilacs*—in *Oklahoma!* As for the setting, that could be easily changed, too. She suggested making Liliom a Creole and moving the action to New Orleans.

Oscar rejected this suggestion, for he was reminded at once of his debacle with Cajun accents, what he called the "zeez, zemz and zoze," in *Sunny River* just two years earlier, but the poignancy and the fantasy in the story still intrigued him. He told Helburn and Langner that he would try to think of a more suitable locale, for he was genuinely eager to find a way to adapt *Liliom.*

His enthusiasm changed from lukewarm to avid when Dick came up with the idea of changing the setting to the New England coastline. When it was agreed to set the time period back to the 1870s, at the time of the Industrial Revolution, Oscar began to see how he could bring in an attractive ensemble. "Sailors, whalers, girls who worked in the mills up the river, clambakes on nearby islands and an amusement park on the seaboard," he was to say later. "I saw people who were alive and lusty, people who had always been depicted onstage as thin-lipped Puritans." Now they tentatively told the Guild they would do it *if* they could work out the motivation and *if* Molnár's permission could be gotten advantageously.[2]

The all-important plot motivation and contractual arrangements took over a year to iron out, but R & H were far from idle during that time. They busied themselves with producing activities while offers of every sort came daily over their desks.

With millions of theatergoers and moviegoers eager to see their show, it was only a matter of time before they would be approached by Twentieth Century-Fox's Darryl Zanuck, who planned to capitalize on *Oklahoma!*'s success in another bucolic romance. Iowa, which he suggested as the setting for a nostalgic look at a bygone Americana, was not too dissimilar from Oklahoma, and the story chosen, *State Fair*, was one Fox already owned and had filmed a decade before starring Will Rogers and Janet Gaynor.

When Oscar and Dick screened the film they were charmed. They decided almost immediately that Oscar would write a screenplay and lyrics for the remake while Dick would provide the score. But they inserted one important stipulation in their contract: neither of them was to be obliged to work in Hollywood. Past experience had taught them that in the cinema capital the writer is the least respected member of the crew, and they hoped their new celebrity would translate into clout. Staying east, they figured, would prevent their work being truncated by a hack out to please a mogul like Zanuck, who would certainly stand over their shoulders. The ploy worked. Soon Rodgers was to crow that "they were so intimidated by the fact that Oscar and I had written *Oklahoma!* they made the picture just the way we wrote it."

Filmed in California, but mostly written on Oscar's Pennsylvania farm and at Dick's home in Connecticut, *State Fair*'s simple plot concerns a naïve country girl and her brother who find romance in the midst of a cattle-and-pie-judging county fair. It was in the popular genre of down home, just-folks family movie musicals, the most famous of which was to be *Meet Me in St. Louis*. The film contained half a dozen songs, three of which deserve mentioning.

In "It's a Grand Night for Singin'" and in "That's For Me," the poet in Oscar who seems to have been hiding just below the veneer in *Oklahoma!* comes to the surface in exquisite lines like

---

[2] Molnár had already rejected Puccini's offer to turn his play into an opera, saying he preferred *Liliom* to be remembered as a play by Molnár rather than an opera by Puccini. As for the contract, the Hungarian playwright received the smallest percentage of all the principal contributors: eight-tenths of one percent.

. . . The moon is flying high,
And somewhere a bird who is bound he'll be
    heard
Is throwing his heart at the sky.

It's a grand night for singing,
The stars are bright above,
The earth is aglow, and, to add to the show
I think I am falling in love . . .

"That's for Me" is this writer's special favorite for several reasons. Not only is the lyric poetic and even, in its ending, genuinely surprising, but the whole song exudes the optimistic dream that was soon to become—for better or worse—a Hammerstein anthem. In this lyric Oscar plays fast and loose with form.[3] The song handily encompasses a whole day in the lives of the lovers— meeting and separation. The separation is handled in the masterful manner Oscar had not used since *Show Boat*'s "Can't Help Lovin' Dat Man." With nothing superfluous, even the verse, this has to be the ultimate "love at first sight" song.

VERSE
Right between the eyes,
Quite a belt,
That blow I felt this morning!
Fate gave me no warning.
Great was my surprise—

REFRAIN
I saw you standing in the sun,
And you were something to see.
I know what I like, and I liked what I saw,
And I said to myself,
"That's for me!"

"A lovely morning," I remarked,
And you were quick to agree.
You wanted to walk and I nodded my head
As I breathlessly said,
"That's for me."
I left you standing under stars—
The day's adventures are through.
There's nothing for me but the dream in my
    heart—
And the dream in my heart—
That's for you.
    Oh, my darling,
    That's for you.

State Fair's biggest hit was "It Might As Well Be Spring," which brought Oscar his second (and Dick his only) Academy Award. Pedants who have ana-lyzed this fine song have attributed much of its appeal and sense of malaise to the obvious "wrong note" and "jumpy rhythm" Rodgers created when he set Oscar's lyric illustrating the laconic heroine who is so out of step she gets "spring fever" in the fall (see Example A). But they quite ignore Rodgers's walking into a strange new key when Hammerstein clues him with "I keep wishing I were somewhere else / Walking down a strange new street." Later, on the masterful lines "Hearing words that I have never heard / From a man I've yet to meet," the listener is transported to yet another new land by a daring shift—about as far as one can go—from C to F♯ minor (see Example B).

---

[3] As does Rodgers. "That's for Me" is a breakthrough 35-bar song. A^1 and A^2 are each ten bars long. The release of five bars is followed by a shortened A^3, now five bars. Then the whole song is tied together with a five-bar coda. For musicians accustomed to working in four- or eight-bar phrases, this is indeed a challenging departure.

Example A

I'm as rest-less as a wil-low in a wind-storm. I'm as jum-py as a pup-pet on a string.

Example B

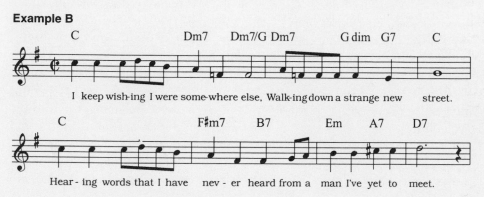

I keep wish-ing I were some-where else, Walk-ing down a strange new street.

Hear - ing words that I have nev - er heard from a man I've yet to meet.

Shortly after he sent off the finished script of *State Fair,* Oscar was hospitalized with what he thought was an attack of appendicitis. Somewhat relieved when he found it was diverticulitis (from which he had suffered previously), he was much cheered to learn that during his confinement *Oklahoma!* had been awarded a special Pulitzer Prize. Even more gratifying to the lyricist still bouncing back from his decade of rejection was the news from Hollywood that no revisions would be needed in the *State Fair* script.[4]

When he was discharged from the hospital, Oscar spent several what he misnamed "lazy" weeks at the farm, his head filled with thoughts of the problems ahead in turning *Liliom* into *Carousel.* Both he and Dick read and reread the Glaser translation of the play, trying to find a clue that might transform their brusque, self-centered protagonist into a sympathetic character.

"Suddenly," Rodgers recalled, "we got the notion for a soliloquy in which, at the end of the first act, the leading character would reveal his varied emotions about impending fatherhood. That broke the ice."

The "Soliloquy," an almost eight minute operatic solo, fleshes out Billy's character and explains much about the angry loafer who strikes out at everyone, even the wife he loves. Here the lyric-libretto—either word alone does not satisfy—does much to replace the laissez-faire attitude Bigelow has evinced

[4] According to an announcement from the R & H Organization, a stage musical based on "State Fair." and using an augmented score is planned for Broadway in 1995

so far with genuine concern for his unborn child. So stirring is this aria that now we are rooting for Billy, hoping he can pull off the robbery in the next act.

Since it begins with "My Boy, Bill, I will see that he's named after me," this writer asked Bill Hammerstein if he was the model for this breakthrough number.

"No, he didn't write it about me. It's just a good name to work lyrically . . . there's only one part of that lyric that's a memory, the line about 'when we go out in the mornings for our swim,' because when we lived in that house in Great Neck which is right on the Long Island Sound we would indeed go in the mornings for our swim."

Whatever inspired it, the scene's—one cannot call it a mere song—pivotal position in the *Carousel* is critical. It is a given that the immature Bigelow would look forward to the birth of a son for the macho companionship it would afford. Reading the lyric, which fairly sings off the page, one is struck by Oscar's poetic sense. Indented and italicized passages are used to let his composer, interpreter and eventually audiences know that here in a separate section are interior thoughts.

Halfway though Hammerstein subtly introduces the idea that his offspring might be a girl. When Billy sheepishly offers his imagined son advice on how to get "round any girl," his utterance triggers his thoughts to the possibility of a daughter and the misty image of his boy evaporates. In its place there comes his vision of his little girl and the fatherly support she will be needing. Out of this comes the decision to rob that eventually leads to his own death. This is more than a musical number, for it points up the validity of the entire book. Perhaps this scene is Hammerstein at the very peak of his powers as a lyric-librettist.

I wonder what he'll think of me!
I guess he'll call me
"The old man."
I guess he'll think I can lick
Ev'ry other feller's father—
Well, I can!

I bet that he'll turn out to be
The spit an' image
Of his dad.
But he'll have more common sense
Than his puddin'-headed father
Ever had.

I'll teach him to wrassle,
And dive through a wave,
When we go in the mornings for our swim.
His mother can teach him

The way to behave,
But she won't make a sissy out o' him—
Not him!
Not my boy!
Not Bill . . .
Bill!

My boy, Bill!
(I will see that he's named
After me,
I will!)
My boy, Bill—
He'll be tall
And as tough
As a tree,
Will Bill.
Like a tree he'll grow,
With his head held high,

And his feet planted firm on the ground,
And you won't see nobody dare to try
To boss him or toss him around!
No pot-bellied, baggy-eyed bully'll toss him
    around!

I don't give a damn what he does
As long as he does what he likes,
He can sit on his tail
Or work on a rail
With a hammer, a-hammerin' spikes.

He can ferry a boat on the river
Or peddle a pack on his back
Or work up and down
The streets of a town
With a whip and a horse and a hack

He can haul a scow along a canal,
Run a cow around a corral,
Or maybe bark for a carousel—
(Of course it take talent to do that well.)

He might be a champ of the heavyweights
Or a feller that sells you glue,
Or President of the United States-
That'd be all right, too.

(Spoken)
(His mother'd like that. But he wouldn't be
    President unless he wanted to be!)
Not Bill!
My boy, Bill—
He'll be tall
And as tough
As a tree,
Will Bill!
Like a tree he'll grow
With his head held high,
And his feet planted firm on the ground
And you won't see nobody dare to try
To boss him or toss him around!
No fat-bottomed, flabby-faced, pot-bellied
    baggy-eyed bastard'll boss him around!

And I'm damned if he'll marry his boss's
    daughter,
A skinny-lipped virgin with blood like water,
Who'll give him a peck and call it a kiss
And look in his eyes through a lorgnette . . .
Say!
Why am I takin' on like this?
My kid ain't even been born yet! . . .

I can see him
When he's seventeen or so,
And startin' in to go
With a girl.

I can give him
Lots of pointers, very sound
On the way to get round
Any girl.
I can tell him—

Wait a minute! Could it be?
What the hell! What if he
Is a girl!
Bill!
Oh, Bill! . . .

(Spoken)
(What would I do with her? What could I do
    for her?
A bum—with no money!)
You can have fun with a son,
But you got to be a *father*
To a girl! . . .

She mightn't be so bad at that—
A kid with ribbons
In her hair,
A kind o' sweet and petite
Little tintype of her mother—
What a pair!
(I can just hear myself braggin' about her)

My little girl
Pink and white
As peaches and cream is she.
My little girl
Is half again as bright
As girls were meant to be!
Dozens of boys pursue her,
Many a likely lad
Does what he can to woo her
From her faithful dad.
She has a few
Pink and white young fellers of two or
    three—
But my little girl
Gets hungry ev'ry night
And she comes home to me . . .

(Spoken)   She's got to be sheltered and fed, and dressed
My little girl!   In the best that money can buy!
My little girl!   I never knew how to get money,
                                            But I'll try—
I got to get ready before she comes,   By God! I'll try!
I got to make certain that she   I'll go out and make it
Won't be dragged up in slums   Or steal it or take it
With a lot o' bums—   Or die!
Like me!

With the crucial problem solved, "everything else," Rodgers was to say, "then fell into place." Moving ahead, casting was announced for January 1945. It was decided to use the same director, Rouben Mamoulian, and choreographer, Agnes de Mille, who had served so well in *Oklahoma!* This time, because neither had anything to prove to the other, they got along.

Carousel opens without an overture in the usual sense, but Rodgers's dark, slightly out-of-tune merry-go-round waltzes coupled with Oscar's stage directions as realized by Mamoulian set out the relationship of Julie, Mrs. Mullin, Carrie, and Billy far better than any opening scene with dialogue could. Moreover, by plunging directly into "The Carousel Waltz," R & H changed the decades-old habit of theatergoers drifting haphazardly into their seats while the orchestra played a medley of the forthcoming hit tunes.

"Sit up and watch. The plot is unfolding already," this first piece with its liberal use of bitonality demanded of the audience. And the length of the pantomime scene (seven minutes) hinted at groundbreaking extended musical sections to come. Rodgers was so adamant (and so prescient) in his feeling that the musical tempo not be interrupted that when de Mille insisted, "You've got to illustrate this a little bit more musically. You can't just go on playing a waltz in three-four time, unbroken without any punctuation from the orchestra," he shot back, "I want those waltzes absolutely the way they were written."[5]

Although by the time of *Carousel*'s creation the waltz was a long outmoded form, replaced early in the century by ragtime, in Rodgers's hands even the many he wrote with Hart never smack of *alt Wien*. They are as fresh as tomor-

[5] Nicolas Hytner, who directed the Royal National Theatre's acclaimed London revival in 1993 (which was greeted with greater hosannas when transferred to New York's Lincoln Center the following year), infused Rodger's continual three/four meter with arresting and swirling circular images.

At curtain's rise we see an enormous clock under which the factory weavers are busy plying their looms. At six, when they are released from their work with a whoop, the scene offers a gigantic low-hanging white-yellow moon. Eventually a swirling carousel unfolds and is now constructed before our very eyes. Hytner's imaginative visual conception makes us feel that de Mille's request for variety and Rodger's caveat of no time change are meeting each other half way.

row and as American as apple pie. "The Carousel Waltz"[6] with half a dozen glorious melodies, is by far the best of his many cheeky incursions into three-quarter time. Brilliantly orchestrated by Don Walker, with themes swirling now from the strings, now from the tuba (which Rodgers insisted be included in the orchestration), it can hold its head up in the concert hall alongside longhair pieces by Gershwin and Bernstein.

This brilliant scene having set the iconoclastic tone, *Carousel* was ready to display its collection of expanded musical sections braided together with dialogue. In its vocal ambition it is far closer to opera than to what it was called, "a musical play." That is why it is less frequently performed today (except in major revivals) than the other members of the R & H blockbuster quintet, *Oklahoma!*, *South Pacific*, *The King and I*, and *The Sound of Music*. There just aren't many actors around who can sing the roles.

The five-minute musical scene between Carrie and Julie is the first of these extended segments as well as Oscar's initial excursion into subplot. In the opening of *Liliom* Julie's friend Marie talks of her beau Wolf, a Jew, now a porter but ambitious enough to become a steward in the forthcoming act. He is the alter-ego for Mr. Snow, the herring fisherman, arch-Presbyterian, and Marie (given the New England name of Carrie) first tells Julie about him to get her "interested in findin' a feller." Later, when we get to meet him, Oscar outlines his character using a square-cut sing-songy rhythm, a counterpart of his rhythmless pragmatic soul:

When I make enough money outa one little boat,
I'll put all my money in another little boat.
I'll make twic't as much outa two little boats,
And the fust thing you know I'll have four little boats!
Then eight little boats, then a fleet of little boats!
Then a great big fleet of great, big boats.

Hammerstein gives all the fun to Carrie as he lightens the scene when Enoch Snow continues with

The fust year we're married we'll hev one little kid,
The second year we'll go and hev another little kid,
You'll soon be darning socks for eight little feet.
Carrie: Are you buildin' up to another fleet?

[6] In 1944 Rodgers, along with twelve other composers (Leonard Bernstein, Igor Stravinsky, Aaron Copland among them) accepted a commission for a concert piece from Paul Whiteman's radio sponsors. Rodgers's piece, called "Tales of Central Park," was never handed in and never performed. Biographer David Ewen is convinced that this piece "so happily caught the gay and colorful feeling of the opening carousel scene that Hammerstein persuaded Rodgers to use it for the play."

Jan Clayton and John Raitt play the long "bench scene" of Carousel, *which climaxes in "If I Loved You." In the 1993–94 revival in London and New York the bench was eliminated, requiring the principals to stand around uncomfortably.* MOCNY.

Then the couple go into "When the Children Are Asleep" (We'll sit and dream / The things that every other / Dad and mother dream), a duet about their quest for respectability. Their middle-class conformity is strong contrast to Billy's lawlessness and Julie's blind, fatalistic acceptance of her husband's being out of step with the society around him. "What's the use of wond'rin?'" she tells everybody later, "he's your fellow, you're his girl, there's nothing more to say."

This contrasting relationship, only hinted at in *Liliom*, becomes the whole crux of *Carousel*'s second half. In a stunning de Mille ballet we see the Snow children sixteen years after Billy's death taunting his daughter Louise, and we witness Louise's alienation and loneliness. Now there is a valid reason for Billy to return to earth and to create whatever empathy this unredeeming character ends up with.

In the last scene, the high school graduation, Billy observes Louise's suffering and reacts to it. He exits proudly seeming to have accomplished his good deed, because he insisted his daughter listen to the godly speaker. The starkeeper–country doctor–orator talks directly to Louise, telling her not to be held back by her parent's failures. "Stand on your own two feet," he says. Billy, whose words are heard only by the audience, urges, "Listen to him. Believe him."

The starkeeper continues with "and try not to be skeered o' people not likin' you—jest you try likin' THEM." Suddenly Louise puts her arm around the uppity Snow girl and rather simplistically we realize she has gotten the message and is on her way to being accepted into society. Now while everyone is singing that if you "walk on with faith in your heart you'll never walk alone." Billy goes over and kisses Julie, saying, "I loved you, Julie. Know that I loved you!"

Great and moving as the song is, it doesn't seem to be what is needed at that moment. Hammerstein the librettist seems to be telling us to embrace others, while Hammerstein the lyricist tells us in song that we must keep the faith.

It makes for a lame ending, one that works while we are in the theater but whose honesty seems to evaporate when we hit the fresh air. In Molnár's morality play Liliom is unreformed, while in Hammerstein's, the leopard who suddenly changes his spots appears rigged. Even Hammerstein noted his own artifice when he said he couldn't let Billy go down unchanged.

"It was not the anxiety to have a happy ending," he told interviewer Arnold Michaelis, "that made me shy away from that original, but because I can't conceive of an unregenerate soul—and to indulge myself I changed the ending."

But self-indulgent or not, Hammerstein's technique is strong enough to make us accept what is happening on stage. What *does* make us go further and weep at the end for the finally articulate Billy is Rodgers's magnificent music here and earlier, "If I Loved You" and "You'll Never Walk Alone," two of Rodgers's greatest achievements, underscoring the dialogue.

Hammerstein, deeply concerned as to how Molnár would react to the changes of venue and story line, was relieved when the Hungarian playwright came up to him with tears in his eyes after witnessing a run-through of the musical. "What you have done is so beautiful," he said, "and you know what I like best? The ending."

But the ending is powerful only when we remember the beginning. Early in the play each of these diffident people gives an oblique hypothesis about what it would be like *if* they were to be in love. Each comes to the conclusion that their shyness would eventually separate them. As the play progresses, we realize Julie's uncomplaining ways do not stop Billy from committing the robbery that will lead to his suicide; Billy's inarticulateness leads only to braggadocio and to the same tragic results. Since both Julie and Billy are operating on different levels of expression, Hammerstein never lets them sing together, and to end the scene Rodgers's orchestra creates a smashing crescendo that tells much more than their tentative kiss of their passion for each other.

"If I Loved You" is never reprised until the penultimate scene when, far too late, Billy is at last able to articulate his love with no ifs but only to the audience. Eventually Billy is even able to communicate his message to Julie, who, unable to hear him, senses his presence. In the reprise Hammerstein alters the last eight lines, simply and masterfully. (The earlier version is printed alongside the reprise.)

. . . Longing to tell you but afraid and shy,	. . . Longing to tell you but afraid and shy,
I'd let my golden chances pass me by.	I let my golden chances pass me by.
Soon you'd leave me,	Now I've lost you,
Off you would go in the mist of day	Soon I will go in the mist of day,
Never, never to know	And you never will know
How I loved you	How I loved you,
If I loved you.	How I loved you.

There is little downright humor in *Carousel*; no jokes, no lighthearted characters like Ado Annie in *Oklahoma!* This seriousness of subject almost put R & H off from tackling *Liliom* and, as mentioned earlier, was mitigated when Richard Rodgers suggested the New England setting. Now only the lightness of the factory girls, the macho of the fishermen, the subplot, and especially the youth were able to save *Carousel* from becoming a dark morality exercise. Hammerstein uses every bit of fun he can find. Even the rite of spring, which comes to Maine a bit later than the rest of the country, is extolled lustily in "June Is Bustin' Out All Over." My favorite verse, the third, is printed below.

June is bustin' out all over	All the rams that chase the ewe sheep
The ocean is full of Jacks and Jills!	Are determined there'll be new sheep,
With her little tail a-swishin'	And the ewe sheep aren't even keepin' score!
Ev'ry lady fish is wishin'	On account a it's June!
That a male would come and grab her by the	June, June, June,
gills.	Jest because it's June, June June!
June is bustin' out all over!	
The sheep aren't sleepin' any more,	

For the second act's opening number, "This Was a Real Nice Clambake," Oscar's daughter Alice was sent up to Maine to discover just what went into a typical clambake. Although some Maine authorities voiced their objections to lobsters being slit down the back, Alice maintained that opinion was divided as to which side of the lobster is called the back. Her father then kept the back-slit lobster line to conform with his daughter's research and incorporated it into his mouthwatering recipe:

Fust come codfish chowder
Cooked in iron kettles
Onions floatin' on the top,
Curlin' up in petals.
Throwed in ribbons of salted pork,
An old New England trick,
And lapped it all up with a clamshell
Tied onto a bayberry stick!
Remember when we raked them red hot lob-
sters
Out of the driftwood fire?
They sizzled and crackled and sputtered a song,
Fitten for an angels' choir.
We slit' em down the back
And peppered 'em good
And doused 'em with melted butter,
Then we tore away the claws,
And cracked 'em with our teeth,
Cause we weren't in a mood to putter!

Then at last come the clams—
Steamed under rockweed
And poppin' from their shells,
Just how many of 'em
Galloped down our gullets,
We couldn't say oursel's.
OH - H - H
This was a real nice clambake,
We're mighty glad we came.
The vittles we et
Were good, you bet!
The company was the same.
Our hearts are warm,
Our bellies are full,
And we are feelin' prime.
This was a real nice clambake,
And we all had a real good time!

Aprés clambake, a relaxed company. "Our hearts are warm, our bellies are full, and we are feelin' prime." MOCNY.

WE HAVE
NO TICKETS
for "OKLAHOMA!"
AND CERTAINLY NONE
for "*Carousel*"
AND DON'T EVEN MENTION
"O MISTRESS MINE"

This sign was kept on display in the windows of most Broadway theater ticket brokers from 1945 through 1947. MOCNY.

But even with the presence of these two delightful numbers, *Carousel* was long and heavy-handed when it reached its tryout performances in New Haven. Oscar and Dick worked furiously to pare it down. They cut five scenes and half the ballet, eliminated Mr. and Mrs. God as a New England minister and his wife entirely, substituting the Starkeeper, and were still cutting songs and eliminating scenes and verses when the show opened in Boston.

By the time *Carousel* was ready to be seen at the Majestic Theatre, directly across from the St. James where *Oklahoma!* was housed and still selling out, the word handed down from Boston was that R & H had created a masterpiece. Almost all the New York critics agreed,[7] and the public kept the show running for 870 performances. In London, though not a critical success, it was nevertheless a popular one and, following on the heels of *Oklahoma!* at the Drury Lane, chalked up a run of sixteen months.

[7] The lone dissenter, Wilella Waldorf, writing in the *Post*, complained that "the *Oklahoma!* formula is becoming a bit monotonous, and so are Miss de Mille's ballets. All right, go ahead and shoot."

Carousel garnered most of the awards of its season, winning the Drama Critics Circle and the Donaldson Award for best musical. Donaldsons were also won by Hammerstein, Rodgers, Mamoulian, and de Mille. Nor were John Raitt, who played Billy, and dancers Bambi Linn and Peter Birch overlooked. They all won awards.

To the end of his life it remained Rodgers's favorite of all his work. "Oscar never wrote more meaningful or more moving lyrics, and to me, my score is more satisfying than anything I've ever written," he confessed. "It's the whole play . . . tender without being mawkish."

According to Oscar's son James, the aforementioned 1993 revival came closer to achieving what his father had imagined. "Here, rather than being a braggart who is unsure of himself, Billy is shown to be a man who's ready to explode . . . But underneath that is someone poetic . . . and it makes Julie excited by the violence, in a sense, and trapped into it. She doesn't know why."

Every work of musical art is capable of differing interpretations. Theater buffs will argue over the layers of meaning in *The Threepenny Opera*, *My Fair Lady*, and *Show Boat* for decades to come. But it is no small barometer of this musical's adaptability to the times that half a century after its creation it can speak meaningfully to a generation aware of violence and masochism.

Seeing or listening to *Carousel* over and over again is like watching a slowly unfolding rose. With every petal that unfurls one glimpses an additional bittersweet beauty beneath. Perhaps that is why it is generally considered Hammerstein's finest work, certainly his most poetic and moving. As its creator, Oscar confessed to "crying all through the rehearsals." Richard Rodgers expressed the added feelings of most theatergoers when he said, "It affects me deeply every time I see it performed."

Hammerstein
1945–1949

ON OCTOBER 10, 1947, *Allegro*, the fourth R & H collaboration (the third one staged), opened at the Majestic Theatre, where *Carousel* had ended its long run five months earlier. Walking through the Broadway theater area that evening, a prospective ticket-buyer might have chosen among several other R & H shows on display. *Oklahoma!*, now in its fifth year, was *still* selling out; *Annie Get Your Gun, Happy Birthday*, and *John Loves Mary*, all of which Oscar and Dick had produced, were running simultaneously.

Theatergoers away from Broadway likewise had a selection of R & H shows available to them. The touring *Oklahoma!* was in Boston, the road company of *Carousel* in Chicago, and the second company of *Annie Get Your Gun*, featuring Mary Martin, was a hot ticket in Dallas. Both *Oklahoma!* and *Annie Get Your Gun* were sell-outs in London, and a clone of the smash-hit R & H production of *Show Boat* would be sent on the road in ten days.

All this simultaneous production plus their film, recording, and publishing deals had turned the R & H office into a hive of activity. And Oscar and Dick were constantly called on to solve theatrical crises (real and imaginary—of which this industry has more than its share) because they took their responsibility so seriously. If they were successful it was because they themselves supervised everything—even to casting replacements for the smallest roles in their various touring entertainments.

One wonders why they kept their plates so full with all these varied projects. William Hammerstein believes it was their mutual lawyer and financial advisor, Howard Reinheimer, who advised Oscar and Dick that they would only be able to keep 10 percent of their vast profits. (At that time United States federal revenue took 90 percent of one's *net* profit in the higher brackets.) They figured they had little to lose and all to gain if the shows they backed were successful. And they knew that the producer's share of a show can be even more lucrative than the earnings of the work's creators.

William Hammerstein adds:

They didn't expect to succeed as well as they did. They didn't expect to lose money, but if they had lost it, it wouldn't have been a great loss. In England . . . because of English law, all the money that was accumulating from their London productions was collecting and they couldn't take it out of the country—so they started producing everybody else's shows. They did *Damn Yankees* and *Pajama Game* abroad. . . . I think all that producing came about as a practical answer to some financial questions.

With all this activity, it was not until early in 1947 that they turned to the creative: writing their new show. Soon the media broke the news that R & H had an unconventional musical planned for the fall. It was announced that this would be a "fluid musical" designed around a unit set.

Oscar's original concept was to write a story of a man's life from birth until death, showing how he faces his individual integrity and copes with the conspiracy of the world.

Oscar explained:

I think I got the idea because I was trying to write a play without much scenery. I was thinking not only of Broadway, but the colleges which need very much properties that they can do. I evolved a play with just props, chairs and tables and so forth. . . . I borrowed the chorus idea from the Greeks, and then found that I could do other things with the chorus to provide the audience with insight into the characters. I intended Dick to write music for it, but we wound up reciting the chorus instead.

But the R & H reputation for a big musical would not be denied, and Oscar and Dick finally wound up with a great deal of scenery. Jo Mielziner's stage designs incorporated treadmills that shifted back and forth from the wings, platforms, loudspeakers, and back projection, and Oscar's original idea soon became subverted in technology. Actually that was the major reason for the musical's artistic as well as financial failure. Its massive scenery coupled with the huge cast assembled used up just about all the margin of production costs. When they took *Allegro* out on the road with a simpler revolving stage and reduced forces, it more nearly broke even.

"I was concerned when I wrote *Allegro* about men who are good at anything and are diverted from the field of their expertise by a kind of strange informal conspiracy that goes on," Oscar was to say. "People start asking them to join committees . . . and the first thing you know they are no longer writing or practicing medicine or law. They are committee chairmen, they are speech makers, they are dinner attenders."

In his concern for these subverted professionals he spoke of "writers, doctors, lawyers, businessmen," but he might as well have added "lyricist-librettist," for *Allegro* was more about himself than about the red-blooded American, Dr. Joseph Taylor Jr., he had chosen as protagonist and alter ego.

"He was so successful, he had so many responsibilities," Stephen Sondheim reported, "that they cut into his artistic life." Always keen to champion humanitarian responsibilities, Oscar had been an organizer of the Anti-Nazi League in the '30s, a member of the War Writers' Board from its inception until after World War II, and recently had been elected president of the Authors League.

"He accepted these responsibilities because he felt they were worth attending to. But at the same time he found himself farther away from his profession," Sondheim explained. "He transformed that into the story of a young doctor who comes to the big city at the behest of his upwardly mobile wife. Oscar turns the doctor into a politician who ends up laying cornerstones at hospitals."

Allegro tried to tell the story of the responsibilities that come with success but audiences—and most of the critics—misunderstood. They thought Oscar's message was simply anti-urban, curable with a return to the Grade A Milk values of the farm. Actually he never told his story very well—probably because Oscar, like everybody else, could find no solution to the dilemma. His sons James and William, both of whom have studied *Allegro* with a view to further performances, disagree on what the eventual outcome should be. "The man has got to leave his wife and go off with his secretary," states James, referring to the unfaithful and money-hungry spouse versus the wise-cracking, in-love-with-the-doctor nurse. The implication is that together they will go back to Dr. Joe's small town and practice medicine gloriously ever after.

"He must succeed *in* the city, not go back home and run away. . . . He must have his own practice now with integrity, and get rid of all those hangers on," William insists. Oscar himself must have sensed his predicament for he said, "I think I know where I got a little wobbly," and thought he had come up with a satisfactory ending for the play. The librettist's solution (for he was rewriting *Allegro* at the time of his death) was different from the hypothesis of either son. "As the doctor is leaving at the end of the play, he gives his wife a chance to go with him, and after a dramatic pause, she goes. In other words, he does not walk out on her. There is no suggestion that there will be a divorce. There are also other lines (in preparation for this change) earlier." Perhaps he was being hypocritical, having been through both these twentieth-century maladies, but divorce and infidelity were not Oscar's bailiwick.

Unfortunately, the validity of the ending matters little, for by the time we reach the denouement, we no longer care about the characters, the lyrics, or Rodgers's melodies. Oscar said he was aiming to create an *Our Town* universality and simplicity, but his concept of Everyman seems closer to a cardboard Norman Rockwell picture. Yet even this might have been acceptable without the pretension that *Allegro* evinces. Perhaps all the faults might have been overlooked had the songs been up to the usual R & H standard.

The decision to use the chorus was a mixed blessing whose novelty intrigues for half the first act until it becomes redundant. It was *Allegro*'s first indication that Hammerstein had boxed himself into a no win situation. The protagonist, Joseph Taylor Jr., does not appear until his college days but is only commented on by the massed groups on either side of the stage. When we finally see him, the chorus who has been telling us what the growing boy has been thinking and feeling through college and in his life as a doctor might as well go home.

Allegro has the weakest score of all the R & H collaborations. With only three semi-extractable songs—two of which are given to secondary characters— and not a hit among them, its music and lyrics don't seem to have been written by the team with a fund of melody and cogent lyrics that turned out *Oklahoma!* or *Carousel.* The wholesome, almost treacly songs in the first act seem out of tune with the quasi-ironic, message-laden ones in the second, but listed as they are below can serve as a synopsis of the *Allegro*'s plot.

"A fellow needs a girl to sit by his side at the end of a weary day," sings Dr. Taylor, Joe's father, to his wife early in the play. "It's a darn nice campus. . . . I like my roommate / And you would like him too / It's a darn nice campus / But I'm lonely for you," Joe writes to childhood girlfriend Jennie. Then later he flirts with the girl his swinging roommate, Charlie, introduces him to. But Joe is too goody-goody to seduce her, so he serenades her with, "We have nothing to remember so far."

The story of Joe's romance culminates in "You Are Never Away" (from your home in my heart), which Joe sings before capitulating to the materialistic Jennie. After a truly purple scene in which Joe's mother dies after a confrontation with Jennie, the act ends with "What a Lovely Day for a Wedding," which incorporates dreadful lines like

> There's a lively tang in the air,
> It's a treat to meet at a wedding
> When fam'lies are letting down their hair.

Act Two opens after the Crash, with Jennie hanging clothes on the washline and consoling the female ensemble with "Money isn't everything . . . as long as you have dough!" The centerpiece of the act is a cocktail scene titled "Yatata, Yatata, Yatata," wherein Hammerstein's alter ego, now using the voice of Charlie, comments

> Broccoli, Hogwash, Balderdash,
> Phoney Baloney, Tripe and Trash.

At last we come to the title song, a pallid rehash of Noel Coward's twenty year old hit, "Dance, Little Lady," it seems like a yet another version of "Yatata, Yatata, Yatata":

Brisk, lively, merry and bright!	Don't stop whatever you do!
Allegro!	Do something dizzy and new!
Same tempo morning and night!	Keep up the hullabaloo!
Allegro!	Allegro!

Joe's nurse, Emily, now comes on with perhaps the best song in the show—certainly the most interesting musically—but one written in such pop style it seems to belong to another play. In love with Joe and aware that his wife, Jennie, is having an affair with a member of the hospital board, she sings "The Gentleman Is a Dope." ("Why am I crying my eyes out, he doesn't belong to me? He'll never belong to me.") The play ends with the ghost of Joe's mother exhorting and finally convincing him to "Come Home" ("You will find a world of honest friends who miss you / You will shake the hands of men whose hands are strong / And when all their wives and kids run up to kiss you / You will know that you are back where you belong"). Joe, Emily, and Charlie will go back to their small town and become dedicated to the practice of medicine.

Agnes de Mille, who was given her first choreographic break with *Oklahoma!*, had by this time parlayed her triumph as Broadway's major force in dance and movement with *One Touch of Venus*, *Bloomer Girl*, *Carousel*, and *Brigadoon*. Now she had been engaged to direct the entire production and so of course was the first to see the script. Here is a transcript from one of the last interviews she gave this writer:

R & H sent me the first two scenes, and I wept because I thought it was so beautiful. I phoned Oscar and said, "If Thornton Wilder wrote a musical it would be this one. I think you're writing pure poetry. It's just lovely." And then when the ending of Act One came, I said, "This is not right, I want to talk to you about it." And he said, "Let me finish the play." Well, he finished the play two days before we went into rehearsal, and said, "Now, now we'll talk." And I said, "Oscar, will you tell me what this play's about?" He said, "It's about a man not being allowed to do what he wants to do in life." I said, "That's a wonderful subject, and very important." And I asked him not to go into rehearsal. "Just drop an iron bar." And he said, "That's not possible." And I said, "You probably won't have a failure, because I don't think it's possible for you two to have a failure, but the other thing could be great." They went in. . . . They didn't listen to me.

De Mille's account shows just how deeply Hammerstein was caught up in the Dr.-Joseph-Taylor-Jr. syndrome. Heavily occupied with the R & H schedule of plays in production and committed to a fixed opening night, he could not conceive of delaying the premiere. Nor, with all his extraneous activities, could he seem to find time to sit back and take a critical look at what he and Rodgers had created. At last, with a deadline upon them, Oscar rallied to help his overworked director by directing the book scenes himself. It must be admitted that the vogue for ballet-in-musical had passed, and Oscar, looking back on what was to be his most personal show, always felt de Mille was not up to the task of directing the entire production. She had passed her peak, and although she was to continue to choreograph-direct, only her choreographic efforts were successful.

Since R & H refused to postpone the opening, there was no time to replace the muddy-colored costumes, repaint the unrelieved blue-green of the unit set, which had what de Mille described as "a worm running up the back" (a path), or rechoreograph dances that had the cast frantically somersaulting and scampering rather than dancing.

But all that might have been forgiven if the two acts had development and consistency. The first act's country doctor, his wife, and his mother turn out to be Grant Wood clones. They have nothing on Joe Jr., the All-American son, and his All-American girlfriend. In the second act, in the city, we have an entirely different play.

"They concluded that I had said that city people were wicked, and rich people were wicked, and poor people were virtuous which isn't true," Oscar said. "That was not in the play, and yet, I had somehow given out that flavor to some people in the audience and some critics, which only means that there was some fault in my writing."

It must also be admitted that Oscar (as Lerner, too) was far more creative in adaptation or collaboration than he was in working out an original story. Most of his major hits—*Show Boat, New Moon Oklahoma!, Carousel*—had been adaptations. His only original story had been *Music in the Air*, which owed much to Jerome Kern. *Allegro*'s story unfortunately allowed him to get on a soap box and preach about abolishing artifice and returning to good honest values, a criticism that was to become more and more frequently leveled at himself in the ensuing years.

Thus, on opening night of *Allegro*, Oscar and Dick, who had been hailed for four years as the most successful theatrical collaboration since Gilbert and Sullivan, were sticking their necks out with what they felt was a new and daring type musical. Since nothing in the theater wants to be brought down like established success, critics of their new show, which had amassed a then unheard-of three-quarters of a million dollars *before* it opened, were certain to be sharply divided.

An exhausted Richard Rodgers watches a dancer rehearse a routine from Allegro *in the lounge of the Majestic Theater while Agnes de Mille and her assistant Dania Krupska lend encouragement.* MOCNY.

Robert Coleman wrote in the *Daily Mirror:*

Perfection is a thrilling thing, be it a Joe DiMaggio catch, a Cushing brain operation, a Rembrandt painting, a Whistler etching, a Markova-Dolin *Giselle,* or *Allegro,* the great new musical by Richard Rodgers and Oscar Hammerstein. Perfection and great are not words to be lightly used. They have become commonplace through misuse. But *Allegro* is perfection, great. It is a stunning blend of beauty, integrity, intelligence, imagination, taste and skill. It races the pulses and puts lumps in the throat. But for its utmost enjoyment, the patron out front must resound with mind and heart.

Robert Garland, writing in the *Journal-American,* seems almost to have been sitting in on a different musical. His critique ran:

To the Messrs. Funk and Wagnalls, publisher of *The Standard Dictionary*, "Allegro" means "brisk," "lively," and even downright "gay." But to the Messrs. Rodgers and Hammerstein the word means nothing of the kind. They seem to have confused "allegro" with, say, "lento." Last night their *Allegro* went its slow, unhurried way, telling with the aid of moving platforms, travelling curtains, lantern slides, Greek choruses, loudspeakers, a huge company of actors, singers, dancers and a symphony orchestra, a simple run-of-the-U. S. A. biography. . . . The could-be heartening narrative of a small town medico who goes to the big town and sees the error of his ways is all but lost in the contrivance of its projection.

It is understandable that critics would be sharply divided. R & H, who had opened a new door with the naturalistic musical in *Oklahoma!* and created a truly integrated musical in *Carousel*, were marching through that door with a theatrical happening of the kind that never had been seen on Broadway. Although the show was swallowed in technology, back-projections, "travelers," a Greek chorus, and a hero who doesn't even appear until half the evening is over, there were enough critics who admired R & H's daring to give *Allegro* the Donaldson Award for book, lyrics, and musical score.

But even that could not induce the public to come. The intrepid who did attend never ran back to tell friends they *must* see *Allegro*. The only thing to marvel at was the technical smoothness. *Allegro* is the show that spawned a take-off on the well-known vitriolic barb, "one walks out humming the scenery—except that there was none!" Without word of mouth the show was doomed to failure.[1] Although it played a respectable nine months followed by a thirty-one-week, sixteen-city tour, because of its huge overhead the show finished in the red.

But financial failure often translates into a *succès d'estime*, and contemporary historians have been kinder to *Allegro* than critics were in 1947. They often cite *Allegro* as the first concept musical, or a show written around a theme. With no character standing out, none being given more than one song, one comes away with the "idea" that Joe's particular story is secondary to the concept that success corrupts. Flawed though *Allegro* may have been, it has the distinction of being the first of a class of splendid new musicals. Plot became secondary to philosophy. *Company*, whose idea is marriage; *Baby*, which is concerned with approaching parenthood; *Starlight Express*, an experiment concerning trains; *Sunday in the Park with George*, how an artist rearranges reality; and *A Chorus Line*, whose concept is auditioning, owe a debt to *Allegro*.

R & H's vision that the show should remain fluid was also responsible for

---

[1] *Allegro* has frequently been revived, with little success (Lambertville Music Circus, 1952; St. Louis Municipal Opera, 1955; Goodspeed Opera House, 1968; Equity Library Theatre, 1978; Lost Musicals, Barbican, 1992; New York City Center, 1994).

the emergence of the choreographer-director. Although *Allegro* was to be her last involvement with R & H, de Mille was the first in a line that was to flower into powerful theatrical personalities, from Michael Kidd, Bob Fosse, Jerome Robbins, and Michael Bennett down to Tommy Tune.

But there is a still deeper implication to *Allegro*'s failure that goes far beyond the innovations that it was to engender. This financial blow to Oscar and Dick's no-flop record hurt them far more than the money they lost on a theatrical venture into unchartered territory. *Allegro* was Oscar's first and last attempt to remold the musical theater, to nudge in a new direction the form he had tried all his life to develop. Had they been younger or had *Allegro* been somewhat more successful, he and Dick would certainly have been daring again. For Hammerstein and Rodgers it remains their only attempt at real innovation. Henceforth for the rest of their collaborative years they would play safe.

Now they knew they needed a hit.

Their next was a hit of such massive proportions that it would be better classified as an out-of-the-ballpark home run. Instead of the few pallid, quickly forgotten numbers *Allegro* produced, *South Pacific* was to have fourteen songs. Four of them—"Some Enchanted Evening," "Bali Ha'i," "Younger Than Springtime," and "I'm in Love with a Wonderful Guy"—reached perennial or standard status. Others, less well known but still familiar, like "A Cockeyed Optimist," "Carefully Taught," "Bloody Mary," "There Is Nothin' Like a Dame," "I'm Gonna Wash That Man Right Outa My Hair," "Happy Talk," "Honey Bun," and "This Nearly Was Mine," became only slightly less popular. Actually there were only two songs that were overlooked in most anthologies, "Dites-Moi" and "Twin Soliloquies," and for this there is good reason. The first was in French and the latter was a duet of thoughts presumably unspoken.

To write a score of such dazzling wealth had not been decided beforehand, but in January 1948, after Broadway director Josh Logan told first the composer and then the librettist to read a moving, somewhat Madama Butterflyish story titled "Fo' Dolla" in a collection called *Tales of the South Pacific* by hardly known author James Michener, they knew they had found their next project and all their plans began to hang together.

"I was in Philadelphia, stage managing *Mister Roberts*," William Hammerstein remembers, "and Dad called me up and said 'I want you to go and buy a book called *Tales of the South Pacific*. I'm going to make a musical out of it.' I bought the book, read it and called him the next day. I said, 'It's a wonderful book, but how the hell are you going to make a musical out of this?' He said, 'Well, we are!' "

The book was full of stories besides "Fo' Dolla," and although the tale of the strait-laced naval lieutenant and his love affair with the sixteen-year-old daughter of Bloody Mary was both moving and appealing, both Oscar and Dick felt it was not original enough to support a full evening. Looking for a subplot they were intrigued by "Our Heroine," the story of an Army nurse and her romance with a French planter. Since this combination represented two *serious* stories (*Oklahoma!* had offset the romantic relationship of Curly and Laurey with the comic one of Ado Annie and Will Parker; *Carousel* similarly used Carrie and Mr. Snow as foils for Billy and Julie), the collaborators decided to add yet another lighthearted story to the cauldron. "A Boar's Tooth," which described comic con man Luther Billis, served as a way to link him amusingly with the harridan Bloody Mary. "Fo' Dolla" remained the moving main story but now there were two subplots, a kind of construction Oscar had not used since *Show Boat.* Other tales in the collection contributed extra color and background or nuances, but the above three supplied the principal ingredients.

It was decided from the outset that Joshua Logan, who had found the property originally, would direct. This was no plum of gratitude for bringing the story to their attention. Josh had brilliantly directed Rodgers and Hart's *I Married an Angel, Higher and Higher,* and that team's last show, *By Jupiter.* After a wartime stint serving in army intelligence, he had recently done the same for the R & H production of Irving Berlin's *Annie Get Your Gun.*

With a strong director chosen, Oscar and Dick were careful to avoid falling into the traps that had strangled *Allegro.* To gain further control of the project, to prevent any producer from casting his latest heartthrob in a key role, and since they had no need to worry about raising money, they would produce *South Pacific* under their own aegis.[2] Turning further away from the mistakes of *Allegro,* there would be no ballet, indeed no choreographer. Josh Logan was entirely capable of staging the dance numbers.

By late May, James Michener's book had been optioned and contracts with Logan and co-producer Leland Hayward had been signed. Now the word on the street was that R & H were looking for a mature singing actor. Edwin Lester, who had signed Ezio Pinza to star in a musical he was producing for the Los Angeles Civic Light Opera Association, called to say that since his musical had fallen through, Pinza's contract was for sale. The availability of the famous Metropolitan Opera singing actor made Oscar and Dick look more deeply into the story of the mature and successful French planter Emile de Becque and his love affair with Nellie Forbush, the Navy nurse from Arkansas. It was soon

---

[2] R & H were reluctant to share producer's credit. Eventually they were obliged to give away minority interest. Above the title the billboards proclaimed: **Richard Rodgers and Oscar Hammerstein 2nd in association with Leland Hayward and Joshua Logan present** . . .

decided that the *main* and *subplots* could happily be reversed. Now the May to
November romance would be fresh in a musical theater that usually confined
love stories to youth.

A few weeks later they lunched with Pinza and signed him. Ever wary of
overuse of his vocal chords, he had stipulated in his contract that he would not
have to do more total singing in his eight weekly shows than what the equivalent vocal parts of two average operas called for. Shortly after that their
thoughts turned to Mary Martin, who was touring in *Annie Get Your Gun.*
Eventually they persuaded her to play Nellie Forbush—knowing the simple naïveté she could muster would make a perfect foil for the elegant Ezio.

Mary, who at first joked by asking, "What do you want with two basses?,"
was terrified of having her vocal quality compared with that of a world-famous
Don Giovanni. "I have thought about it very carefully," Rodgers assured Mary.
"You will never have to sing in opposition to Pinza. You'll sing in contrast
to him."[3]

They had broken the initial pledge of their altruistic alliance—that they
would never write for stars—but this combination of operatic respectability coupled with the down-to-earthiness of Martin's show-business glitz was a commercial brainstorm. Rodgers wrote "Some Enchanted Evening," with its tender
coda ending on a pianissimo high E, with Pinza's voice in mind. Martin's belting or chest range was also considered when Dick composed "Cockeyed Optimist" and "I'm in Love with a Wonderful Guy." Soon they began choosing
some of the other key roles as well. Juanita Hall's rich voice easily won her the
role of Bloody Mary even before the book was finished.

Writing for Pinza, who didn't want to sing too much, forced another decision. It would preclude the long musical passages wherein the story is told.
Indeed, the songs in *South Pacific* sit like jewels on the water. They give the
audience a good time but hardly help to create the integrated musical Oscar
and Dick had now become famous for, represented by *Oklahoma!* and *Carousel.*

But the decision as to what *kind* of musical they were about to create allowed
Oscar to plunge into the libretto and come up with one of his finest construc-
tions (one can hardly call it an adaptation). In his previous two musicals, although he had moved scenes and changed motivations, he hewed to the outline
of *Green Grow the Lilacs* and *Liliom.* But here, since there was no play, not
even a novel's thread to adapt, he was more in a position of originator working
with already established characters.

Fixing the desires and insecurities of these contrasting individuals in the audience's mind at the outset of the play was paramount. Oscar and Dick recog-

---

[3] Rodgers broke his promise and had his stars join voices near the end of Act I in a reprise of
"Cockeyed Optimist."

nized the chemistry that Martin and Pinza would be able to exude on stage and early in the play created their unspoken twin soliloquies, a daring musical scene that tells much about each one's character. The quasi-operatic inspiration was found in Michener:

> "I was looking at the cacaos," Nellie said in a sing-song kind of voice. To herself she was saying, "I shall marry this man. This shall be my life from now on." . . . To himself de Becque said, "This is what I have been waiting for. All the long years. Whoever thought a fresh smiling girl like this would climb up my hill? . . . "

If the inspiration here is clearly Michener's, the concept and execution are certainly Hammerstein's. He knows when to let the music take over. Now he throws the scene to Rodgers, who quickens the action by shifting from three-bar phrases to two-bar ones. This leads to a musical postlude with a near or-gasmic orchestral climax when Emile and Nellie look into each other's eyes and clink brandy glasses. The scene that sets up the entire play to come is played in less than three minutes!

Nellie:
Wonder how I'd feel
Living on a hillside
Looking at an ocean
Beautiful and still.

Emile:
Younger men than I,
Officers and doctors,
Probably pursue her,
She could have her pick.

Emile:
This is what I need,
This is what I've longed for,
Someone young and smiling
Climbing up my hill!

Nellie:
Wonder why I feel
Jittery and jumpy!
I am like a schoolgirl
Waiting for a dance.

Nellie:
We are not alike.
Probably I'd bore him
He's a cultured Frenchman,
I'm a little hick.

Emile:
Can I ask her now?
I am like a schoolboy!
What will be her answer?
Do I stand a chance?

With his main characters introduced, Oscar's job was to elongate and to create suspense in a simple story, and he did it brilliantly. He came up with the idea to have de Becque agree to go on a dangerous military mission after Nellie's break with him. Now the mission prolongs the separation between the lovers and allows time for a show-within-a show. In the original story, the break and reconciliation occur in the same night. Oscar's treatment is vastly more theatrical.

Further, the libretto's use of Luther Billis as part of this particular tale (in which he fits admirably), making him Bloody Mary's complement, is brilliant.

After singing the "Twin Soliloquies," Ezio Pinza and Mary Martin drink a toast to friendship, mostly with their eyes. This is followed with a passionate kiss. MOCNY.

By being paired with Luther, Bloody Mary becomes more developed, and in addition to her original qualities she also becomes a major comedienne.

The dovetailing of these elements with the poignant story of Liat and Lieutenant Cable completes the triangular plot, making the audience deeply aware of the marvelous contrast between mature and very young love. Finally, in the libretto it is Lieutenant Cable who accompanies Emile on the dangerous mission. The knowledge of Cable's death is in itself deeply moving, but it also serves to conclude the Liat-Cable plot satisfactorily without Cable's having to refuse to marry her on racial-social grounds as he does in the original *Tales*.

Oscar did a great deal of soul-searching and at first wanted to have Cable and Liat remain together at the final curtain, but he eventually realized that the Lieutenant's prejudice was too deeply ingrained for a Broadway à la Hollywood "happy ending." Satisfying his knowledge of human nature, he achieved an additional lagniappe, for by having Cable killed offstage he was able to satisfy censors who might object to the liaison as fostering miscegenation.

But even though Oscar, who was violently anti-discrimination, was able to wiggle out of Cable's misuse of the native girl by killing him off, he insisted on fleshing out Cable's character in his second song, "You've Got To Be Carefully Taught." This courageously honest song, which originally compared hate and love, has none of the banner-waving excesses of agitprop. The final, parenthesized section of the song, while interesting in itself, mitigates the song's message and was cut from the score.

You've got to be taught to hate and fear,
You've got to be taught from year to year,
It's got to be drummed in your dear little ear,
You've got to be carefully taught.

You've got to be taught to be afraid
Of people whose eyes are oddly made
And people whose skin is a diff'rent shade,
You've got to be carefully taught.

You've got to be taught
Before it's too late
Before you are six,
Or seven or eight,
To hate all the people

Your relatives hate.
You've got to be carefully taught!
You've got to be carefully taught!

(Love is quite different.
It grows by itself.
It will grow like a weed
On a mountain of stones;
You don't have to feed
Or put fat on its bones;
It can live on a smile
Or a note of a song;
It may starve for a while,
But it stumbles along,
Stumbles along with its banner unfurled
The joy and the beauty,
The hope of the world.)

"Carefully Taught" was written and remained in the show in spite of many who advised both Oscar and Dick to delete it. Rodgers, however, as though trying to soft-pedal the song's hard-hitting message, couched Oscar's diatribe in a gentle, almost innocuous melody. The song was sung only once in the show and never reprised.

"But it was retained. My father was adamant about its inclusion," says son James,

> because he believed it's true—that kind of discrimination. But the US State De-
> partment took his passport and he had a helleva time. My father had belonged to
> the Anti-Nazi League which became a total communist front organization, but he
> left it as it was becoming one. So he was in an absolutely clear position. They
> didn't have anything on him because he was as clean as a whistle, and he was a
> patriot through and through. He was told to write what he believed about
> America, and he wrote a little dissertation on being an American, and showed it
> to me very proudly. I was young and I said, "You can't do this. This is humili-
> ating!"
> He said, "Why? I believe what I'm writing. Why shouldn't I say this?" I said,
> "Because it's on demand." He said, "That makes no difference to me, I still believe
> it." He got his passport . . . I guess he knew this kind of thing was going to pass.[4]

Oscar had no fear of saying what he felt was appropriate to his play and characters. Nor did he seek any help beyond his talking out the story with Dick, for he was secure in his knowledge of play construction. But throughout the summer of 1948, when he began writing the libretto, he felt insecure writing the language and humor of a group of men and women in wartime on an island in the South Pacific.

He was brave enough to admit it and to give an urgent call to Josh Logan, who in his youth had been educated at Culver Military Academy. Logan, co-author and director of *Mister Roberts*, had spent the war years in the U. S. Army. "I hate the military so much," Oscar told Josh, "that I'm ignorant of it."

Logan and his wife Nedda immediately arranged to come down to Highland Farm, and with the aid of a dictaphone—the early tape recorder—he and Oscar improvised many of the show's saltier scenes. Collaboration had always inspired Oscar, and there is no doubt that much of the authentic ring *South Pacific* had come during playacting extemporization in Doylestown.

Unfortunately the collaboration was to turn into a *cause célèbre*, for Logan

---

[4] It is noteworthy that the tenor of the times was such that a sheet music copy of "Carefully Taught" in its truncated version was not published until a decade later, when the movie of *South Pacific* was released. It is also interesting to recall that in 1953, after a road company performance of this show in Atlanta, Georgia legislators issued a vehement protest against the inclusion of this song. They introduced legislation to outlaw entertainment works having "an underlying philosophy inspired by Moscow." Oscar responded immediately on a more liberal note, saying these lawmakers could not represent the views of the people of Georgia, adding, with his tongue clearly in his cheek, that he was surprised by the idea "that anything kind or humane must necessarily originate in Moscow."

rightly wanted co-librettist credit. This he was eventually given, but since R & H reserved author's royalties, Logan was denied the financial bonus. He might have sued, but had he done so it is certain he would have been replaced as director. It was a bitter pill for Josh, who sparked so much of the comedy and had invented so much of the dialogue. His input gave him the authority to change many of the lines once the show was in rehearsal.[5] It rankled especially when *South Pacific* received the Pulitzer Prize and Logan's contribution was completely overlooked.

Collaboration of another sort was offered by Mary Martin, whose freshness Dick and Oscar said inspired them as much as Nellie Forbush's naïve Arkansas quality. They wrote "I'm in Love with a Wonderful Guy," an open-hearted show-stopper with which she would ever after be identified, for her voice. Perhaps Oscar was showing off his rhyming prowess to Mary in the release, which seems overly sophisticated for the "corny as Kansas in August" Nellie, but the overall effect of the song is unadulterated joy.

I'm as trite and as gay   I'm bromidic and bright
As a daisy in May,   As a moon-happy night
A cliché coming true!   Pouring light on the dew!

When they were in rehearsal Martin offered yet another inspiration. Having recently cropped her hair, she suggested it might be amusing to wash it onstage as part of the play. Oscar, who had been searching for a song to solidify his heroine's decision to break with her French lover, came up with "I'm Gonna Wash That Man Right Outa My Hair," and she washed Emile de Becque outa her hair 1,866 times.

Martin's and Pinza's voices were to serve as inspiration, but they mattered far less to the mood of the play than the lyric and song that would encapsulate the main thrust of the story. Oscar went back again to the *Tales*, believing, as with *Show Boat*'s "Ol' Man River," that a unifying theme would have to be found. Now Hammerstein lighted on the paradisiacal island of Bali Ha'i and reread Michener's exotic description:

Like a jewel it could be perceived in one loving glance. . . . It had a jagged hill to give it character. It was green, like something ever youthful, and it seemed to curve itself like a woman into the rough shadows formed by the volcanos of the greater island of Vanicoro. . . . Like most lovely things, one had to seek it out.

---

[5] One he *should* have changed because it is always incomprehensible in performance is to be found in "Bloody Mary": "Her skin is tender as Di Maggio's glove." Audiences invariable looked at each other and said "What?" (Later editions of the score clarified it as "Her skin is tender as a leather glove.")

Oscar then turned Michener's description into one of his most philosophical and poetic poems. He made the island a lure, a seductive siren, a Shangri-La in the middle of the Pacific as extolled by the wisest native of them all, Bloody Mary:

Most people live on a lonely island
Lost in the middle of a foggy sea.
Most people long for another island,
One where they know they would like to be.

Bali Ha'i
May call you
Any night
Any day.
In your heart
You'll hear it call you.
"Come away,
Come away."
Bali Ha'i
Will whisper
On the wind
Of the sea.
"Here am I,

Your special island!
Come to me,
Come to me!"

Your own special hopes,
Your own special dreams,
Bloom on the hillside
And shine in the streams.
If you try
You'll find me
Where the sky
Meets the sea
"Here am I,
Your special island!
Come to me,
Come to me!"
Bali Ha'i
  Bali Ha'i
    Bali Ha'i.

Showbiz denizens who knew how quickly Dick worked soon spread the story that when Hammerstein handed the completed lyric to Rodgers after a solid week's work, Dick set it in ten minutes. Rodgers denied this, pointing out that all he came up with was the opening motive: an octave skip followed by a descending semitone. But then, the composer added, once he got the motive, the rest of the composition came easily. The imagined timbre of Juanita Hall's rich mezzo, Rodgers said, had been running through his head since the days he cast her as Bloody Mary.

Good as the A section of Rodgers's descending melody is, with its liberal use of the diminished chord (Example A), it must be compared with the exotic ascending release. Here the composer achieves extreme contrast. He has used the diminished chord liberally in the A section; now he uses its diametric opposite—the augmented (Example B).

When Jo Mielziner heard the song he was terribly moved and made a sketch

---

It was Mary Martin's idea, and here she soaps up and rinses "that man right out of her hair." She and her followers shampooed to music for 3,727 performances in New York and London. MOCNY.

Example A

Ba-li Ha'i may call you, An-y night, an-y day.

Example B

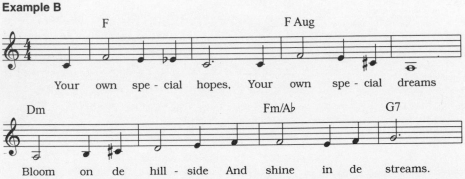

Your own spe - cial hopes, Your own spe - cial dreams

Bloom on de hill - side And shine in de streams.

of the mountain, its crater surrounded by mists, for the show's backdrop. Seeing the completed drawing inspired Oscar to write the following interlude.

Someday you'll see me
Floating in the sunshine,
My head sticking out
From a low-flying cloud;
You'll hear me call you,

Swinging through the sunshine,
Sweet and clear as can be.
"Come to me,
Here am I,
Come to me!"

"Bali Ha'i" is a prime example of the total collaborative art that a musical play represents. Announced in the orchestra at the very outset of the overture in a shimmering arrangement by Robert Russell Bennett, the song only emerged as we know it after input from Hammerstein, Rodgers, Mielziner, Hall, and Bennett.

If "Bali Ha'i" is an example of the mysticism and mystery that are to be found in *South Pacific*, the nurses and lusty army personnel stand for the sheer fun and frenzied existence, à la M*A*S*H, trying to crowd so much living into young lives on the brink of death in wartime.

Josh Logan, who created this naturalism, did so by avoiding what Oscar, who had cut his teeth in early musicals, called "the ensemble." Every member of the cast became an individual, just as all these strange bedfellows are actually put together in the army. There was no choral singing—merely separate lines given to various characters. This helped extract all the laughter inherent in the lines—especially the bawdier ones from "There Is Nothing Like a Dame" ("We get letters doused wid poifume / We get dizzy from de smell. / What *don't* we get? / You know damn well!") or "Honey Bun" (". . . she's broad where a broad should be broad!").

Beyond a few carps about inconsistency of character, this writer finds very little in *South Pacific* that needs changing. Once the show got into rehearsal a year after R & H had first heard of the *Tales* in January 1948, "Now Is the Time," a dramatic aria planned for Emile before he goes off on his dangerous mission, was deleted in favor of the rueful waltz that Dick wrote after Oscar thought of the title "This Nearly Was Mine." Another similar case occurred when the collaborators couldn't seem to come up with a song that might express Cable's reserved ardor. Oscar remembered Dick's lovely melody called "My Wife" that they had tossed out of *Allegro*. Writing a new and poetic lyric transformed it into one of their most ravishing songs, "Younger Than Springtime." Note how stating the title at the very outset adds vastly to the song's intensity.

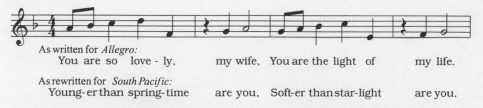

As written for *Allegro:*
You are so love-ly, my wife, You are the light of my life.

As rewritten for *South Pacific:*
Young-er than spring-time are you, Soft-er than star-light are you.

As soon as ticket availability was announced, it became clear that R & H's luster had not been dimmed by *Allegro's* failure. The box office at New York's Majestic Theatre quickly amassed a million-dollar advance. That was only the beginning, for *South Pacific,* even though it did not run as long as *Oklahoma!,* achieved more publicity and grossed far more than any of the team's predecessors.

After opening night the critics brought out their most opulent adjectives; "Magnificent . . . glowing . . . tenderly beautiful," claimed *The New York Times.* Not to be outdone, the *New York Herald-Tribune* gushed, "Pearls . . . enchantment . . . eloquent," while all the other papers, without a single naysayer, followed suit.

Two years later, after *South Pacific* garnered *all* the awards of the season including the prestigious Pulitzer Prize (which hadn't been awarded to a musical since *Of Thee I Sing* in 1931), Mary Martin inaugurated the London run at the Drury Lane. Perhaps the most gracious tribute to the show was offered by Kenneth Tynan, who wrote: "There really nothing more to do except thank the composer and the lyricist and climb up off one's knees, a little cramped from the effort of typing in such an unusual position of gratitude."

Lerner

1948–1951

IT SEEMED ILLOGICAL THAT COLLABORATORS who had written two shows that were total disasters followed by one that had mild success would separate after their first smash hit. So when Lerner announced six months after the opening of *Brigadoon* that he was teaming with Kurt Weill to write "An American Vaudeville," rumors as to *why* the most promising pair since Rodgers and Hammerstein were now splitting up ran through the theater community faster than money through a drunkard's pockets.

There were many real causes for the split. It was well known that Loewe wanted to go back to serious composing. He had already made a concert version of the songs from *Brigadoon*. It has also frequently been said that the older man was bridling at Alan's arrogant bossiness. This had increased with *Brigadoon*'s success.

On Alan's part, the new attitude stemmed from a deeper insecurity: that the success of *Brigadoon* was a fluke. He needed to write at once to prove to himself that his talent might not suddenly dry up, but Fritz, who had no such fears, insisted on a long holiday before they began another show. Alan simply couldn't wait.

Once *Brigadoon* opened, Lerner moved in with Marion Bell. They were married shortly afterwards. But Loewe had no such idyllic relationship, and he wanted to be out of reach of his estranged wife, Tina. He was often behind in support payments. The couple had not yet entered into a fixed separation agreement. With the success of *Brigadoon*, Loewe had reason to believe that Tina would demand excessive alimony. Loewe, who often boasted about his sexual prowess, never tried to hide his penchant for young girls. He had told many friends that he would like to settle on a tropic isle, have enough money, and enjoy the rest of his life. Now that he was in his late forties and had the MGM nest egg and a hit show, it seemed a propitious time.

Like the reasons for the split, there are variations as to who put the new team of Lerner and Weill together. According to archivist Miles Kreuger, Cheryl Crawford, having produced the Weill-Ogden Nash *One Touch of Venus*, mentioned that she would be interested in seeing what Alan and *Venus's* composer Kurt Weill could come up with. (Nash, a specialist in light verse, was not a true lyricist with knowledge of what was singable in the way Lerner was.) If the idea of the show suited her, Crawford promised to produce it.

Maurice Abravanel, who had conducted *The Day Before Spring* and was a friend of Lerner, Loewe, and Weill, had lunch with Loewe shortly after *Brigadoon* opened. When Loewe swore, "I will never work with that son of a bitch again," Abravanel revealed he said, " 'If you're serious, Fritz, I'm going to tell Kurt because I think they could hit it well together.' So I told Kurt and he came to see *Brigadoon*. Afterwards he said, 'Oh, Maurice, it's not really up to my level.' But later he began working with Alan."

Weill, a Jew who had come to the United States to escape Hitler, was the famed composer of *The Threepenny Opera* and *Mahoganny*, both written with Bertold Brecht. Since he had emigrated to the United States he abandoned his acerbic Germanic style and, superb composer that he was, had embraced the prevailing Broadway sound (as well as everything American.) He followed his first New York show, *Johnny Johnson* (a sort of New World *Good Soldier Schweik*), with *Knickerbocker Holiday*, which had book and lyrics by Maxwell Anderson. Although the show was not terribly successful, Weill's royalties from its only hit, "September Song," were considerable.

In the late '30s he and his wife Lotte Lenya rented a farmhouse in Suffern, New York, not far from Anderson's home in New City, about thirty miles north of Manhattan. With the Andersons as gurus, the Weills were introduced into a kind of Greenwich Village North atmosphere. Milton Caniff, the cartoonist of *Terry and the Pirates*, painter Henry Varnam Poor, actor Burgess Meredith, and author Marion Hargrove were among those who would meet at the home of the Andersons. It was there in 1939 at a birthday party honoring Walter Houston, the star of *Knickerbocker Holiday*, that Weill had met Moss Hart, with whom he would eventually write his first American success, *Lady in the Dark*. Weill had followed this with *One Touch of Venus*; *The Firebrand of Florence*, a flop; and a gorgeous opera that was too suave for Broadway, *Street Scene*.

It was after *Street Scene* opened that it became obvious that now Weill was adamantly "American" when he coruscated a reporter from *Life* who gave him a favorable review but referred to him as a German composer. "I do not consider myself a German composer," he wrote. "The Nazis obviously did not consider me as such either, and I left their country (an arrangement which suited both me and my rulers admirably in 1933)."

In addition to the 67th Street apartment into which he and Marion had recently moved, Alan Lerner bought a home in Pomona, New York, not far from the Andersons, and joined the group. He was honored to be collaborating with so respected a composer as Weill, and, once the preliminary arrangements were out of the way, he began scraping around for an idea. It was during one of his walks up South Mountain Road in New City, he told Miles Kreuger, that he was struck with the vernal beauty of the countryside and tried to imagine what the area must have looked like before the industrial revolution. To the idea of contrasting country life to life in an urban center, which he knew Oscar Hammerstein was preparing in *Allegro*, he added another dimension— that of a continuing marriage.

Kurt Weill welcomed the idea, probably because it had certain similarities to his generational saga, *The Eternal Road,* a pageant he had written with Franz Werfel in 1935 that had been called "a résumé of the Old Testament." But Lerner's idea went farther in that it encompassed some of the agitprop of Weill's works with Brecht.

Alan was no Brecht. He was a liberal and an idealist, not a revolutionary. And Weill, now three years before his untimely death in 1950, had long ago abandoned the biting, angular melodic lines he chose in the peak of his German period.

Lerner's story, which he called *Love Life* (he said because his characters have such a "lovely life" they don't grow older as the generations pass around them), begins around the time of the American Revolution when Sam and Susan Cooper move into a new town. The libretto travels though the decades highlighting the changes that industrialization, economic panic, women's suffrage, the Great Depression, and modern times wreak and incorporates the couple's separation and their eventual tentative reunion. Between scenes, so that the two leading players might have a chance at least to change costumes, there would be a sort of chorus, a minstrel show. Throughout each incarnation the protagonists were not to change. Sound familiar? Certainly. It was *Brigadoon* all over again but with a difference. To foil death and to escape the encroaching witches of society, instead of moving the town and the love affair into the sheltering safe braes of Scotland, Lerner had the Coopers face the buffeting that progress brings with it. Sam and Susan are far less believable than *Brigadoon*'s Fiona and Tommy. The Coopers gain nothing by being so fragmented, and they lose the romance—Alan's stock in trade. The ills of society caused by industrialization was hardly a subject Americans wanted to think about so soon after World War II. But Lerner, using almost a Soviet technique, rammed his message down his audience's throats. *Love Life* was to have a song called "Economics" as well as one called "Progress." A few stanzas of the latter are quoted below.

One time this was a very quiet planet.
The reason was nobody was around.
But then one day, Jehovah, who began it,
Got bored and clamped a couple on the
    ground.

And right away when man and woman came
    here
They took a peek and nature took its course.
He said "I love you," and she answered "same
    here";
And love became the greatest human
    force . . .

. . . For suddenly the mind of man was
    changing;
He started moving on in greater haste
And perspectives took a drastic rearranging
For love and home had fin'lly been replaced.

What is this thing
That's better than spring?
What thing is this
That's greater than a kiss?
What is the "X"
That's bigger than sex?
What could it be?
What could it be?
What could it be?
It's progress!

Where ev'ry man can be a king,
Why next to progress
Love's a juvenile thing.
Yes, with progress
Your chance to hit the top is great.
One year you need a loan
And the following year you own
New York State.

Lerner's lyric technique had by no means deserted him. If only he had not worn his message on his sleeve, *Love Life* might have succeeded, but he refused to disguise it and therefore didn't display his ability to bring romance to a song.

"Here I'll Stay" is one of the most enchanting romantic lyrics Lerner ever wrote, and Weill's music, too, is exquisite, with an enormously moving descending bass line. Yet the song, whose lyrics are printed below and which by now has become a standard, never set the show soaring. One wonders why.

There's a far land, I'm told
Where I'll find a field of gold,
But here I'll stay with you.
And they say there's an isle deep with clover
Where your heart wears a smile all day
    through.
But I know well they're wrong,
And I know where I belong,

And here I'll stay with you.
For that land is a sandy illusion;
It's the theme of a dream gone astray.
And the world others woo
I can find loving you,
And so here I'll stay.

The answer is because the song comes so early in the first act, before the audience has had a chance to empathize with the characters, their intense love song is wasted. More important is Alan's language, lush and almost over-rhymed (but eminently singable), more the clever Larry Hart than the carpenter, Sam, who has just been introduced as a character.

Lerner gets no help in believability from his composer. Weill's gorgeous melody is one of the best and most up-to-date show tunes that 1948 offered, while the characters are supposed to be set in Colonial New England.

"Green-Up Time" was the title of the other fairly popular song from the show. "There was a minor strain in everything Kurt Weill wrote," Lerner told

The chorus of eight who sing "Progress" and bring out the buck-and-wing vaudeville aspect of Love Life. *MOCNY.*

Moving from the past to the present, as Lerner would attempt again in On a Clear Day, *this plot follows the many lives of an American couple. MOCNY.*

interviewer Tony Thomas, "so I said, 'Let's write a really happy song about spring,' and he came back and played that . . . with the minor in it again. But I suppose in every joy there is a little sadness." Actually the minor strain gives the song a modicum of charm. (See Example A below.)

With lines like "Then I began to look around / And in ev'ry field I found / Greens were a-pushing up through the ground," the song's lyric is not very imaginative considering the plethora of past spring songs. This one, needing an oblique point of view, always struck me as a first draft for a fresh way of invoking the vernal season. Lerner would hit on it from two decades later when he wrote "Hurry It's Lovely Up Here" in *On A Clear Day*.

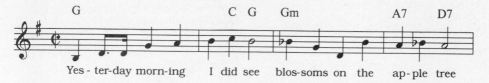

Yes-ter-day morn-ing I did see blos-soms on the ap-ple tree

Later in his life when talk of reviving *Love Life* surfaced, Lerner always turned thumbs down on the idea, saying he had stolen so much from it to use in other plays that only a skeleton remained. As we have seen, he did borrow the idea of eternal life from *Brigadoon* and would develop the continuing generation theme portrayed by the same characters again (with diminishing returns) in *1600 Pennsylvania Avenue*. But the most specific pilfering is his idea for "I Remember It Well." Except for Cole Porter's two quite different songs called "Just One of Those Things," I know of no other case where a lyricist has copyrighted two songs with identical titles.) Here it is sung by young Sam and Susan. I have set beside this lyric its more masterful and infinitely more moving version written for the movie *Gigi* and sung by an older Honoré and Mamita, unforgettably personified by Maurice Chevalier and Hermione Gingold.

Sam: Remember? I do indeed, Susan, I remember every detail of that evening just as if it were yesterday. Just as if it were yesterday.
(*He* sings)
It was late at night.
Susan: It was six-fifteen.
Sam: You were dressed in white.
Susan: I was all in green.
Sam: That's right! I remember it well.
It began to pour.
Susan: It was warm and clear.
Sam: We were in the store.
Susan: We were both out here.
Sam: That's right! I remember it well.
We were all alone.
The kids had flown.
Upstairs to bed for the night.

Honoré: But I've never forgotten you. I remember that last night we were together as if it were yesterday.
(*He* sings)
We met at nine
Mamita: We met at eight.
Honoré: I was on time.
Mamita: No, you were late.
Honoré: Ah, yes! I remember it well.
We dined with friends.
Mamita: We dined alone.
Honoré: A tenor sang.
Mamita: A baritone.
Honoré: Ah, yes, I remember it well.
That dazzling April moon!
Mamita: There was none that night.
And the month was June

Susan: When you brought the chair
They both were there,
But outside of that, you're right.
Sam: But the moon was low.
Susan: Yes, the moon was low.
Sam: And I loved you so,
Yes, I did love you so.
And it seemed even more,
More than ever before;
Am I right?
Susan: Yes, dear, you're right.
Sam: I remember it well.

Honoré: That's right! That's right!
Mamita: It warms my heart
To know that you
Remember still
The way you do.
Honoré: Ah, yes! I remember it well.
How often I've thought of that Friday . . .
Mamita: Monday . . .
Honoré: . . . night
When we had our last rendez-vous,
And somehow I've foolishly wondered if you
    might
By some chance be thinking of it too.
That carriage ride . . .
Mamita: You walked me home.
Honoré: You lost a glove.
Mamita: I lost a comb.
Honoré: Ah, yes! I remember it well
That brilliant sky . . .
Mamita: We had some rain.
Honoré: Those Russian songs.
Mamita: From sunny Spain?
Honoré: Ah, yes! I remember it well.
You wore a gown of gold.
Mamita: I was all in blue.
Honoré: Am I getting old?
Mamita: Oh, no! Not you!
How strong you were.
How young and gay!
A prince of love in ev'ry way.
Honoré: Ah, yes! I remember it well.

Other songs in this vaudeville were a polyglot, sometimes encompassing a whole scene. In the second act, Lerner writes his obligatory male soliloquy, and this self-revelatory one (for Alan and Marion's marriage was to end shortly) might pass for a '50s night-club sketch. It is an ironic threnody called "This Is the Life." Sam has left Susan. He has checked into a hotel room and talks to the audience about how although he misses the kids, he is determined to enjoy his freedom. He orders a huge dinner (all in song) and mentions how he must forget Susan. Eventually he cancels the dinner, and shouting, "This is the way it ought to be! I'm free! I'm free!," he goes out to paint the town.

It is not very far from Higgins's tirade at the end of *My Fair Lady*. Fortunately there we have Eliza to throw his slippers at him, and by that time Lerner had learned not to go over the top.

The show opened on October 7, 1948, to the most mixed reviews in memory. Half the critics raved, the other half panned. Brooks Atkinson, writing in *The New York Times*, ventured, "Although billed as 'a vaudeville,' *Love Life* is cute, complex, and joyless—a general gripe masquerading as entertainment," while Robert Coleman of the *Daily Mirror*, a normally reserved critic, brought out all his hosannas: "Cheryl Crawford established a new high standard for Broadway musicals last evening. Author Alan Jay Lerner and composer Kurt Weill have fashioned a superlative entertainment—a song and dance show with great heart, soaring imagination, welcome novelty and keen observation."

Kurt Weill's biographer Ronald Sanders ascribes the show's curtailed run of 252 performances to faults in Lerner's script, which he says contained "a certain intellectual pretentiousness." Maurice Abravanel blames Kurt's score, and Cheryl Crawford hedges by saying that the script had major shortcomings which affected Weill's work. "Because Kurt's score served the style of the writing," she added, "it didn't have the warmth of his best ballads."

Love Life might have been able to limp along to a respectable run attended by half-filled houses of the intelligentsia had it not been a question of "bad economics"—as Lerner's song in the show by that name may have prophesied. At the time of the show's opening James Petrillo, the president of the American Federation of Musicians, had imposed a recording ban, and so no original cast album was made. Perhaps even more important was the fact that *Love Life* had hardly gotten a foothold when Frank Loesser's *Where's Charlie* opened. And before the season was out two smash hits—Cole Porter's *Kiss Me, Kate* and Rodgers and Hammerstein's *South Pacific*—would vie for theatergoers' dollars and add to its box office woes.

Today, echoing what the critics wrote at the time of its debut, *Love Life* is deemed to be one of the great lost musicals, the forerunner of the concept musical by some; others (myself included) find it a pretentious and predictable bore with one superb song. It is simply that the songs *are* the show. Since they are general rather than specific, they don't help flesh out the characters. Worse, the story is not allowed to develop beyond a series of catastrophes. Each time we meet the characters in another set of costumes we are forced to rebuild our interest. Beyond that, the minstrel show, with all its stereotypical ramifications, has no place in today's theater.

Even before the troubles with *Love Life* began, Alan was having trouble with his private love life. Marion Bell wanted to play the part of Susan, but because of her frequent absences from *Brigadoon* and largely because of Alan and Cheryl's negative opinion of her reliability, she was passed over. Nanette Fabray, a vivacious, snub-nosed singing actress, veteran of many musicals and lately the star of *High Button Shoes*, was given the role.[1]

---

[1] But Kurt Weill did not forget the sound of Marion Bell's voice. It was she he chose to portray Jenny in the recording of his folk-opera, *Down in the Valley*.

There is no doubt that Lerner kept a firm hold on the proceedings even to the point of firing director Robert Lewis and hiring, then bossing, director Elia Kazan, whose first musical this was. Alan demanded strict attendance and attention at rehearsals from Marion and Kurt Weill's wife Lotte Lenya, even though Love Life's plot contains episodes of concern for women's rights. "During rehearsals, Lenya and I would sometimes go for coffee." Marion said. "Alan did not like that. He wanted us in the theater at all times." Although her husband involved her in discussions about his play, "Alan and I argued about the writing of the woman's part. Then he often went off to the theater."

Marion and Alan shuttled between their Manhattan flat and the country house, and talked of having a child. Marion told this writer she feared she might be incapable of conceiving a child at that time.

It was to become Lerner's lifelong habit that whenever things romantic became impossible, he would look for a new partner or a new habitat. In this case, with Love Life safely launched at the 46th Street Theatre, it was the latter. Now Alan contacted Lillie Messinger to see what work she could round up for him on the West Coast. Soon Lillie put him in touch with Arthur Freed, the head of the musical production unit at MGM, who urged Alan to come west so they could work out the story of a film for Fred Astaire.

Freed headed "The Freed Unit" under whose aegis Metro had filmed their finest musicals and become preeminent in the industry. Now he was not only able to spark Lerner's ideas into new and filmic areas but was able to supervise technically and artistically Lerner's first foray into the incredible maze of screenwriting. Freed and Lerner hit it off at once, for besides being an erudite and tasteful man, Freed was a highly professional lyricist.[2]

Alan and Arthur together came upon the idea of plotting their musical around a thinly disguised and heavily romanticized story involving Astaire and his sister Adele. Having danced together since they were children, the Astaires had come to London in 1928 to repeat their New York success Funny Face. It was then that Adele had met and fell in love with Charles Cavendish, a son of the Duke of Devonshire. Charles proposed marriage at once, but Adele re-

---

[2] Arthur Freed (1894–1973) wrote lyrics for many of the films produced under his aegis. Among his hit songs were "Temptation," "You Are My Lucky Star," "Would You" and "Singin' in the Rain." The list of films he produced reads like a catalog of landmarks in the history of the musical film. Some of the highlights are Babes in Arms (1939); Cabin in the Sky (1943); Meet Me in St. Louis (1944); Till the Clouds Roll By (1946); Good News (1947); Easter Parade, The Barkleys of Broadway, On the Town (1949); Annie Get Your Gun (1950). Besides Royal Wedding, An American in Paris, Brigadoon, and Gigi, which involved Lerner, Freed produced Show Boat (1951); Singin' in the Rain (1952); The Band Wagon (1953); Kismet (1955); Silk Stockings (1957); and Bells Are Ringing (1962).

jected him, preferring to continue her career. At last in 1931, after the Astaires' great success in *The Band Wagon*, Adele accepted the proposal and retired from the stage, leaving her brother partnerless. Fortunately, despite the lack of enthusiasm Fred's subsequent screen test ("Can't act. Slightly bald. Can dance a little") provoked, he was given a small part opposite Joan Crawford in *Dancing Lady*.

The outline of Alan's script closely followed the facts of the dissolution of the Astaires' partnership. Fred (Astaire) and Adele (Jane Powell) became Tom and Ellen Bowen, *Funny Face* became a musical retitled *Every Night at Seven*, and Cavendish became Lord John Brindale (Peter Lawford). To round out the foursome, Astaire was provided with an inamorata as well in Sarah Churchill, the great statesman's daughter, who fortunately could dance a bit and perhaps more fortunately resembled her father not at all. At the film's close the foursome, arms linked, are seen marching to the church anticipating a double wedding ceremony that very afternoon. Lerner leaves the point of whether or not Tom and Ellen will continue their theatrical careers after marriage unresolved at the final frame. But then, the picture is about love and weddings, not about careers.

The plot is ordinary but charming. What gave it an extra brilliant dimension was the setting—London during the Royal Wedding of Princess Elizabeth and Prince Philip—and the clever fashion the event and the story were intercut. MGM had access to the considerable footage of the marriage two years earlier in November 1947, and Alan skillfully intercut them with the Bowens' romances.

As for the score, Freed chose a true professional to provide the music for Alan's first screenplay. Burton Lane had a notable career that shuttled between Broadway and Hollywood. Born Burton Levy in New York City, he began writing songs as a child and broke through working for a music publisher while still a teenager. With "Yip" Harburg he contributed to music and lyrics of *Earl Carroll's Vanities of 1930* and *Three's a Crowd* when he was only eighteen. Soon Freed brought him out to the West Coast to write the score for *Dancing Lady*, which produced his first hit song (with Freed), "Everything I Have Is Yours." For the rest of the '30s Lane went on to write a string of mediocre musicals, but in 1941 he produced another big hit, "How About You?," for Freed's *Babes on Broadway*. Near the end of that decade he came back to Broadway to collaborate once more with Harburg on the smash hit *Finian's Rainbow*, which reached the stage only two months earlier than Lerner and Loewe's *Brigadoon*.

Alan and Burt, who were eventually to write two musicals, *On a Clear Day You Can See Forever* and *Carmelina*, besides the current film and an aborted one, had known each other slightly in New York. Perhaps more valuable to the collaboration was the fact that each respected the other's talent. Both men were well known for trying to run the show, with Lane living up to his reputation of "being difficult to work with."

Alan's script is full of sunny good humor and the score is full of bright up-tempo songs in spite of Lane's somewhat somber nature. At one point as the collaborators were driving to the studio Alan mentioned that the film was altogether *too* charming and that they needed something to take the curse off it. Lane said, "Well, give me something ricky-tick, a title," at which point Lerner came up with what has since been verified as the longest title in ASCAP history, "How Could You Believe Me When I Said I Loved You When You Know I've Been a Liar All My Life?" Fortunately the title only appears at the beginning and end of the song. Lerner had the wit to rhyme the title lines with "How could you believe me when I said we'd marry when you know I'd rather hang than have a wife."

Other bright songs were "Every Night at Seven," the torrid dance number "I Left My Hat in Haiti," and the martial "It's a Lovely Day for a Wedding" (in spite of rain that is seen falling on Princess Elizabeth's coach), but the best song in *Royal Wedding* was the ballad "Too Late Now." Coming after Lord Cavendish notes the complications in the way of their romance and his suggestion that he and Ellen might be better off with different partners, Lerner's simple lyric moves the plot forward in a way that was rare in movie musicals. It is reprinted below.

Too late now to forget your smile,
The way we cling when we've danced a while;
Too late now to forget and go on to someone
    new.

Too late now to forget your voice,
The way one word makes my heart rejoice;
Too late now to imagine myself away from
    you.

All the things we've done together
I relive when we're apart.
All the tender fun together
Stays on in my heart.
How could I ever close the door
And be the same as I was before?
Darling, no, no I can't any more—
It's too late now.

With Astaire as star, Lerner was called upon to invent fresh situations for dancing. Since the story placed the Bowens on a liner crossing to England, Alan put Fred into the ship's health club to dance with and around the equipment for one sequence. Later he chose amusingly to have the couple do their dance during rough tides on a shifting ballroom floor, a familiar feeling to anyone who has ever crossed the Atlantic in rough weather. But by far the most distinctive and mystifying moment in the film is the scene where a lovelorn Astaire dances on the floor, walls, and ceiling of his room. Lerner said he imagined the scene in a dream, and asked the studio men the next morning if it was feasible.

Throughout this project Lerner and Lane worked relatively smoothly together, although Lane as recently as 1990 said that with the exception of "How Could You Believe Me . . ." and "Too Late Now" he was "not happy with Alan's lyrics. I had some tunes that deserved better, including 'Every Night At Seven' and 'I Left My Hat in Haiti.'" In Lerner's defense, both these lyrics seem superior to the their run-of-the-mill tunes. Moreover, this writer finds it ignoble of Lane (or any other collaborator, for that matter) to decry a partner's contribution. For the record, Lane's melody for the ceiling dance was rescued by Lerner and shorn of its mundane Harold Adamson lyrics which were sung in 1934 by Eddie Cantor in *Kid Millions*. A few bars juxtaposed with the original will show the superiority of the Lerner version.

I WANT TO BE A MINSTREL MAN	YOU'RE ALL THE WORLD TO ME
We always love a minstrel man	You're like Paris in April and May,
He thrills us like nobody can	You're New York on a silvery day,
The way he dances sure is dandy,	A Swiss Alp as the sun grows fainter,
And sings his songs about sugar candy . . .	You're Loch Lomond when autumn is the painter . . .

The minor difficulties encountered by the collaborators had nothing on *Royal Wedding*'s casting problem. Not only was the picture to feature Fred Astaire but it was planned to feature June Allyson teamed with Peter Lawford, in the hope they would repeat their success in *Two Sisters from Boston* and more especially in *Good News*. When Miss Allyson discovered she was pregnant, Judy Garland, who had starred opposite Fred Astaire in the enormously successful *Easter Parade*, was chosen to replace her. Garland even recorded one of the songs, but eager to get in shape for the film, she had been dieting severely. After a few days shooting she became ill and was fired from MGM.

Jane Powell, her replacement, was the perfect choice for the lighthearted role of the vivacious and charming Ellen Bowen. Possessed of a voice that goes easily from coloratura to a chest belt, she even played tough moll to Astaire's cad in the "How Could You Believe Me . . ." number, milking every line for humor. It was certainly her success in *Royal Wedding* that led to her being cast in the movie musical that peaked her career, perhaps the only successful Broadway-type musical ever *written* directly for the screen, *Seven Brides for Seven Brothers*.

Although critics gave the film excellent reviews, Lerner generally chose to denigrate his script, saying, "My contribution to *Royal Wedding* left me in such a state of cringe that I could barely straighten up."

For many years Arthur Freed had wanted to make a musical set in Paris. When he saw a *Life* magazine article featuring photographs of a great number of Americans who had gone there on what was known as the GI Bill—veterans' tuition paid by the U.S. government—it sparked what was a timely idea. Why not a film about an American in Paris? Travel abroad had increased substantially since the war, and now that the '40s were almost over, the filmgoers would surely be eager to see a musical set in what was called "the world's most romantic city."

Since Freed was a poker-playing buddy of Ira Gershwin's, one Saturday night Arthur offered to buy the title and the use of George's tone poem, "An American in Paris" as a ballet to climax the film. Ira agreed to the idea only on condition that all the songs in the musical be by himself and George.

"That's the object of the picture. I wouldn't use anything else," Freed said.

Ira, who was in charge of his brother's estate, was paid $300,000 for the use of his and George's music and lyrics and an additional consultation fee of $50,000, and once the contracts were signed in August 1950, Freed began to confer with his director, Vincente Minnelli, who had directed several brilliant musicals including *Cabin in the Sky, Meet Me in St. Louis,* and *The Pirate* for MGM. Minnelli, an acknowledged master of films with lavish visual style, was an obvious choice. Both Fred Astaire and Gene Kelly were considered, but once it was decided that the plot should revolve about a young American who has stayed on in Paris after the war to study painting, Kelly, a dozen years younger than Astaire, won out. Now it was decided that Kelly would also choreograph the entire picture and that balletic dancing would be the film's main thrust.

Freed also wanted to use gravel-voiced pianist Oscar Levant, who had been a friend and disciple of George Gershwin's and one of the notable interpreters of his work. "Including Oscar in the film," said Vincente Minnelli, "lent the enterprise a sort of legitimacy." Kelly as painter and Levant as mercurial concert pianist were to be sidekicks, the latter being always ready with an acerbic comment.

Freed conceived a vague plot concerning three men and their involvement with a girl. As for the girl, Kelly had spotted seventeen-year-old ballerina Leslie Caron in Roland Petit's Ballets de Champs-Élysées. He was delighted to discover she spoke passable English, since she had an American-born mother. After a screen test, it was further determined that she was an effective actress. It was a brave or perhaps foolhardy thing to do—to build an expensive musical around a newcomer—but Freed and Minnelli were convinced that Caron, while not beautiful, was star material. More than that, they wanted the film to be authentically gallic, not what was known as Hollywood-French—American actors affecting a fake French accent.

To that end, since Kelly played so well opposite children, one of the film's most charming scenes would be written: Kelly teaching the children to say-sing a few English words. The scene was built around "I Got Rhythm": a child shouts, "I got," and Kelly finishes with "rhythm," or "music," or "my gal." Irresistible. For authenticity, MGM scoured the area around Los Angeles to find fifty French-speaking kids. They hoped to get Maurice Chevalier to play the role of Caron's protector and Kelly's competition. When Chevalier was not available, Georges Guetary, an entertainer largely unknown to American audiences but popular with Europeans, was chosen.

Now it was time to procure a writer. Freed mentioned Alan Lerner, and Minnelli concurred heartily. "I didn't think I'd get him," Freed remembered, "because Alan likes to write songs. But when I told him the idea and the characters, he said, 'I'd love to do it.' "

Lerner had to write the screenplay backwards, as Minnelli mentions, "like a Chinese puzzle. The elements of the script and song were pre-supplied. So was the approach. The script had to be tailored to Gene Kelly's ingratiating personality, as well as reflect his love of Paris."

Although shooting began in August, it was not until November that Alan came up with an idea for the plot of a kept man who falls in love with a kept woman. Freed cautioned him that "the girl could not be kept per se, so with this idea in mind," Lerner said, "I began to write and by Christmas I had the first act completed." Freed and Minnelli had to stand by patiently waiting for Alan to finish the script. Three months later, in mid-March 1950, Alan got his ideas together and finished the photoplay.

Softening his approach somewhat (approval from a censoring board called the Hays Office was essential for motion picture release in 1950), Kelly's "sponsor," rich, elegant Nina Foch, becomes merely a patron of the arts who has an unconsummated yen for Gene. From the dreadful quality of his oils that are shown in the film, Foch indeed has very little artistic taste, but no matter. Kelly in his turn is smitten with Caron, whom he first sees in a café.

The young gamine returns his feelings but rejects his advances because she is engaged to Guetary, her protector since the death of her father, who was killed in the Resistance.[3] As their romance deepens, neither Kelly nor Caron can come to confess to each other of the alliances in their lives. At last, when Kelly avows his love, Caron reveals that she is planning to marry. Kelly, in pique, rushes to Foch's apartment, kisses her passionately, and whirls her to the wild Beaux Arts Ball.

At the ball he learns that Guetary plans to marry Caron and take her to America that very evening. His car is waiting below, and Caron tearfully confesses she will join him. Now comes a moving scene in which Caron says, "Paris has ways of making people forget," and Kelly responds with, "Not this city! It's too real, too beautiful. It never lets you forget a thing. It reaches in and opens you

---

[3] The younger girl and older man would be a lifelong pursuit and plot device for Lerner. Besides *My Fair Lady*, the same thing happens in *On a Clear Day*, *Gigi*, and *Lolita*.

wide . . . and you stay that way." The lovers tearfully embrace. Caron runs off, followed by a pensive Guetary, who has overheard their conversation.

This leads to the musical highlight of the film, a seventeen-minute ballet which tries to show in an oblique manner the whole story of the film. The ballet, although charmingly danced and photographed, is basically a mistake. It does, however, serve some purpose. It gives us another chance to see Kelly and Caron in *pas de deux,* and it allows the audience time to watch Kelly reflect on the loss of his love.

After the ballet, of course, Guetary, noble Frenchman that he is, turns around the car in which he and Caron were eloping. He releases Caron from her vow, and the lovers are united on what looks very much like an authentic staircase street leading up to Sacre Coeur but which was in reality a Hollywood sound-stage.

The film received so-so reviews, with most critics praising the way the Gershwin score augmented the visual beauty of Kelly's choreography while denigrating Lerner's contribution. *Newsweek* flippantly called Alan's script "a silly story." Bosley Crowther, the *New York Times* reviewer, went farther when he reported: "Mr. Kelly is the one who pulls the faint thread of Alan Jay Lerner's peach-fuzz script into some sort of pattern of coherence and keeps it from snapping into a hundred pieces and blowing away."

In spite of that, *An American in Paris* won Academy Awards for best picture, best color cinematography, art direction, set direction, costume design, and musical scoring, and despite the critics' pans, best story and original screenplay. A special award was presented to Kelly for his "contribution to the art of choreography on film." It was the first time in fifteen years, since *The Great Ziegfeld* captured the best picture Oscar, that a musical won Hollywood's highest accolade. Although Alan was very proud to have won his awards for best story and original screenplay—the first ever for a musical—one must admit that it was hardly a contest. The other nominees in this category—*The Big Carnival, David and Bathsheba, Go for Broke,* and *The Well*—were dreadful and except for De Mille's biblical behemoth have vanished into moviedom obscurity.

Nonetheless, one should not underestimate Alan's considerable contribution to *An American in Paris*—for creating a plot that never interfered with the musical, choreographic, or visual beauties of the film. Additionally, Alan's scenario included many tricks that were particularly theatrical. The film's very opening, panning the *quartier,* having each of the three men in the film introduce himself to the audience in a witty speech, was a stage device not common to the realistic medium of cinema. Beyond that Lerner tailored the script admirably to the personality, the talent—as Minnelli says—the "jauntiness" of Kelly, as he was later to tailor *My Fair Lady* to Rex Harrison.

Although Alan picked up the Oscars rather than the producer, part of his awards belong to Freed,[4] whose vision set the project rolling and who was in on every script conference. Unsung too by the Academy was Ira Gershwin's contribution. He helped choose fresh songs and even wrote some additional lyrics to his brother's music. And certainly the film would have been second-rate without the remarkable performance of Caron. She epitomized the romance and fragility of the movie. Perhaps the most overlooked of all the collaborators was Vincente Minnelli, who singlehandedly supervised everything and whose exquisite taste shows in every frame. Of all the MGM musicals about Paris, this one seems to have the most authentic ring, although except for the opening stock footage of the city, the entire film was shot in Hollywood. Incomprehensibly, Minnelli was beaten out for the Oscar by George Stevens, who routinely directed A Place in the Sun.

Alan was not in Hollywood at award time. He had reconciled with Fritz Loewe, their problems at least temporarily put aside, and was hard at work on their next project, Paint Your Wagon, which was scheduled to open on Broadway in less than six months. But there was another reason why he would not leave Manhattan. He had been called back to New York to be near his father. Joe Lerner was in the midst of his seventeen-year-long bout with cancer, and he lay in a coma at Memorial Sloan-Kettering as the awards were announced.

After the ceremony Alan went immediately to the hospital and asked the nurse to tell his father the good news, should he awaken.

"Oh, he knows," the nurse said.

"How?" Lerner asked.

"Well, at five minutes to eleven he came out of his coma," she explained, "reached over and turned on the radio. After hearing that you won the Academy Award, he turned it off again and went back into the coma."

Joe Lerner was a valiant fighter. He came out of that coma, and even though half of his neck and jaw had been cut away and the power of speech totally denied him, he was in remission for a full year before being struck down again.

Back in the hospital, shortly before the operation from which he would never recover, he wrote Alan, who was at his bedside, a query on his pad: "I suppose you're wondering why I want to live?" Alan nodded, and before they wheeled his father into the operating room, Joe scribbled a note and stuffed it into his son's hand. Lerner was too benumbed by emotion to look at his father's answer until many hours later. When he did, he read, "To see what happens to you."

---

[4] After An American in Paris garnered so many Academy Awards, Freed signed Lerner to a three-picture open-ended contract. The first, Brigadoon, released in 1954, was not very satisfactory. Lerner felt that was because it was shot on the Hollywood lot rather than in Scotland. The second, Huckleberry Finn, was never filmed although Alan and Burton Lane wrote half a dozen charming songs. The third was another Academy Award musical, Gigi.

Hammerstein

1950–1951

ALTHOUGH IT IS TITLED *The King and I*, the "I" was always the more important character in Oscar and Dick's next musical. From its genesis, when the "I," Gertrude Lawrence, sewed up the rights to the story, to the libretto, with this magnetic star's mellifluous speaking voice ringing in Oscar's ear, to the score, which Dick tailored to Lawrence's wavering soprano, the entire show revolved around Gertie. That was, at least, until Yul Brynner came onto the scene.

It need not have been so had Hammerstein and Rodgers heeded the advice of their wives and come in on the project at the beginning. Long before Margaret Landon's best-selling novel, *Anna and the King of Siam*, had been filmed as a dramatic work starring Rex Harrison and Irene Dunne, each Dorothy had urged her husband to adapt it as a musical.

The impetus was to come from another side. Fanny Holtzman, Lawrence's friend, agent, and lawyer, realizing that her client was no longer being offered the plum musical roles Mary Martin or Ethel Merman were, decided to *create* a project for her star.

Anna and the King of Siam had been published in 1943 and the film came out three years later, but by 1950 the story had fallen into limbo. So it was by sheer coincidence that the William Morris Agency, representing Landon, sent Holtzman a copy of the book, hoping to resurrect their property and suggesting that an adaptation would make a smashing play for Lawrence. After rereading the story of Anna Leonowens, the British widow who went to Siam in the 1860s, both Fanny and Gertie agreed that a perfect role had been found. But it certainly cried out to be a musical.

Fanny first approached Cole Porter, who was uninterested, and was about to approach Gertie's longtime friend, total man-of-the-theatre Noël Coward, when she ran into Dorothy Hammerstein on a Manhattan street. Immediately she thought of Oscar and sent the book to R & H for consideration. When the team still could not see their way to adapt the book, she asked them to screen the film.

"That did it," Rodgers was to write. "It was obvious that the story of an English governess who travels to Siam to become a teacher to the children of a semibarbaric monarch had the makings of a beautiful musical play." Both Oscar and Dick could see the contrast in the story between Eastern and Western cultures. In that area it possessed the same tension as *South Pacific*. But here there was an important added theme: democratic teaching triumphing over autocratic rule. It was as though they had found a wedge into solving the problems that they posed in *South Pacific*. Almost more important, the two protagonists besides representing East and West were far stronger than any they had yet dealt with—a totalitarian bully with cracks in his armor who has a thirst for scientific knowledge versus a strong-willed, independent woman who approaches her teaching with dignity. This promised fireworks.

"Here was a project," Rodgers was to add, "Oscar and I could really believe in, and we notified Fanny that we were ready to go to work." Fanny was as crafty in hammering out contracts as she was in finding a property for her star. She wangled an unheard of 10 percent of the gross *as well as* 5 percent of the net for Gertie while she appeared in the musical. (This was twice what Dick and Oscar earned.)

Gertie needed (or wanted) no co-star, but R & H felt someone with theatrical magnitude would help insure the production's success. Alfred Drake, who since *Oklahoma!* had become Broadway's most serviceable leading man, was the initial choice for the role of the King, but since he only wanted to sign for a limited run he was not acceptable to R & H. Then they thought of Noël Coward, who could easily have been made up to look Oriental. Another plus for Coward was that he always played so movingly opposite Lawrence. But when Coward rejected the part, R & H approached Rex Harrison, who had starred in the movie version opposite Irene Dunne.

At the time, Harrison was having financial, artistic, and amorous troubles. Worse, he had fallen in love with some property on the Italian Riviera and had signed to do a film of far less than top quality *simply* because the producer advanced him enough money to buy the house.

"If I hadn't gone on that holiday in Santa Margherita," Harrison says, "if I hadn't found that plot of land and committed myself to *The Long Dark Hall* I might have chosen to do *The King and I* and stayed with it for five years, and gone on to do the film."

He allows that he might never then have been chosen for *My Fair Lady* and says he would "now be known as the man with dozens of wives and ninety children, rather than as an irascible misogynist."

The choice and identification of Yul Brynner as "The King" was one of show business's fortunate accidents. Yul Brynner was the King incarnate and his gravelly voice, half singing, half shouting, was the template for Rex Harrison's *sprechstimme* in *My Fair Lady*. The happenstance was not unlike Mary Martin's being the first to be offered the role of Laurey in *Oklahoma!* Rodgers wrote her afterwards: "So glad you chose not to do it. With you in it would have been an entirely different show."

When the show opened with Yul Brynner, who immediately came to personify the King of Siam with such strength and canny charm that he totally erased Harrison's cinema portrayal in the public's mind, Rex sent Yul a telegram: "The King is Dead, long live the King."

Oscar's first task in "going to work" was to organize what was essentially a collection of disconnected incidents (not unlike *Tales of the South Pacific*) into a cogent libretto. He offered Josh Logan co-authorship, but Josh, who was still smarting from his treatment in their former venture, refused. "It was a decision I will regret for the rest of my life," Logan was to say ruefully.

Hammerstein perforce then tackled the book alone and turned out what was to become his masterpiece. He attacked Landon's book and the scenario for the movie with a bold hand, leaving hardly any detail of either story unchanged. In the end he created a true distillation of the novel's original intention, although it was greatly different in episode.

His first plot change was to make the King Mongkut attractive, where the book describes him as "of medium height and excessively thin." Although this was barely hinted at in the film, Hammerstein felt an emotional attachment, a bit more than mutual admiration, between the King and Anna was possible. It was a difficult line to walk, especially in 1950, and made more so because the play was set in 1860 when there would have been no contact between the "proper Englishwoman and the Oriental barbarian."

The subplot involving Tuptim and Lun Tha goes far beyond Landon's story as well. One of the dozens of horrifying tales in the novel and scenario occurs more than three-quarters of the way into the book; the man involved with Tuptim is a priest whose teachings she idolizes. Her religious fervor makes Tuptim leave the palace, dress as an acolyte, and secrete herself in the monastery simply to be near her spiritual leader. Once the deception is discovered, the King, furious at her disobedience and Anna's pleas for clemency, orders a pallet and scaffolding for the couple's eventual torture and hanging set up directly under Anna's window. Anna thus is involved but powerless against the King's brutal and unrelenting punishment for the guiltless couple.

Hammerstein's libretto alters the story to make the King's two victims lovers even prior to Tuptim's "arranged" marriage to the King. Now the subplot, which begins when Tuptim is delivered as a "present" from the King of Burma, extends throughout the musical, with Tuptim an outspoken champion of freedom. This subplot climaxes with Tuptim brought before the King and threatened with torture. Now Anna's intervention is effective and leads to perhaps the most dramatic scene Hammerstein ever wrote.

In it Anna pleads with the King, telling him that this girl has "hurt your vanity, not your heart." As the King lifts the whip to beat the cowering slave, Hammerstein's stage direction reads: "His eyes meet Anna's and hold for a few moments of mental battle. Slowly the whip drops. Then, the battle won by Anna, the King gives her an agonized look and runs off." His prime minister the Kralahome, representing the old guard, addresses the schoolteacher with broken heart and cold hatred: "You have destroyed him. You have destroyed King."

The Kralahome is right, of course, but the confrontation has brought to the audience the principles Anna (Hammerstein) has been espousing throughout the play. Now in the final scene at the King's deathbed, there is nothing left but for her pupil, the young Prince Chululongkorn, to announce his precepts for modernizing Siam. With the underscoring of "Something Wonderful," one of Rodgers's most moving melodies, the Prince, still a child, decrees his reign will have fireworks, boat races, and "no more kow-towing," but we know he will gradually come to espouse his teacher's principles of dignity and freedom.

This makes for the most ennobling and moving finale in all musical theater.

But Hammerstein was too clever a craftsman not to realize that the other side of emotion is humor, and although it was sorely lacking in Landon's book, he supplied it—a vastly different kind from the guffaws of South Pacific—with the King's dilemma song, "A Puzzlement." He was even able to find some lightness in Anna's diatribe against the King, "Shall I Tell You What I Think of You?"

The dancing, invented by Jerome Robbins, was geared to ritual and pageantry. Robbins, who originally was hired only to choreograph "The Small House of Uncle Thomas," treated his work with loving humor. Gradually, as his input spread to encompass all the musical's dancing, he brought an enchanting lightness to the rest of the evening. "The March of the Siamese Children," "Western People Funny," and "Getting to Know You," as well as the fifteen-minute ballet, were all amusing show-stoppers.

The King and I is the second musical that needs a scene by scene run-down for further understanding.

ACT ONE

Overture

Rodgers avoided the pentatonic and other Oriental scales in favor of more bizarre chordal juxtapositions, unorthodox voice-leading, open fourths and fifths to create his Siamese atmosphere. The liberal use of gongs, wood-blocks, and other percussion from the pit helped as well. The overture begins with these arresting hollow sounds and is followed immediately by "Something Wonderful," whose clever shift of voice-leading creates an exotic sound. "I Whistle a Happy Tune," "I Have Dreamed" and "Hello, Young Lovers" bring the medley to a close.

Scene 1: The Deck of the Chow Phya

Anna and her young son Louis arrive in Bangkok and are met by the King's prime minister, the Kralahome, and his retinue. Faced with these forbidding characters, the terrified boy asks his mother if anything ever frightens her. "Sometimes," she replies and then adds that when that happens, she whistles. The scene ends with Anna and Louis singing "I Whistle a Happy Tune," whose entire range is contained within a single octave.[1]

Scene 2 : The King's Library

The King receives Tuptim into his court and indicates he is charmed by her beauty. Tuptim sings "My Lord and Master," whose first words are "He is pleased with me," tying the song into the plot. At the end of the song she lets us know she loves another man.

The song's rhythm, lyric, and style are in contrast to the Western-type song ("I Whistle a Happy Tune") Anna has just sung. Hammerstein's contribution, with words like "painted cheek, tap'ring limb," enforces Rodgers's use of an exotic melody and stop and go accompanying rhythm (see Example A).

Example A

---

[1] Rodgers was careful to give Lawrence, who habitually sang flat and whose range was embarrassingly hooty near the top of the staff, "personality songs." He tailored his rangier melodies to the voices of his two other principals—the quicksilver soprano of Tuptim (Doretta Morrow) and the lush mezzo of Lady Thiang (Dorothy Sarnoff).

This background rhythmic differentiation will be used in the very next song ("Hello, Young Lovers") to accent the East-West separateness (see Example B). "My Lord and Master" is a rangy song (from low D-sharp to high A-sharp,) which creates further Eastern effect by leaning heavily on I and IV chords. Rodgers built the melody largely on the hexatonic scale but used a sophisticated, almost Ravelian harmonization underneath.

The plot then introduces Anna's desire for a house of her own, head wife Lady Thiang and Tuptim's quest for knowledge, and the latter's special eagerness to read *Uncle Tom's Cabin*. As soon as the King leaves, all the wives crowd around Anna. They think she wears her full-skirted dress because she is shaped that way. Then they ask her about her husband, and she sings "Hello, Young Lovers."

This lyric, an exquisite one and as important to the whole play as the "Soliloquy" was to *Carousel*, was one over which Oscar agonized for five weeks. He wrote several different versions before sending it over to Dick. It nettled Oscar that after he had dispatched his typewritten pages to Dick he heard nothing from his collaborator beyond, "It works fine." So upset was he that he called in Josh Logan and poured out his hurt and anger. It was to put further strain on what may have been a glowing collaboration but a shaky friendship.

As for the song itself, the lyric of the verse is reminiscent of an English recitative and very personal when Anna talks lovingly about her late husband, Tom. But the point of view changes dramatically, as though Anna were seeing love through Siamese eyes, in the refrain which moves into 6/8 time, "Hello, young lovers, whoever you are / I hope your troubles are few, / All my good wishes go with you tonight," she sings to an Oriental background. Then, slipping back into her nostalgic three-quarter-time English waltz world, she adds, "I've been in love like you."

Example B

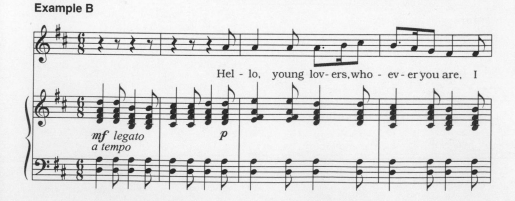

hope your trou-bles are   few.   All   my good wish - es go

with you to-night   I've been   in   love   like   you.

poco rit.   *a tempo*

The delicate scene is interrupted by the arrival of the children for presentation to their schoolteacher. The music of their "March," in which Rodgers uses
the same "wrong note" to intrigue the ear as he did in *Oklahoma!*'s "Oh What
a Beautiful Mornin'!'" but to totally different effect, is bewitching and would
seem to call for matching imaginative movement on stage. It should be mentioned that John Van Druten, playwright and director of drawing-room comedies, was totally miscast as director of this musical. Luckily, besides choreographing the show, Jerome Robbins was available to supply necessary
movement to many scenes.

The almost ritualistic entrance of the seventeen children to Rodgers's
"March of the Siamese Children" is a typical Robbins contribution. His youngsters come in singly or in pairs. The music crescendos to a smashing forte timed
to the entrance of Crown Prince Chululongkorn and then diminishes to a whisper as the smallest child is carried on stage, tugs at Anna's dress, and backs
into position with the rest.

Scene 3: In Front of Curtain

This short scene serves to show the insecurity of the King and the reasons for his bluster, while giving the stagehands behind the "traveller" a chance to ready the set for the upcoming schoolroom scene. R & H were creating operetta here and were not interested in the fluid staging they maintained in *Allegro* and *South Pacific*.

The Crown Prince elicits from his father the confession that he is "not sure of anything." The song, "A Puzzlement," written to be half sung, half spoken (not unlike Henry Higgins's numbers in *My Fair Lady*), was a late addition to the score and written with Yul Brynner's voice in mind. The song is amusing, but at its heart is the unsure monarch as exemplified by the excerpts below:

When my father was a king
He was a king who knew exactly what he
  knew
And his brain was not a thing
Forever swinging to and fro and fro and to.
Shall I, then, be like my father
And be willingly unmovable and strong?
Or is better to be right,
Or am I right when I believe I may be
  wrong . . .

. . . There are times I almost think
Nobody sure of what he absolutely know.
Everybody find confusion
In conclusion he concluded long ago.
And it puzzle me to learn
That tho' a man may be in doubt of what he
  know,
Very quickly he will fight,
He'll fight to prove that what he does not
  know is so!

Scene 4: The Schoolroom

"Getting to Know You," Anna's song to the wives and children, turned out to be the show's most popular number. It was added when the show was previewing in Boston and the general feeling was that the first act was too slow. Gertie suggested a number with her and the children might "brighten things."

As "Suddenly Lucky," the tune had been rejected for *South Pacific* and might have been overlooked even this time had Mary Martin not reminded Oscar of its existence. With its juxtaposed accents (see Example C) it seems far more Siamese than Polynesian. Using tutorial words like "pupil," "teacher," "expert," "subject," and "knowing," and ending with "because of all the beautiful and new / Things I'm learning about you / Day by day," the song is specific to the teacher and this play while remaining general enough to be applicable to anyone.

Example C

After the song and dance Anna reverts to harping on the house the King promised her. This small episode in Landon's book has been expanded in Hammerstein's libretto into a running feud which vexes the King and dominates the entire first act.

Anna's request and the King's denial lead to a standoff ending with her lines: "A land . . . where there is talk of great change, but where everything still remains according to the wishes of the King!" This almost breaks her relationship with the monarch. After both she and the King storm off, Tuptim and her lover Lun Tha enter and explain that their secret rendezvous will no longer be possible if Anna, who has been a party to their meetings, leaves Siam. They sing "We Kiss in a Shadow," whose delicate lyric against Robert Russell Bennett's haunting orchestration has the proper Oriental flavor.

We kiss in a shadow,   We speak in a whisper,
We hide from the moon,   Afraid to be heard:
Our meetings are few,   When people are near,
And over too soon.   We speak not a word . . .

Intermediate Scene: In Front of Curtain

Another late addition. This gives the author a chance to reprise the King's "Puzzlement" number, this time led by Anna's son, Louis, in duet with the Crown Prince. They want to fight but like their elders, Anna and the King, they don't know what they are fighting about. This vignette leads directly into

Scene 5: Anna's Bedroom

The scene begins with Anna's soliloquy (the interior monologue, an R & H innovation, having by this time become an expected part of every show). She pours out her anger at being considered the King's servant and affirms her decision to leave Siam. But then, halfway through, a born teacher, she remembers moments when "I've heard an occasional question / That implied at least a suggestion / That the work I was trying to do / Was beginning to show with a few . . ." Wavering now, she concludes with, "I must leave this place before they break my heart."

Her reverie is interrupted by Lady Thiang, who implores Anna to go to the King. "He is deeply wounded man," she says. "No one has ever spoken to him as you did today in schoolroom." The head wife then adds that the King needs help in protecting Siam. "Lady Thiang, please don't think I'm just being stubborn," Anna says, "but I cannot go to him. I will not."

Then Lady Thiang pulls her trump card, the exquisite song "Something

Wonderful." (I have quoted below the opening lines of the verse and the closing lines of the refrain.)

This is a man who thinks with his heart,
His heart is not always wise.
This is a man who stumbles and falls,
But this is a man who tries . . .

. . . You'll always go along,
Defend him when he's wrong,

And tell him when he's strong,
He is wonderful.
He'll always need your love,
And so he'll get your love.
A man who needs your love
Can be wonderful.

Scene 6: The King's Study

This final scene of the act has no singing until its conclusion but because of all the plot activity and heavy underscoring one does not miss it. From the charming moment of Anna's entrance where the King says, "You come to apologize?"

ANNA: I am sorry, your Majesty but—
KING: Good! You apologize.
ANNA: Your Majesty! I . . .
KING: I accept!

it is clear that this is a crafty manipulator. When Anna suggests the King entertain in grand manner the British envoys who are coming to Siam, he takes up her suggestion at once as *his* idea, sounding the gong, waking up the whole palace. He puts everyone to work to prepare for their imminent arrival. "Just as it is done abroad," the King reminds Anna. She in turn tells him that they can entertain the guests after dinner with a play that Tuptim has written.

Before the act ends the King invokes Buddha to "help prove to the visiting English that we are extraordinary and remarkable people."

"And Buddha," he adds before the curtain comes down, "I shall give this unworthy woman a house—a house of her own—a brick residence adjoining the Royal Palace, according to our agreement, etcetera, etcetera, etcetera."

ACT TWO

Scene 1: The Reception Room of the Palace

The King's wives, dressed in European-type "swollen skirt," are ready for presentation to the British. In a charming moment when they prostrate themselves as the King enters, Anna discovers they have practically no underwear. With the British already in the palace, it is too late to remedy the situation. As the King pooh-poohs the value of undergarments, Anna removes her smock, revealing her dress of the evening, bare shoulders, and cleavage. The King now perceives her, perhaps for the first time, as a desirable woman. He shows some pique at Anna's familiarity with Sir Edward Ramsay, an old beau, insisting she abandon Sir Edward and take charge of seating the guests as they go in to dinner.

Scene 2: In Front of Curtain

Lady Thiang, knowing of Lun Tha's love affair with Tuptim, has ordered him to leave Siam this very night. When Lun Tha comes onstage, he tells Tuptim of the edict and of his plans to secrete her on his boat with him. Together they sing "I Have Dreamed," one of Rodgers's finest musical achievements. Oscar, beyond choosing the unusual form[2] for this song—$A^1A^2A^3B$—hardly contributed a very original lyric, but no matter. The song is a tour de force and hits its emotional mark by spending eight bars in F, eight in G, eight in A, and finally modulating back to F.[3]

The lovers are interrupted by Anna who comes to remind Tuptim that the play is about to begin. Tuptim bids a tearful good-by to Anna who says, "God bless you both," and then reprises "Hello, Young Lovers."

Scene 3: The Ballet

Tuptim's ballet, which was actually created by choreographer Jerome Robbins to Trudy Rittman's[4] music and rhythms, was based on Hammerstein's scenario. Robbins studied authentic Thai dancing, and then, as Yuriko, who danced Eliza in the ballet, remembered, "Hammerstein told him he was more interested that the ballet be effective than that it be authentic." Costumed with jeweled headdresses, masks, swirling silks, and ribbons, this ballet was a visual delight

[2] This form, apparent in songs from "Tea for Two" to "Gigi," usually modulates only in its A^2 section.

[3] Hammerstein, always hypercritical of his work, was dismissive of this scene and its Act I counterpart. His son James recalled that every time he would go the theater with his father, Oscar would say, "Let's go outside and take a walk" during this scene. James also told this writer that "Lun Tha is a terrible part; Tuptim is saved by the ballet and 'Lord and Master.' "

[4] Ms. Rittman is the doyen proponent of an unsung group—rehearsal pianist. It fell to the "rehearsal pianist" (sometimes called "dance arranger") to adapt the music for dancing. Ms. Rittman, who worked on several R & H shows, generally used Rodgers's music—cutting, splicing, changing rhythms, etc.—according to the choreographer's design. In the case of the ballet "The Small House of Uncle Thomas," she created her own rhythms and melodic lines using woodblocks, cymbals, and rhythmic speech. Of Rodgers's score she used only snippets from the composer's "We Kiss in a Shadow" and "Buddha" themes.

The main thrust of its scenario is Eliza's race to join her lover George who has been sold "into the far away province of Oheeo." Tuptim's entertainment parallels her own situation when she refers to "King Simon of Legree who has decided to hunt Eliza with scientific dogs who sniff and smell, and thereby discover all who run from King." Fleeing from the King, Eliza comes to the river. "Who can save her?" asks Tuptim.

"Buddha make a miracle. Buddha send an angel down. Angel make the wind blow cold. Make river water hard, hard enough to walk upon."

When Simon comes to the river, Buddha has called out the sun, and Tuptim tells us that now the river water has become soft, drowning wicked Simon and his slaves.

Before the ballet ends, Tuptim loses her composure and blurts out, "I too am glad for death of King. Of any King who pursues a slave who is unhappy and wish to join her lover." Then regaining her poise she leads to the finish of the entertainment with praise to Buddha.

Scene 4: The King's Library

After Anna and the King congratulate each other on the evening's success, the monarch rewards her with the gift of a large emerald ring. Anna hardly has time to thank him when the secret police report that Tuptim is missing from the palace. This, and the King's insistence that all his females remain faithful to him, provokes a discussion of the King's double standard. "It is like old Siamese rhyme," maintains the King:

A girl must be like a blossom
With honey for just one man.
A man must live like honey bee
And gather all he can.

To fly from blossom to blossom
A honey bee must be free,
But blossom must not ever fly
From bee to bee to bee.

Anna has a more courtly concept in mind when she tells about how Englishmen are introduced to women, ask them for a dance, and may end up falling in love with them. First she sings "Shall We Dance?" while moving about the library. When the King asks her why she stops, she replies, "Your Majesty . . . in my country a girl would not dance while a man was looking on." "Good!" he says. "We dance together. You show me. You teach!"

This leads to the most expansively joyous moment in the play. The King in red penang and sari jacket puts his arm around Anna's waist. She is wearing the exquisite oyster-pink satin ballgown Irene Sharaff designed for Gertie. He then leads her into the dance. So great is the emotion coming across the footlights that I have never seen a production of this musical when applause did

In Jerome Robbins's stylized ballet, a Siamese version of Uncle Tom's Cabin, *an exotic Yuriko as Eliza carries her baby "George" as she flees from wicked King Simon of Legree.* MOCNY.

not spontaneously break out at the moment when Anna and the King, exhilarated by this infectious polka and almost overwhelmed by their newfound relationship, spin around the library.

The energetic dance is encored but hardly begun when it is interrupted by the announcement that Tuptim has been found. This leads to the aforementioned scene where the King is forced to retreat from his barbaric ways. Is

Hammerstein telling us that knowing love at last the King is unable to punish his slave for that emotion? If so, he carries it further into a *liebestod*. As she is taken offstage, Tuptim, learning that Lun Tha has been executed, kills herself. Anna, left alone with the Kralahome, returns the emerald ring to him and announces that she plans to leave Bangkok on the next boat. "I wish you have

"Shall We Dance?," the exhilarating polka that is perhaps the most romantic single moment in all the R & H canon. Yul Brynner in penang and red jacket leads Gertrude Lawrence, who wears an exquisite ball gown designed by Irene Sharaff. MOCNY.

never come to Siam," the prime minister says with heartbroken rage. "So do I," Anna answers, running off. "So do I!"

Intermediate Scene

A short explanatory moment advising the audience that the ship captain has arrived and that Prince Chululongkorn is in charge of a procession because his father is ill. At one point an assistant comes from the palace to tell the Crown Prince to return at once because his father's condition has worsened.

Scene 5: A Room in Anna's House

Lady Thiang and Chululongkorn have come to persuade Anna to come to see the King. The King has written a letter which his wife delivers. Anna reads it:

> While I am lying here, I think perhaps I die. This heart, which you say I have not got, is a matter of concern. It occurs to me that there shall be nothing wrong that men shall die for all that shall matter about man is that he shall have tried his utmost best. But I do not wish to die without saying this gratitude, etc. etc. I think it is strange that a woman shall have been most earnest help of all. But, Mrs. Anna, you must remember that you have been a very difficult woman, and much more difficult than generally.

Anna breaks off with, "I must go to him."

Scene 6: The King's Study

The dying King tries banter, insisting Anna take back the emerald ring. All the children and the wives try to keep Anna from going. One of the young students who cannot write has made up a moving letter which she recites. The rest of the children say they are afraid without their teacher, whereupon Anna reprises, "I Whistle a Happy Tune" as her prescription to pretended bravery. At last, when the boat whistle sounds and Anna starts to leave, the children crowd around her. "Do not go, Mrs. Anna. Please do not go," they repeat.

Anna can take no more. Asking Louis to go down to the dock and tell Captain Orton to have their things taken off and put back in their house, she starts to remove her hat and gloves.

The musical ends as Chululongkorn with the underscoring of "Something Wonderful" announces to the assemblage his decree against crawling and crouching. "The gentlemen this way, only bending from the waist. The ladies will make dip as in Europe." Here, at the finale, Oscar's stage directions, always masterful, are perhaps the most patrician and moving in all the play.

He tries to show them a curtsey, but cannot. In the manner of his father, he signals to Lady Thiang with his left hand. She crosses to the bed. [Music off. "And so he'll get your love." [5]*] With great pride she turns and faces the wives and children and drops a low curtsey before them. As the music swells all the children and women imitate her, sinking to the floor as the curtain falls in final obeisance to the dead King and with respect for the new one.*

By the time *The King and I* went into rehearsal in January 1951, many of the preliminary technical problems of mounting so large and expensive a show had been solved. The costumes under Irene Sharaff's supervision were being custom made of specially dyed Thai silk; Jo Mielziner had designed his imposing sets. (It took six railroad cars to get them to the Boston previews.) Most importantly, Oscar's book, including the aforementioned complete stage directions, had been written. Unlike any former R & H show, only three songs had to be cut.

Once on stage, things went less well. Oscar, always polite, had to contend with Van Druten's flaccid direction without seeming to take over. "Wouldn't it

[5] Hammerstein's stage directions always correlated what he wanted the actors to do or feel to the exact words of the song the orchestra was playing. He could clue the actor to the lyric. In this he was rather like Puccini, who indicated where his complete ideas for staging fell in his scores.

In the moving finale during which the King dies, young Prince Chululongkorn (Johnny Stewart) takes over with an assist from his mother, Lady Thiang (Dorothy Sarnoff), standing, left. MOCNY.

be better to do it this way?" he would suggest, hopping onto the stage. What he couldn't fix he would instruct Jerry Robbins or Gertie to look at. But everyone involved with the show had to admit that Yul was a powerful actor who instinctively knew what to do with the role and no shrinking violet in the directoral department. Much of the urgency the play possessed came from him. He even directed the mercurial, often improvisatory Gertrude Lawrence.

The show opened at the St. James Theatre on March 29, 1951, and was to run almost three years, chalking up fewer performances than *Oklahoma!* or *South Pacific* but far more than *Carousel.* It ran almost as long in London[6] as it did in New York, capping a string of R & H hits that had monopolized the Drury Lane for nine years.

Critics were spotty. *Time* critic Louis Kronenberger said that this "may not quite be Rodgers and Hammerstein at their best, but it is musicomedy at its most charming." "Some interesting music, but the most important work in this big, long show is the work of Mr. Hammerstein," said the *Daily News,* while the *Journal-American* sprinkled further salt onto Rodgers's music by complaining that it was "not a great score according to Rodgers. There are no musical moments which made you almost jump out of your seat as in most of his successes."

Although the public paid no mind to the critics, the reception did not help the collaboration. Each man went his separate way for a while, Rodgers writing a massive thirteen-hour score for an NBC television series, *Victory at Sea,* while Oscar busied himself with a revival of *Music in the Air.* Dick's score was to take four months longer to play out in half-hour segments than the total run of the Kern-Hammerstein remounting, which barely eked out 56 performances.

Perhaps Oscar felt after *The King and I,* which was pure operetta, that musicals were due for a return to that form, but his actions show he was shaky about the revival. Adding and then deleting his now classic "All the Things You Are" and moving the locale from Munich to Zurich did not help the creaky plot.

Bert McCord, writing in the *New York Herald-Tribune,* put most succinctly the reasons why Oscar had murdered all chances of pre-R & H revivals: "One of the ironies is that Hammerstein is perhaps more responsible than anyone else for the fact that the show can't finish in the running today. By his work with Rodgers, he has been instrumental in educating audiences to more developed musical theater."

---

[6] One of the saddest footnotes to the glorious London opening was the absence of Gertrude Lawrence, who had hoped to bring the production "home" to the Drury Lane. A great musical star, Ms. Lawrence died after a valiant struggle with cancer the year before. She had played the role of Anna in New York for seventeen months.

Lerner

1951–1955

FAR MORE IMPORTANT THAN THE RESPECT AND FINANCIAL rewards his Hollywood sojourn was to afford him, Lerner's almost two years on the West Coast were to introduce him to two people who were paramount in what might be called the "middle period" of his life. The first was Nancy Olson, to whom he was married for seven years. She became Lerner's third wife, mother of two of his daughters, and a constant adviser during the writing of My Fair Lady. The other was Stone "Bud" Widney, who might be called Lerner's production associate, although friend and advisor would be a better title. Bud was introduced to Lerner by Nancy Olson and become deeply involved in every show for the rest of Lerner's life.

It was while they were working on Royal Wedding, Burton Lane recalled, that Lerner met Nancy Olson. When he told his collaborator that he was very interested in the film actress, a decade his junior, and planning to leave Marion Bell, Lane advised him to slow down, pointing out that he had already been twice married and was not yet divorced.

But Lerner could not slow down any more than he could face being alone. In this case he had asked friends for introductions to any attractive women with whom he would have something in common, and they brought out a roster of possible candidates. Nancy, they said, was one of the nicest of the "nice girls," a part in which she would eventually be typecast by the press and in cinema.

Nancy Olson was born in the Midwest in 1928. Her father, a well-known obstetrician and gynecologist, moved the family to Los Angeles when he was offered a professorship at the University of Southern California. Nancy withdrew from her drama course at Northwestern University and enrolled in an acting course at UCLA. Her family was quite wealthy, and student classmates remember her coming to class in a chauffeur-driven limousine.

Bud Widney, who was a directing student in the same program, recalled that when a revue with music and lyrics by Billy Barnes, *Footprints on the Ceiling,* was done on UCLA's main stage, Nancy, who had quite a good singing voice, was given the leading part. It was in this revue that she was spotted by a talent scout from Paramount. "We all knew it would take about thirty minutes for *her* to become a star," Widney said.

But in typical Hollywood fashion, although under contract and salaried, she was allowed to continue her college course. Ambitious for her own career but refreshingly oblivious to Lerner's success, she told Billy Barnes, "I've met the most wonderful fellow—and you must get to meet him too. You'd have so much in common for he's a songwriter, but not quite in your league because he only writes the words."

Months later she was given a part opposite Randolph Scott in *Canadian Pacific* and then the role of the wholesome ingenue who tries to help extricate her boyfriend William Holden from his relationship with harridan Gloria Swanson in *Sunset Boulevard.*[1] Neither picture had been released at the time she met Lerner.

Known on the lot as "wholesome Olson," she was five-foot-five (the same height as Lerner), with high Scandinavian cheekbones, fair skin, honey-blond hair, and blue eyes. It was through Nancy that Alan came to know Bud Widney.

"Nancy and I were not going together," Widney said recently.

> I had a steady gal at the time. We did acting scenes together . . . and became good friends. After that period she had begun to date Alan. . . . It had to have been '49 or '50. I was developing a master's thesis on the Mexican-American for the motion picture department of UCLA curiously enough in conjunction with the sociology department.

Lerner, who was working on several projects at the time, had recently outlined the plot of *Paint Your Wagon* and had changed the leading character's ethnic background from Swedish to Mexican. The change from stoic Norseman to high-born Latin allowed Lerner's libretto to indulge in the author's favorite extra dividend—injustice. Julio Valveras's first speech to the miner's daughter with whom he will soon fall in love sets him up. "This is gringo land now. Not Mexican. One time all this was part of Mexico. I'm a citizen. Suddenly, a few years ago they start fighting in some place called Texas. (He snaps his fingers.) I'm a foreigner."

---

[1] In 1950 Nancy's *Sunset Boulevard* performance was nominated for an Academy Award as best supporting actress. She lost out to the rotund Josephine Hull's dithering interpretation of James Stewart's aunt in *Harvey.*

Alan knew nothing of the history or sociology of Mexican-Americans and so when Nancy said, "Bud knows something about Mexicans," Lerner was anxious to meet him.

"It was a good meeting because we respected each other instantly. He wanted to talk about the Mexican-American and I had a lot to say on the subject, and he appreciated it," Widney said. Lerner, always insecure about the quality of his writing then handed Widney the scripts of *Royal Wedding* and *An American in Paris*.

"Read these and give me your reaction to them." I guess it was some kind of a test. . . . I went through them both. I told him exactly how I felt, in quite a lot of detail. Where I thought he was short-changing himself and where the thing didn't materialize. He was delighted that he could get some reactions with love—without in any way feeling that he was challenged . . . or that somebody had an axe to grind.

From the start of their relationship Widney understood Lerner and respected his dependence on an intellectual friendship.

Alan's real need was to express things in the presence of somebody else. When he got an idea for a song or a plot or a musical, he could tell right away whether it floated. That's usually what you do with your collaborator. In Alan's case collaboration was such a long and exhaustive kind of thing that, except for Fritz, his collaborators were generally too impatient to deal with it. The idea of having somebody to toss things around with—as long as it takes to iron the goddamned thing out . . . that's what he might have thought when he suggested I come to N.Y.

He called Cheryl Crawford [the show's producer] and said, "I've got a guy I'd like you to hire as assistant stage manager," and she said "Have him write me a letter." I wrote her a joshing, fun letter and she said come on. Alan made it possible for me to come back and learn the theater . . . and didn't restrict that to the musical theater.

Six months later Alan asked Marion for a divorce, selflessly she agreed and soon applied for a California decree. "I loved him too well not to want to see him happy," Bell said. But Lerner became impatient because Marion's California application would not be acted upon for a year. He soon began making regular trips to Nevada so that he could build up the six-weeks' residency necessary to obtain a divorce in Las Vegas. After working on his films during the day he would fly there in the evening, check into his hotel and come back the next day.

"I did not contest it, I knew he wanted to marry Nancy Olson, and I loved him enough to let him go," Marion Bell said. "There was no alimony. Alan put $6,000 in the bank for me to get started on my [solo singing] career."

Alan with Marion Bell, his second wife, in one of the rare photographs of him wearing glasses with flesh-colored frames. These were soon discarded in favor of his signature heavy black-rimmed spectacles. Photo courtesy Marion Bell.

Shortly after his divorce from Marion, Alan married Nancy Olson on March 19, 1950. He finished the script of *An American in Paris* the night before the wedding. "I sat down at eight that evening and wrote sixty pages. Somehow I ended it and I never changed anything," he was to tell musicologist Donald Knox.

When they returned from their honeymoon they went directly to New York so that Alan, who had reconciled with Fritz, could begin to work full time on *Paint Your Wagon*. Although the score came easily this time, as Bud Widney noted, "It became apparent early on that the book wasn't working." The idea was noble, using the Gold Rush as a metaphor for the American Dream. Both Lerner and Hammerstein hoped their stories would somehow ennoble their audiences. But Alan's gold miners who are looking everywhere for something to change their lives around are not very far from a high school version of Voltaire's *Candide* as combined with *The Girl of the Golden West*.

Once the show was put on stage, the plot, which concerned a town where gold is found, an uneducated miner who sends his daughter back east to get an education, and her love for an illiterate Mexican, seemed sparse and malconstructed. Even though at the end the couple and townsfolk decide to irrigate their land and no longer go off after the pot of gold—specifically their "Wanderin' Star"—the majestic thought was somehow lost.

Agnes de Mille, who had contributed so much to the success of *Brigadoon*, agreed to involve herself in the project because she had "liked the beginning of the libretto and thought the score was first class." A competent overall director of musicals, she would confine herself, as she did in *Brigadoon*, to choreography because of pressures of other work. The directorship of the entire production fell on Daniel Mann, who, like Elia Kazan, the director of the ill-fated *Love Life*, was an Actor's Studio alumnus.

"But Alan couldn't write the story," de Mille confided. "We got on the road with it and he would do a different ending to act one and act two—every day." Bud Widney, who was now assistant stage manager, recalled that

soon director Danny Mann's advice and opinions [this was his initial musical] were no longer sought. Agnes de Mille was in the driver's seat instead of him. That was my first lesson in Broadway showbiz. We rehearsed for two or three weeks and did a run-through. The powers met on stage. I crawled into the theater and listened to them sitting round the round table. My vivid impression was Danny on one side of the table and Agnes on the other. Everyone was talking to Danny, but as the conversation went on, and as the message was "it ain't working," all the heads went over to Agnes. Body language alone told me that's how it works. Danny stayed on as director, but all of the attention was being paid to Agnes. Danny was filled with incredible frustration. He fired me but the stage manager then threatened to quit so they kept me on.

As the out-of-town preview date got closer it became a search for the culprit. Alan thought it must be James Barton, the star . . . Barton wasn't bad. The book was bad. Anytime you're in trouble it's the book. Then it was decided to get Agnes to do more dances.

De Mille said, "It was insanity. I did dances, I did scenes, I did this, I did that, and it was thrown out and another thing put in. And finally, in a meeting of the staff one day sitting on the stage he turned to me and said, 'Agnes, which is more important to you, the book or your dances?' And I said, 'What book?' He turned purple and from then on he hated de Mille. In *Brigadoon* he loved me. We used to have breakfast in the hotels every morning and scream with laughter."

It was during those tension-filled days readying *Paint Your Wagon* that de Mille lost her respect for Lerner. "Alan was a liar," she told this writer. "I don't think he could keep track of his lies. Now I think Loewe was more gifted than Alan—I think Loewe was a real melodist. He was a pretty good musician, he had been trained. But Alan couldn't write a plot. He couldn't end *Paint Your Wagon*. We staged it. It ends. You see, it was bits and pieces of a superb thing. He simply couldn't put it together."

History has borne de Mille out. *Paint Your Wagon* as a musical play is simply not revivable. The ennobling element that he wanted the musical to have was

obfuscated by Lerner's chauvinistic attitude towards women. With a female auc-
tion as the highlight of the first act and the whooping arrival of a carload of
whores as the cornerstone of the second, Lerner's grizzled prospectors seemed
more like high schoolers with permanent erections.

But the score, as de Mille pointed out, contains some song gems that cannot
go unmentioned. In his first attempt to write a lyric for the uneducated, Lerner
followed Hammerstein's axiom: "Heavy rhyming equates with sophistication
and schooling." *Paint Your Wagon*'s most important ballad, introduced by the
hero and heroine, Julio and Jennifer, both supposed illiterates, is "I Talk to the
Trees." It has only one rhyme in its refrain, yet its poetic language helps it get
the sensitive nature of the musical's young lovers across. Unfortunately Fritz
Loewe's lovely but most sophisticated melody marked "Tempo di Beguine"
works at cross purposes with the lyric.

I talk to the trees   But suddenly my words
But they don't listen to me,   Reach someone else's ears
I talk to the stars   Touch someone else's heartstrings, too.
But they never hear me.   I'll tell you my dreams
The breeze hasn't time   And while you're list'ning to me
To stop and hear what I say.   I suddenly see them come true.
I talk to them all in vain.

Loewe was somewhat more supportive in Lerner's other nonrhymer, the rousing "I'm on My Way."

Where'm I goin'?   When will I get there?
I don't know.   I don't know.
Where am I headin'   When will I get there?
I ain't certain.   I ain't certain.
All I know is I am on my way!   All I know is I am on my way!

Melodically and often lyrically, *Paint Your Wagon* contains several other
haunting gems. "I Still See Elisa" is an old curmudgeon's recollection of his
late wife who "talked softer than a leaf hittin' the water." It has an exquisite
release: "Her heart was made of holidays / Her smile was made of dawn / Her
laughter was an April song / That echoes on and on . . . ."

This writer's special favorite is the rueful "Another Autumn," coming as it
does at a time of the year when the prospectors can no longer go up north in
search of a fresh vein of gold. The lyric of the refrain is reprinted below:

Another autumn . . .   For you can dream in spring
I've known the chill before.   When ev'ry hope is high,
But ev'ry autumn   But when the fall comes in
I feel it more and more.   They all begin

To fade and die.
Another autumn . . .
So sweet when all is well;
But how it haunts you when all is wrong.

For one thing time has shown,
If you're alone
When autumn comes
You'll be alone all winter long.

And the score has joy aplenty with songs like "Whoop-Ti-Ay" and "How Can I Wait." In the latter the young heroine, anticipating a meeting with her lover, sings "How can I talk, can I breathe, can I eat? / What can I do with my hands and my feet? How can I wait till tomorrow comes?"

Perhaps the most unusual song in the score is a beautiful ballad of lonely prospectors hungering for their women, "They Call the Wind Maria"—not chauvinistic in this case, for each man is yearning for his own girl.

Away out here they got a name
For wind and rain and fire;
The rain is Tess, the fire's Jo,
And they call the wind Maria.
Maria blows the stars around
And sends the clouds a-flyin'

Maria makes the mountains sound
Like folks were up there dyin' . . .
And I'm a lost an' lonely man
Without a star to guide me.
Maria, blow my love to me;
I need my girl beside me.

Paint Your Wagon won far more approval for Loewe's music than for Lerner's contribution. It received one rave, two favorable, and three unfavorable reviews from the New York critics after its Broadway opening on November 12, 1951. Robert Garland in the *Journal-American* summed up the feelings of the critical fraternity by noting that "it is a lush and lusty score, more lush and lusty than the book and lyrics." In spite of its cool reception it was able to play close to 300 performances, although with its large cast and many empty seats near the end of its run, it ended up losing money.

In an ironic twist, *Three Wishes for Jamie*, a fantasy about a barren woman who eventually bears a mute son, was the very next new musical to come to Broadway after *Paint Your Wagon*. Its star while it was previewing on the West Coast was Marion Bell. During her marriage to Alan, Marion too had been barren, but after the divorce, Alan paid for an operation which allowed her to conceive. In short order she married, was selected for the role in *Three Wishes*, discovered she was pregnant, and withdrew from the stage. She was replaced by Anne Jeffreys. By the time the musical opened, bombed, and closed three weeks later, Marion's son Thomas Charlsworth was born. Marion said recently that she never regretted exchanging motherhood for a stage career.

In 1968, more than fifteen years after *Paint Your Wagon* completed its Broadway run, Lerner himself undertook to produce a screen version. He had learned much about librettos by this time and according to Bud Widney "knew that his stage version had a bummer of a book and felt there must be a way to save the thrust of the plot. So he got Paddy Chayefsky to write a new book." It was too verbose, but Lerner reedited and rewrote Chayefsky's story of men who meet when each is chasing a dream. Five forgettable new songs, this time with music by André Previn, were added. But now the musical, which featured popular actor Lee Marvin, a young Clint Eastwood, and a rewritten story of male bonding, was a total disaster. Whatever little charm and sensitivity the Broadway show had was replaced in the cinema version by macho crudeness.

Paint Your Wagon was still running in the spring of 1952 when Alan returned to Hollywood for an extended time so he could work on the screenplay of *Brigadoon*. Nancy had never stopped making pictures and since she had gotten an Academy Award nomination for her role in *Sunset Boulevard* had been much in demand. Now both she and Alan shuttled back and forth. Olson starred in *Union Station, Mr. Music, Force of Arms,* and *Submarine Command,* all of which were released in 1950 and 1951.

She had taken time out to have a baby, but by the time they returned to Hollywood she was ready to resume work. Liza Lerner had been born in October 1951. In an interview with a screen columnist, Nancy reported that it was tempting "when you don't have to be a breadwinner" to retire from the screen. She would be returning to a grueling schedule at Paramount and deplored the fact that she'd have less time for Liza, but, she added, "Movies are fun and especially when one can be independent."

Alan too could be independent, for with the Broadway success of *Brigadoon* and an Academy Award for *American in Paris* in his pocket, Arthur Freed, his producer, had no trouble convincing the powers-that-be at the studio to keep *Brigadoon*'s Scottish locale. (They had wanted to change it to Ireland because there were more Irish moviegoers than Scottish in America.)

One day Alan got an invitation to lunch from Gabriel Pascal, the producer-director who was in Hollywood converting Shaw's *Androcles and the Lion* into a film. Pascal, a Romanian who happened to have been born in Transylvania but whose mother tongue was Hungarian, had directed the film versions of *Major Barbara* and *Caesar and Cleopatra.* He had produced those films as well as a version of *Pygmalion,* and it was the musicalization of the latter that he wanted to talk to Alan about.

Pascal, a mountainous man with a basso-profundo voice and an accent which, according to Lerner, "defied any known place of national origin," spoke only in simple declarative sentences without articles. "You will come to restaurant." "Tonight I go to theater." He was well known in Hollywood circles for his stinginess and bluntness. It was also rumored that he was broke after his Technicolor fiasco *Caesar and Cleopatra,* an extravaganza for which he actually brought sand to Egypt to get the right color.

Everyone also knew the story that had allowed him to astonish the motion picture community, of how he acquired the film rights to many of Shaw's plays. For years, in many Hollywood living rooms, he was fond of expanding on the narrative, by now apocryphal. It went something like this.

He appeared at the door of Shaw's cottage in Ayot St. Lawrence and rang the doorbell. When the secretary answered, he said he wished to see Mr. Shaw. "May I ask who sent you?" the secretary inquired, to which Pascal retorted, "Fate sent me." Shaw, who happened to be in the stairwell at the time, overheard this exchange and invited him in.

"Who are you?" asked Shaw.

"Gabriel Pascal. I am motion picture producer and I wish to bring your works of genius to screen."

"How much money do you have?"

Pascal dug down into his trousers. "Eight shillings."

"Good. Come in. You're the first honest film producer I've ever seen."

Pascal boasted that when he left the cottage his pockets were bulging with contracts giving him the film rights to many of Shaw's works, including *Pygmalion.*

Alan's first meeting with Pascal can best be described by this excerpt from *The Street Where I Live.*

Seating himself in a buddha-like squat at the table, he ordered four plates of spaghetti, one for me and three for himself, and four raw eggs. "I will show you," he said, "only way to eat spaghetti." He cracked one raw egg and spilt the contents over my spaghetti and used the other three for himself. To my surprise it was quite good. As he swooshed the spaghetti into his mouth he said, "I want to make musical of *Pygmalion.* I want you to write music." I hastened to explain to him that I did not write music. "Who writes music?" he asked. "The composer," I replied, "Fritz Loewe." "Good," he said. "We will meet again and you will bring man who writes music." I told him Fritz was in New York. "Good," he said. "You will tell him to come out here."

Lerner was fascinated. Fritz was on his way out to Hollywood to discuss a project of his own, a musical he was planning to write with Harold Rome. When Fritz arrived, Alan intrigued him as they both seriously considered the possibilities of a musical on the transformation of Shaw's flower seller. After several meetings with Pascal at the unkempt pseudo-Spanish house he had rented in Westwood, and in spite of the producer's unctuous flattery saying, "You are only men who can fulfill my dream," they could not see how it could be done. *Pygmalion* was full of interminable talk. Most of it takes place in drawing rooms or Higgins's studio, which would allow for no choral writing.

During their meetings Alan came up with the idea of making Higgins a professor of phonetics at Oxford so that a chorus of undergraduates could be brought in, but that seemed clumsy and was discarded. Then other stumbling blocks presented themselves. The play contains no comedy except for Eliza's father Doolittle, and most of his lines were preachy. Moreover, there is no subplot, an absolute essential for a musical. As Alan wrote, "One story and only one, and any author who thinks he can add characters to a play by Shaw is exhibiting behavioral evidence of the first signs of acute paranoia."

Worst of all, *Pygmalion* is a comedy of manners that concerns several heartless men—Higgins, Pickering, Doolittle—who preyed on an ambitious but vulnerable woman, and is totally anti-romantic. Unlike the legend of Pygmalion, the ancient Greek sculptor who brought his creation to life because his love for her was so strong, Shaw's sculptor brings his Galatea to life by ignoring her. He storms and carries on when Eliza leaves him to stay with his mother, but when he finds her there and confronts her, she answers him back in his own terms and he is delighted. He realizes he has succeeded in transforming her from what he called "deliciously low" into a full-blooded woman.

Although Pascal never mentioned it, Lerner and Loewe were very aware that Gaby Pascal had approached Rodgers and Hammerstein years before. After much consideration, they had realized Shaw's play was more of a treatise on social behavior than a musical, and turned the project back. After that, Pascal had tried to negotiate with the team of Howard Dietz and Arthur Schwartz; they, too, threw up their hands. The determined Romanian then requested lyricists "Yip" Harburg and Fred Saidy to consider it with music by Cole Porter. They were last on his list before L & L.

Yet, in spite of its difficulties, the story intrigued Alan and Fritz, and when Pascal asked them a few weeks later what they thought, they did not say no. In their own minds they put the project on hold. Meanwhile, Pascal made a co-production agreement with The Theatre Guild to produce the play at the Westport Playhouse in Connecticut, and it was, as Lerner wrote, "a joy to see it again, but the joy only increased our frustration." Soon both Lerner and Loewe began to look for other stories to adapt while waiting for an idea to break through on *Pygmalion*.

During the summer of '52 Alan, always a political thinker, involved himself in what he felt was an urgent opportunity for change. Adlai Stevenson, a brilliant intellectual, was running on the Democratic ticket against Eisenhower, and since it was also the most shameful time in American history, the era of the demagogue McCarthy, most artists, singers, dancers, comedians, and writers felt they had to drop everything and come to the aid of their party. Nancy, strongly motivated too, dropped all filming and joined Alan in stumping for Stevenson.

It was during a final rally at Madison Square Garden, when Lerner was on the same program as Oscar Hammerstein, that they talked afterwards about

Pygmalion. Oscar asked candidly how Alan was coming with the project. "Slowly," was Lerner's guarded answer, but then he poured out to Hammerstein his feelings about the stumbling blocks he encountered. Oscar nodded and said he and Dick had met the same snags. "Dick and I worked on it for over a year," Oscar said "and we gave it up. It can't be done."

Four weeks later Lerner and Loewe met with Pascal and reluctantly told him they too were giving up any thoughts of making a musical out of *Pygmalion.*

Now in Alan's search for his next project, he asked Bud Widney, whose opinion he greatly respected, to look over some of the possible scripts that had been piling up on his desk.

"One of the things that was on his desk was a paperback collection of Li'l Abner comic strips, and reading through them they suggested a marvelous musical." After Lerner gave Widney the go-ahead, Bud synopsized thirty years of Al Capp's social satire. Now Lerner, who had had another serious argument with Fritz, enlisted Burton Lane to do the score and inveigled Al Capp's good friend Herman Levin, who had hit the jackpot with *Gentlemen Prefer Blondes* and *Call Me Mister,* to produce the show. Alan began to work on the plot with such intensity that after a year he developed encephalitis which developed into spinal meningitis. After a spell of delirium and several weeks in the hospital with a paralyzed left leg, he was discharged at last.

It was at that time in early 1954 that he received a call from composer Arthur Schwartz, who had split with Howard Dietz after a long association to write the successful *A Tree Grows in Brooklyn* with lyricist Dorothy Fields. Schwartz, who was a producer of musical films as well as a composer, had been approached by the principals of a big-screen process called Cinerama to make an epic of *Paint Your Wagon.*

Even though both Lerner and Schwartz had worked successfully with others, they considered themselves "partnerless." Now, because Schwartz had been a lawyer, the two entered into an agreement to be "sole collaborators" (in the manner of Rodgers and Hammerstein) and opened an office on East 62nd Street in New York. In essence it would have been a felicitous collaboration, for Schwartz was an expansive and sensitive composer. Alan, as usual, was notoriously slow in delivering lyrics, and Schwartz, who needed money, eagerly took up again with Dorothy Fields when she came to him with an idea for a bright musical that turned out to be *By the Beautiful Sea.* Now the *Paint Your Wagon* film was stillborn, and no songs of Lerner and Schwartz have ever been published.[2]

---

[2] In his third novel, *The Man Who Knew Cary Grant,* Jonathan Schwartz quoted a fragment of the lyric Lerner wrote for his father's music. Called "Over the Purple Hills," the balance of the lyric is the work of Carly Simon.

Returning to the Dogpatch project he had been struggling with for months, Alan now began talking through the various episodes with Widney—always looking for the serious "message," the search for the American dream that might be hidden in Capp's simple Yokum characters. Composer Lane, showing remarkable forbearance, stood by patiently, for he could not begin writing music until the libretto had at least been sketched out. "I talked through the book with Alan," said Widney, "to help him frame up the beginning, middle, and end of the story. We couldn't find it in two and a half years and all the people around began to get desperate. You have to know that the thing is frameable and that it can be done in two and a half hours, have a first act curtain and a climax."

Then, in the summer of 1954, Pascal died.

Ghoulish as it sounds, Lerner admitted that he immediately began thinking again of *Pygmalion*. At first he merely wondered who had the rights, Pascal's estate or Shaw's. His interest was piqued by the fact that in the two years from 1952 to 1954, since he had last worked on *Pygmalion*, the "rules" of musical construction had changed greatly. Experimental shows like *The Golden Apple, Me and Juliet,* and *The Pajama Game* had opened; there was even a successful and hard-hitting revival of *The Threepenny Opera*. Audiences were now accepting musicals without the choruses he and Fritz had deemed so necessary. One could even get away without a subplot—in fact, that might give a musical added freshness, a turning away from cliché.

"As Fritz and I talked," Alan was to write, "we began to realize that the musical did not require the addition of any new characters. . . . There was enough variety in the moods Shaw had created . . . and we could do *Pygmalion* simply by doing *Pygmalion*, following the screenplay more than the play." Then they began to find ways they could get production values, moving forth and back to Covent Garden, Ascot, all leading to the Embassy Ball where Eliza is presented. Fritz offered that they could even create a sort of mini-chorus by using Higgins's servants.

Herman Levin had been in Europe on holiday. When he returned, Alan told him that he and Fritz were once again committed to *Pygmalion*, that instead of producing the Dogpatch musical, they wanted him to produce Shaw's parable.[3]

Levin's first task would be to unlock the rights. It was soon discovered they belonged to Pascal's estate but were even now being fought over by the woman

---

[3] When Lerner, Lane, and Levin gave up on *Li'l Abner* they gave the fruits of their extensive research to Norman Panama and Melvin Frank, who used none of it. They fashioned their book simply around Dogpatch, Abner, Daisy Mae, and Sadie Hawkins Day. With Michael Kidd's energetic dances applied to Gene de Paul and Johnny Mercer's bouncy score, the show ran for almost 700 performances.

to whom Pascal was married and the woman to whom he was *not* married, but living with. In addition, any transfer of rights from Pascal's estate would have to receive approval from Shaw's British executors.

Into this rabbit's warren plunged Irving Cohen. When Lerner was informed by the Chase (now Chase Manhattan) Bank that MGM, one of their largest depositors, intended to acquire the rights to *Pygmalion*, Alan pulled rank. He pointed out that the very same Chase held more than three million dollars of this late father's money. This stopped the negotiation with MGM and gave the partners cause to hope, but this hiatus did not guarantee that Lerner, Loewe, and Levin would get the rights.

Now they did a foolhardy thing. They decided to write their musical on "spec." The chance was always there that the rights might go elsewhere and they would be stuck with a stillborn musical. Fritz argued that if they went forward, they would be so far ahead of any other team when the rights did eventually come up for assignment that they would have a better chance of being the team selected.

Since the whole project was based on intangibles, Alan and Fritz decided to indulge in utter fantasy by enlisting ideal collaborators. It was not as chimeric as it sounds, for if they could talk top theater people to come into the un-founded dream with them, they would be in that much stronger a position when the rights were eventually awarded.

Oliver Smith, who had done the atmospheric sets for *Brigadoon* and *Paint Your Wagon,* was chosen first. Beyond his talent, Smith had the other not-so-small assets of personal wealth and the experience of having been Herman Levin's co-producer in *Gentlemen Prefer Blondes.* Cecil Beaton, about whom Alan said, "It is difficult to know whether he designed the Edwardian era or the Edwardian era designed him," was an ideal choice for the costumes. Al-though Beaton generally only accepted shows for which he fashioned both sets *and* costumes, it was hoped he would find the project irresistible enough to break this code.

With sets, costumes, libretto, music, lyrics, and producer thus idealized, it was now time to imagine a perfect cast. In the musicals he had written formerly Alan had never been allowed this luxury. An original story has to be completed before it can be presented to a star, but with names like Shaw, *Pygmalion,* Smith, Beaton, Levin, Lerner, and Loewe to bandy about, the collaborators could approach anyone.

Although *Pygmalion* had been written for Mrs. Patrick Campbell (who at forty-nine played the eighteen-year-old Eliza), and notwithstanding Wendy Hiller's charming impersonation in the film adaptation, there is no doubt that the star role is Higgins. In spite of Leslie Howard's rather frenetic performance in Pascal's film, the male role is far more interesting and complex. Scholars have come to agree that Higgins is Shaw. Certainly the Irishman was an inordinately shy man, one whose love affairs exist only on paper—sent through the mail or embodied on stage.

Alan always maintained that Rex Harrison was his first choice, but that is another of his inventions. Michael Redgrave, who had a number of distinct advantages over other competitors, was the original choice. He was splendidly handsome in a fresh, open way and had a real singing voice. He had actually sung Macheath in a revival of *The Beggar's Opera.* The generous offer of 10 percent of the gross of a large-scale musical appealed strongly to this father with three children to support, but the Americans' insistence that he sign for a two-year stint made him shudder. They offered a compromise of a one-year contract, but when Redgrave insisted he could not be away from England for more than six months, negotiations fell apart.

Then Noël Coward, fresh from a triumph in Shaw's *The Apple Cart,* was approached. When Coward, busy with his adaptation of Wilde's *Lady Windemere's Fan* (eventually called *After the Ball*), said no, George Sanders and John Gielgud, it has been reported, were also sounded out. The *Toronto Globe and Mail* drama critic Herbert Whittaker reported that Harrison was chosen for the part "only after every other actor in England had been asked to play it."

Yet, just as Yul Brynner was the ideal King in *The King and I*, there is no doubt that Harrison was the perfect choice for Shaw's misogynist. His detached, imperious manner, his arrogant charm, his suave manner, the high-bred way he walked and talked, and even his reported insensitivity to women[4] made him a perfect Higgins. As to Rex's singing voice, Alan remembered asking Kurt Weill, who had considered him for Macheath in a proposed revival of his *Threepenny Opera,* if he could sing. Weill simply answered, "Enough." Yet it is safe to say that had Harrison been a true singer, the talk-sing style that made the show so groundbreaking (and that Lerner was to use to such good advantage in *Camelot* and *Lolita*) would never have been invented.

Although they were not close friends, Alan had met Rex frequently at Maxwell Anderson's house in the country during the Broadway run of the playwright's *Anne of the Thousand Days.* Rex and his wife Lilli Palmer often came out to spend the weekend, and Alan, an inveterate poker player, usually organized a penny poker game at the Anderson's on Saturday nights.

By the autumn of 1954, the collaborators' concentration shifted from the role of Higgins to his sparring partner, Eliza. A young Julie Andrews was starring in a pastiche of '20s musical comedy called *The Boy Friend.* Even though the role called for her to be "over-the-top," Lerner, Loewe, and Levin, by now known

---

[4] Harrison's ex-wife (he had six marriages) Rachel Roberts and his ex-mistress Carole Landis both committed suicide.

as "the three L's," were enchanted by her flawless pitch and the empathy she created in the posturing role and asked her to audition by reading scenes from various plays. They instantly agreed she passed with talent to spare and made her promise to make no commitments until they had obtained the rights.

Now, in typical Hollywood four-flushing manner, Alan called Rex from New York announcing that he had signed a perfect ingenue to play opposite him, had a producer of hit musicals (Levin), and some of the score to a musical based on *Pygmalion* (temporarily called *Lady Liza*). Without admitting that they did not actually own the rights, he asked if he, Fritz, and Herman Levin could come over and talk to him about starring in it.

With so much riding on those elusive rights, Alan and his lawyers realized it was foolhardy to go any further. Now they tried to force the hand of the Chase Bank and propelled the bank into action. Timid of making the wrong decision and thus alienating Lerner and MGM, both big depositors, the bank aimed to steer clear of any decision. The board appointed literary agent Harold Freedman to arbitrate the matter and placed any resolution as to rights in his hands.

Lerner and Loewe instantly signed on with the Harold Freedman Agency and although it looked fishy to some of the higher-ups at MGM, rights were quickly transferred from the motion picture company to the three L's. This important award was, of course, reported in all the showbiz papers.

Shortly after the project was announced, Alan received a call from Mary Martin's husband Richard Halliday. Mary, who was starring in *Peter Pan*, had read the news and said she would like to hear the score. They had young Julie Andrews on hold but Mary was perpetually young at forty-one and both Fritz and Alan agreed that there would be tremendous insurance in hiring a star of the first magnitude like Martin.

The Hallidays brought along Mary's designer, Mainbocher, after a *Peter Pan* performance and sat stoically at attention through the five songs Lerner and Loewe had completed. No one commented except the designer who said he liked them very much.

When Alan didn't hear from Mary for a week, his curiosity was aroused. At last he called. Richard asked to have lunch with him, at which time, he assured Lerner, he would explain why he hadn't telephoned.

Lerner's strong sense of insecurity was stimulated further when Richard reported, "Mary walked the floor half the night repeating 'those dear boys have lost their talent.' " As Lerner put it: "My memory may be a little hazy about some of his words, but 'those dear boys have lost their talent' is forever engraved on the walls of my duodenal lining." When he reported the meeting to Fritz, the stoic replied, "Well, I guess they didn't like it."

Alan, who always sought another's opinion (more likely approval), was particularly devastated by Mary's trashing what he and Fritz had written so far.[5] After all, she was probably the most experienced and brightest star now playing Broadway. Her words prompted an immediate decision. He and Loewe must go abroad, ostensibly to talk to Cecil Beaton about the costumes and to absorb local color, but imperatively to talk with, to audition, and, if they liked what they heard, to sign Rex Harrison.

Alan's lavish life style and two divorces had used up the considerable profits he had earned with *Brigadoon*,[6] so he asked Fritz to help defray the cost of the trip. But whether his gambling losses had eaten up *his* profits or if he simply didn't want to invest in *Pygmalion*, Loewe declined.

Alan's share of over a million dollars from his father's estate had not yet been divided and distributed. Under the terms of Joe Lerner's will, the principal had to remain intact to insure Edith her $60,000 annual alimony. But Joe had left each of his sons gold mine stock in their own names, and although had Alan held onto the shares they might have been worth far more, he disposed of this legacy in order to finance the trip.

Thus on a rainy night in January 1955, Alan, Nancy, Fritz, and Levin set out to spend some weeks in London, to visit what Alan termed, "the original scenery." It was the first return for Lerner since the summer during his schoolboy days.

But this time he was full of hope and as secure as he ever would be. His name was well known in London; *Brigadoon* had a successful run in the West End back in 1949; and Bobby Howes and his daughter Sally Ann had made a rousing hit out of *Paint Your Wagon* just a year ago.

Flying the Atlantic with his friends and colleagues nearby, expressly in hopes that he could secure a star, Lerner brought two other new assets abroad with him: first, a commitment from the Pascal estate allowing him and his collaborator to musicalize perhaps Shaw's most popular play; second, and perhaps more important, the renewed sense of self given to him by his companion in the seat next to him, the beautiful mother of his two daughters, Nancy Olson Lerner, his very own fair lady.[7] But they were balanced out by two enormous doubts. Would the Shaw estate be as easy to win over as Pascal's had been? And would Harrison hate the songs as much as Mary Martin had?

[5] In her autobiography, *My Heart Belongs*, Martin states that she was offered and turned down the role of Eliza; in *his*, Lerner implies Martin was too old to be considered.

[6] Although *Paint Your Wagon* ran successfully in both New York and London, there were no profits from it until the film version was made, nearly a dozen years later.

[7] The score, when completed as *My Fair Lady*, was to be dedicated to Nancy.

Hammerstein
1951–1955

WITH FOUR SMASH HITS OUT OF FIVE TIMES AT BAT, the only R & H failure having come from Oscar's *original* idea, *Allegro*, it is only logical that showbiz mythology would report that *Me and Juliet*, their next show, must have been suggested by Rodgers because "Hammerstein owed him one." Not so. The idea for a backstage musical had been percolating on the back burner of Oscar's mind for years. But he never brought it forward because he hadn't quite figured out how to avoid all the greasepaint clichés. He didn't want his story to fall into the let's-put-on-a-show mold, featuring the usual ingenue, leading man, heavy, comic, star, and understudy.

Veteran of many long runs, Oscar frequently spoke of wanting to write about how a group of strangers become a family during the months a successful musical is on the boards. Recalling his own super-hits, he always kept in touch with and frequently rehired actors and members of his extended family or moved them from show to touring company. In his decades in the theater he had seen love affairs, marriages, divorces all happen to the people in his shows.

This clever idea might have been stillborn had not a revival of Rodgers and Hart's *Pal Joey* opened to great acclaim in January 1952, only a few months after Oscar's fiasco with *Music in the Air*. *Pal Joey*, certainly the team's most acerbic score, with a no-punches-held book by John O'Hara, was far ahead of its time when it was originally produced in 1940. Set largely backstage in a garish nightclub, Chez Joey, it concerned the misadventures of a sleazy heel and a philandering wife (not his own). In its original debut it prompted the thumbs-down *New York Times* critic to ask if "one could draw sweet water from so foul a well."

Could the critics' sudden about face be a harbinger of the public's demand for a return to musical comedy? Now Oscar mentioned his idea to Dick—who became equally enthusiastic, but, maddeningly, was obliged to postpone work on the new backstage musical until the *Victory at Sea* music was out of the way.

In hindsight, the R & H involvement with *Me and Juliet* seems a logical foray. Oscar and Dick were only being true to their sworn credo (not unlike Stephen Sondheim's): "Once you've done it, don't go back." They had done folk-play (*Oklahoma!*), quasi-opera (*Carousel*), concept musical (*Allegro*), musical play (*South Pacific*), and operetta (*The King and I*). What was left? Only to return to the form they had cut their teeth on. Rodgers had succeeded time and again with Hart in musical comedy, Oscar less so; but *Mary Jane McKane, Sunny,* and even *Very Warm for May* were examples of the form.

When they began to work in the spring of 1952 their enthusiasm was boundless, for research—an important element in all their previous shows—was unnecessary. It would not be set in the Far East, South Pacific, or Industrial Revolutionary New England but would use a background both men practically slept in since they were teenagers. Rodgers was especially thrilled, for he could write song in forms he had been avoiding—boogie, Latin, pop. Hammerstein's libretto even suggested an improvising trio on the corner of the stage.

Like any original story, the plot Oscar and Dick talked out was to come in for many changes. But no matter how different or trivial the story would become, two things were to remain: its theatrical setting and the avoidance of the show-must-go-on theme.

Me and Juliet takes place in various parts of the theater—the dressing rooms, the lobby during intermission, rehearsal stages, a back alley, the orchestra pit, and of course the full stage as the audience sees it—while its current occupant, a hit show, is going on. The hit show, *Me and Juliet,* was to be a stylized *Everyman,* starring prototypes of Juliet, Don Juan, and Carmen, but one might describe the result as a kind of *Kiss Me, Kate* as seen through the eyes of *Rashomon.*

The theatrical venue was an inspiration, but his determination to avoid the show-going-on idea plunged Hammerstein into a trap he couldn't escape. *Without* the urgency of a nightly performance, his people, already shallow, became very ordinary in mundane situations. Additionally, since this was planned as a lavish musical with a large cast, necessitating five songs in the *Me and Juliet* portions, the remaining seven numbers divided among as many principals were unable to flesh out character.

Me and Juliet tried to tell too many stories for its own good. Besides a subplot concerning an uptight stage director whose credo is never to date a dancer in his own show (a dilemma which Oscar solved by having the girl he falls for transferred to another show) and a farcical sub-subplot concerning the orchestra leader and an unknown fan, its plot concerns a chorus singer, Jeanie, in love with an assistant stage manager, Larry, whom she secretly marries. Her former boyfriend, Bob, is a stage electrician and a big burly bully. When he sees Jeanie and Larry in an offstage kiss, he tries to drop a sandbag from the light bridge. He misses the loving couple and only succeeds in forcing the curtain to be brought down frantically to close the first act.

In the second, Bob tries more drastic means and provokes a fistfight with Larry, his opposite—small, retiring, and intellectual. Oscar comes close to his murder-in-self-defense syndrome, but this time Bob is only knocked out by hitting his head against a radiator. Six-foot-one-and-a-half-inch Oscar was always being referred to as "the big guy" as contrasted with Rodgers's diminutive size. After *Me and Juliet* opened he told a reporter he was trying to write about physical misconceptions. To avoid stereotype he must have reversed his images, for he was far less pugnacious than his partner.[1]

Once the libretto was begun, R & H took the trouble to assemble a typical musical comedy team which would be headed by "Mister Musical Comedy" himself, George Abbott, veteran director of a dozen backstage musicals. Don Walker, who built the brassy sound of Broadway into his arrangements, was scheduled to do the orchestrations; Robert Alton, who had been choreographing musicals for thirty-five years, even before most hoofers knew what choreography meant, was chosen. As R & H had brought in the first dramatic musicals, it seemed they would be the first to lead the revival of this much beloved form. Oscar and Dick were pointed in a new-old direction.

The concept was a splendid one, but in typical fashion, once preliminary plans were made (had they learned nothing from *Allegro?*), the collaborators marched straight on into Armageddon. The St. James Theatre was reserved, bumping its reigning hit *South Pacific* into the Broadway. Opening dates were chosen. RCA happily put up the $350,000 production cost in exchange for a half interest in the show and the rights to the original cast album.

Aware that these days shows tour or play previews sometimes for months before being subjected to a critical barrage, William Hammerstein deplores R & H's rigid scheduling. "They would announce that they were going to do something and then they had to do it . . . and in quick time. Along about April you read that R & H had been working on a show about showbiz. Came September and they had to get the damn thing on the stage. That had a lot to do with the failure of *Me & Juliet* and probably with *Allegro*."

Jo Mielziner performed the miracle of creating a set that moved the entire

[1] Ethan Mordden in his excellent book *Rodgers & Hammerstein* mentions Oscar's penchant for deploring bullying in addition to the size differential. "The bully figure," he writes, "runs through Hammerstein's career in *Rose-Marie*, *The New Moon*, *Golden Dawn*, *Rainbow*, and *High, Wide and Handsome*, not to list *Oklahoma!*, *Carousel* (Billy really is something of a hoodlum), and *The King and I* (the King obviously.) To Hammerstein, facing down the aggressor was a fundamental democratic act."

proscenium so one could see the action on-stage and in the wings at the same time. He proved to be the evening's only hero, for the rather cinematic technique that Oscar's libretto foisted upon the set was also its most interesting aspect. The audience would witness, for example, a song being sung by the electricians on the light bridge and then sung a moment later on stage and after the drop of the curtain, the same number performed by the full cast, now in rehearsal clothes (because the director thought their performances were getting flaccid).

The show opened in Cleveland's Hanna Theatre rather than at the accustomed Shubert in New Haven partly because Mielziner's sets were too large for the latter's stage but mainly because the Shubert lacked an orchestra pit, which was germane to the plot. Even there, Rodgers knew they were in trouble when he "heard people raving about the sets without a word being said about the rest of the show." Early in the rehearsal period Hammerstein had given Mr. Abbott the right to cut ruthlessly anything that was not working in the script. "Treat it as if it were your own," he said.

Abbott felt that unfortunately Oscar had never figured out what the "hit show-within-a-show, *Me and Juliet*," was about. Although Abbott never said it, he implied that they might have gotten away with the stylization had Robert Alton been clever enough to choreograph the dances to an amusing climax the way Jerome Robbins had done in *The King and I*'s "Small House of Uncle Thomas."[2] But instead the public was left wondering what this nebulous musical within was about and incredulous that they were being told they were witnessing a long-running hit. Oscar remained sphinxlike and never explained the inner story to press or public; Dick was supportive and equally buttoned up.

The trouble went far deeper than the supposed musical. Hammerstein was never so good at writing an original as he was at adaptation. The script was pure contrivance, and although it presaged freedom not to be obliged to write for an English schoolmarm or an Oklahoma cowboy, Oscar at least had been able to crawl into those characters. These faceless showbiz types in *Me and Juliet* eluded him and remained lacking any sense of personality. Although Oscar's libretto went behind the scenery, it never scratched below the surface.

Oscar was incapable of one-liners (no Abe Burrows or George Kaufman was he), so where his script should have sparkled, it fizzled. The generally amusing or witty lyrics that he was entirely capable of were absent. None of the songs (remember, these are the pioneers of the integrated musical) moved the action forward or elucidated character. And none, except "No Other Love" (actually lifted from *Victory at Sea*), was a hit. Most were downright banal.

---

[2] John Fearnley reported that when they were down to the wire, R & H called in Robbins to satirize the dances. "I can do it, but I won't," snapped Robbins. "It would kill Bob Alton."

Those that tried to be witty tripped over their own feet with embarrassing lines on the order of "Like a snake who meets a mongoose, / That young lady was a gone goose," or sophomoric ones like "I'm your pigeon / Through with roaming / I am homing / Marriage type love." Oscar was not alone in lacking inspiration here. Rodgers's contribution too lacked verve; this score is certainly his dullest.

Perhaps the best and most personal song in the show for both Dick and Oscar is "The Big Black Giant," a threnody directed at and describing the mass of the audience. The only serious song in the score that tries to probe below the veneer, it does show the team's continuing awe of audiences. While not as well executed as its splendid concept would warrant, some of the lines quoted below have a theatrical appeal and flair that in the days before Edward Kleban's honest lyrics to the songs in A Chorus Line might have intrigued.

The water in a river is changed every day
As it flows from the hills to the sea.
But to people on the shore the river is the
    same
Or at least it appears to be.
The audience in a theatre is changed every
    night.
As a show runs along its way.
But to people on the stage the audience looks
    the same,
Every night, every matinée.
A big black giant
Who looks and listens
With thousands of eyes and ears,
A big black mass
Of love and pity
And troubles and hopes and fears;
And every night

The mixture's different,
Although it may look the same.
To feel his way
With every mixture
Is part of the actor's game.
One night it's a laughing giant
Another night a weeping giant
One night it's a coughing giant.
Another night a sleeping giant.
Every night you fight the giant,
And maybe, if you win,
You send him out a better giant
Than he was when he came in.
But if he doesn't like you, then all you can do
Is pack up your make-up and go.
For an actor in a flop there isn't any choice
But to look for another show.

After frantic work, to little avail, in Boston, Me and Juliet kept its commitment to open in New York on May 28, 1953. But by this time, all involved knew the show was unfixable. Too little thought had gone into it; too little communication between Rodgers and Hammerstein had brought in an entertainment that couldn't decide what it wanted to be. Starting as a vaudeville, shifting in the middle to satire, moving to a romantic story, and ending with a brawl capped by more and more frenetic dance, Me and Juliet could not even please the weary businessman.

"Far from a peak achievement," was the guarded critical consensus. Although the show eventually turned a small profit after playing for ten months, which, under any other aegis would have been considered a mild hit, Me and Juliet was not even considered a succès d'estime by the Broadway fraternity. There was no

London production and only a seven-week extended run in Chicago. Both doyen critics Walter Kerr and George Jean Nathan seemed to have found *les mots justes* when they called it "a show *without* a show."

None of the more successful Rodgers and Hammerstein shows seems formulaic. At least not on the surface. But if you looked closely, you'd notice that while it is not quite a formula, each uses an older female as a recurring motif. Each of these women becomes a kind of dea ex machina who pulls the strings that control the essence of the plot.

Oklahoma!'s Aunt Eller singlehandedly manages to interest Laurey in Curly, and it is through her insistence that the hero is acquitted of the charge of murder. Aunt Nettie, while more comforting than controlling in *Carousel*, nevertheless gives Julie the rocklike support and faith she lacks. In *South Pacific*, Bloody Mary engineers the course of Lieutenant Cable and her own daughter Liat. Although exerting less influence on the principals Nellie and Emile de Becque, Bloody Mary and her island Bali Ha'i are the symbolic heart of the musical, representing Eden and, in microcosm, the sanction of intermarriage that pervades the story. *The King and I* is more complex, for the older woman is also the love object. Both Lady Thiang and Mrs. Anna run the show— always making sure the King *thinks* he alone sets the course.

This maternal syndrome, coming understandably from two men who were known to heed the advice of their spouses, was to hold sway until the end of the collaboration. It is seen perforce in *Cinderella*, whose only two males are a weak king and a weaker prince, and in their last collaboration, *The Sound of Music*,[3] where the Mother Abbess makes Maria (even here impersonated by a much-too-old Mary Martin) face the truth, thus turning the plot around.

R & H were thus creating, one might say, the matriarchal musical as distinguished from *Fiddler on the Roof*, *Fiorello*, *Zorba*, *Brigadoon*, or *Finian's Rainbow*. Nowhere is this more evident than in *Pipe Dream*.

John Steinbeck's novel *Sweet Thursday*, from which this R & H 1955 musical was drawn, has a raft of characters. But the one Oscar chose to star was Fauna, the madam-with-a-heart-of-gold of a Cannery Row house of prostitution. Fauna is a typical musical comedy fabrication and quite different from her sister, Dora, the knock-down drag-out hardhearted whore who appeared in *Cannery Row*, Steinbeck's prequel to *Sweet Thursday*. The new madam is a former social worker, so elegant of manner she could easily be confused with Perle Mesta. She teaches her hookers, for example, which fork to select from a barrage of

---

[3] Although the libretto of *The Sound of Music* was written by Lindsay and Crouse, R & H aggrandized the part of the Mother Abbess by giving her "Climb Ev'ry Mountain," perhaps their only song whose impact nears *Carousel*'s overwhelming "You'll Never Walk Alone."

silverware at a formal dinner they will certainly never be invited to, instead of giving them lessons in the amorous arts. Fauna is as patently phoney as the rest of the cartoon characters who appear in *Sweet Thursday* and later *Pipe Dream*. No wonder R & H thought of Helen Traubel, the wholesome Wagnerian soprano late of the Metropolitan Opera, to portray her.

Whitewashing Fauna and *Sweet Thursday*'s other characters was a grievous Steinbeck fault, but one that was accepted eagerly by Oscar. At the time *Pipe Dream* was conceived and written, America was deeply enmired in McCarthyism. Not only did Hammerstein have peering over his shoulder a State Department that had recently taken away his passport, but he had Dick Rodgers, who had always tried to avoid controversy. Dick even announced to the press that the team only wrote "family shows."

One can understand Steinbeck's dilemma as well. He was now playing it safe after having been called before the House Un-American Activities Committee because of his dismally realistic depiction of the hardships in American life during the Great Depression (*Of Mice and Men, The Grapes of Wrath,* and *East of Eden*), which were incomprehensibly considered pro-Communist.

The new novel, far different from Steinbeck's 1945 venture into the disenfranchised, *Cannery Row,* pussyfooted around the same characters: drifters, the unemployable, drunks, dropouts, and hookers. This time, because the novel was *written* to be adapted into a musical, everyone was a benevolent eccentric. *Cannery Row* has three suicides, several fights, and a healthy dash of profanity. *Sweet Thursday* lacks all of these. Unbelievably, it climaxes instead in a masquerade where the denizens of the flophouse and whorehouse arrive dressed in homemade costumes to represent characters from *Snow White and the Seven Dwarfs*. Seriously!

Rodgers and Hammerstein were not the first to be offered the chance to musicalize the adventures of these zanies who live by their wits close to the fish canneries in Monterey, California. The *Cannery Row* stories were optioned by Feuer and Martin, producers of *Guys and Dolls,*[4] in the hopes of interesting Frank Loesser to repeat his stupendous success. It was a good idea, for the West Coast crazies, with their preoccupation with whoring and boozing, were comparable to Runyon's stylized bunch, who were equally obsessed with crap games. But Loesser, not wanting to repeat the past and having loftier ideals, was determined to follow his Broadway fable with a full-fledged opera—*The*

[4] Of course, this show won all the Tony and Donaldson awards, but McCarthy's power had so infiltrated the theater community that *Guys and Dolls* was denied the Pulitzer Prize even though the committee had voted for it. Abe Burrows, its co-librettist (insiders ignore the co-), had been called before the House Un-American Activities Committee. Joseph Pulitzer's will gave the trustees of Columbia University the right of veto, and so, to save face, they awarded no prize that year.

Most Happy Fella. Feuer and Martin were delighted that R & H agreed to take over the Steinbeck opus.

By the summer of 1953, after Oscar had decided there was nothing more he could do to improve *Me and Juliet* and had left for England to supervise the production of *The King and I* at the Drury Lane, Steinbeck began sending chapters of his forthcoming work. Obviously he was writing a novel to be musicalized, for *The New York Times Book Review* found the published book "so gaily inconsequential that it might serve as the working script of a musical comedy, on the order of, say, *Pal Joey*."

Oscar took Steinbeck's characters practically verbatim, only occasionally making them more buffoonish or cartoonish than what had been handed him. It must be admitted that he was able to give some dignity to the leading man, marine biologist Doc, the major character in both books—an escapee from society. Doc, deeply immersed in his scientific research of sea creatures, is only interested in making enough money for necessities—booze, peanut butter, and women. He is revered by the flotsam of their small community for his kindness, patience, advice, and free help in curing all their ailments.

Into the nearby whorehouse comes Suzy, painted by both Steinbeck and more especially Hammerstein as a Chaplinesque waif. The madam (who although christened Flora, prefers to be called Fauna), decides Suzy has no talent as a whore but might have some as a bride and sets about pairing her charge with Doc. She is abetted in this by Doc's oafish friend Hazel. (No, dear reader, Hazel is a man whose mother's brain became addled after having produced eight babies in seven years. By the time she discovered her error it was too late to change the name.) And by the end of the first act, on the heels of the dullest curtain Oscar ever wrote, we find Doc and Suzy having an intimate dinner in a seaside restaurant.

After the aforementioned masquerade, which is treated with the elegance of a *bal masque*, Suzy realizes that Doc was put up by Fauna to take her dining. For some reason, although she has fallen in love with Doc, she decides she must get self-esteem before they can have a successful relationship. The only way to do this, she decides, is to leave Fauna's, get a job as waitress at the local greasy spoon, and move into an abandoned windowless boiler, what she prefers to call a pipe (hence the musical's cutsey title).

Now Hazel, noticing his revered idol Doc's despondency and having heard Suzy confess she would look after him if he had, say, a broken arm, sneaks into Doc's laboratory with a baseball bat and does the deed. The musical ends as Doc with his arm in a cast and Suzy at the wheel drive off to Salinas in search of further marine specimens.

One would assume that with such a superficial plot and characters that *Pipe Dream*'s score would be equally inconsequential. Not so. Hammerstein wrote several heartfelt and very moving songs, and Rodgers's melodic gift, which had deserted him in *Me and Juliet*, seems to have returned. Sometimes both get pretentious, perhaps uncomfortable (because of the collegiate raunchiness of these characters) would be a better word: Rodgers, when he sets the good luck wishes of the prostitutes into conservatory three-part counterpoint, or Oscar, with the precious lyric to "The Happiest House on the Block" ("Where nothing's too good for a guest / Our parlor is cheery / There's rest for the weary / The weary who don't want to rest").

But the ballads for the romantic pair are heartfelt. "Everybody's Got a Home But Me" is Suzy's introductory song. I quote the A^2, the release, and A^3 sections below.

. . . I rode by a house	But once in a while,
Where the moon was on the porch,	When the road is kinda dark
And a girl was on her feller's knee,	And the end is kinda hard to see,
And I said to myself	I look up and I cry
As I rode by myself,	To a cloud goin' by
Everybody's got a home but me.	Won't there ever be a home for me, some-
I am free and I'm happy to be free,	where?
To be free in the way I want to be,	Everybody's got a home but me.

Perhaps the number that carried the action furthest is a sensitive duet instigated by Fauna, intent on building Suzy's self-image before that all-important dinner alone with Doc. She sings the lines and asks her young disciple to repeat after her.

Fauna first, then Suzy:	Suzy's eyes are searching eyes
Suzy is a good thing,	The world they seek is new.
This I know is true	Suzy looks for love
Suzy is a good thing.	As other people do
She may make mistakes	Suzy will find hers, too . . .
As other people do—	Someone is looking for Sue
Everybody makes a few.	

Doc's song, "The Next Time It Happens," which he sings to Suzy before they break up, is one of Hammerstein's most imaginative lyrics:

The next time it happens	"The next time it happens,"
I'll be wise enough to know	What a foolish thing to say:
Not to trust my eyesight when my eyes begin	Who expects a miracle to happen every day?
to glow.	It isn't in the cards
The next time I'm in love	As far as I can see
With anyone like you,	That a thing so beautiful and wonderful
My heart will sing no love song till I know the	Could happen more than once to me.
words are true.	

Within the dilapidated walls of The Bear Flag Cafe, Suzy (Judy Tyler) gets advice from Metropolitan Opera soprano Helen Traubel, the most miscast "madam" in showbiz history. MOCNY.

To this writer, the most popular number in the score, "All at Once You Love Her," seems a dud. It tries to capitalize on the success of *Me and Juliet's* "No Other Love." Both use the same form, start with the same three pitches, use linear diatonic melodies, and have a pseudo-Latin beat[5] (see Examples A and B below).

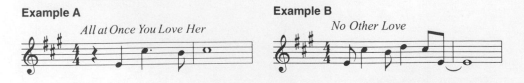

Example A

All at Once You Love Her

Example B

No Other Love

[5] To make matters worse, "All at Once You Love Her" is first introduced in the restaurant scene with a dreadful Spanish lyric.

R & H's reputation coupled with the popularity of "All at Once You Love Her" were enough to ensure that *Pipe Dream,* when it opened on November 30, 1955, broke the record for the largest advance ($1.2 million) of any of the team's shows. But critical reception was less than enthusiastic, and after it exhausted the money in the till, the show broke another record by closing after giving the fewest performances in the R & H canon.

William Hammerstein says the whole project was a mistake. He has no love for the lackluster performances of Suzy (Judy Tyler) and Doc (William Johnson). Director Harold Clurman, lately of the Group Theater, Hammerstein feels would have been more at home with Odets and O'Neill. He deplores the score, the story, and especially the miscasting of Helen Traubel, never an actress, who squabbled constantly with her director and was unfortunately the only singer in the cast who had to be miked.

Why wasn't Traubel replaced?

Rodgers had developed serious cancer of the jaw requiring surgery just about the time *Pipe Dream* went into rehearsal. Oscar, who could never fire anyone, was forced to muddle through the most sensitive rehearsal period alone. Being the eternal optimist, he believed that with work and adjustments to the script he could help Clurman get some intensity out of the diva. Hoping to create a showstopper, he even added two children to one of her numbers. All this to no avail. By the time Rodgers had recuperated enough to attend rehearsals, it was too late to seek out a new star.

William's brother, James, was close to the project too. He spoke recently of why he feels *Pipe Dream* could not possibly have worked. "It goes far beyond Traubel's ineffective performance. R & H were afraid of sleaze. Dad thought of theater as being a very moral place. When you walk into a theater, if you've just screwed your partner out of a million dollars and his wife, you will still cry if you see somebody on stage victimized by someone."

Although moralists might preach that the characters in *Pipe Dream* all were society's victims, it would be hard to cry for any of them. What was needed was a libretto with the depth of, say, a *Carousel.* But, as the critic for *Newsweek* noted, "While no one may ever write another *Carousel,* Rodgers and Hammerstein miss that goal by a mile."

The Broadway season 1955–56 was so lackluster that one might have expected *Pipe Dream* to be adjudged at least an esteemed flop or a noble try. Not so. When award time came around, only Alvin Colt, the costume designer, was given a Tony.

Oscar and Dick must have felt some of the magic of their collaboration had dissipated when they realized hardly four months after their musical's debut that while their show was playing to half-empty houses the *very next* musical to open on Broadway would turn out to be one of the great theatrical experiences of all time. To make matters worse, it was a property that had sifted through their hands. They had given it a try and come away feeling it was "impossible to musicalize." It was *My Fair Lady*.

Lerner
1955–1958

NO SOONER HAD ALAN AND COMPANY checked into Claridge's than they arranged to see Rex Harrison, who was starring in the West End opposite his wife, Lilli Palmer, in *Bell, Book and Candle*. Harrison received them cordially in his dressing room shortly after the curtain went down, but Alan wondered why Palmer was curt to him when he popped in to ask Lilli to join them for supper. It was because she and Harrison were not speaking. Harrison was deeply involved with British actress Kay Kendall. It is a tribute to their mutual acting ability that although on stage Harrison and Palmer were able to play convincing love scenes, when the curtain descended it was goodbye until the next performance.

Alan soon set up an audition around the piano that had been moved into Fritz's suite, at which time, Harrison recalled in his autobiography, *Rex*, "They played me the music of a preliminary version . . . and the music then was not up to the standard of the final version."

The songs included "Just You Wait 'Enry 'Iggins," which Mary Martin had called a copy of Cole Porter's "I Hate Men" from *Kiss Me, Kate*, and the "Ascot Gavotte," of which she had heartily disapproved, saying, "It is just not *funny*." Then they played him two songs written especially for Harrison. One was "Lady Liza," which (minus the words) would be turned into the sweeping "Embassy Waltz." The other was "Please Don't Marry Me," which Rex didn't like at all and which he sensed "Alan and Fritz didn't care for very much either." (The song was later discarded.)

For the next three weeks they persisted, meeting Harrison socially, taking long walks during which Alan convinced Rex of his respect and devotion to Shaw. He vowed that he would use only the master's lines in the musical's dialogue (a promise he broke). Eventually he appealed to Harrison's considerable vanity by assuring him that he could outdo Leslie Howard's cinematic performance, which Harrison revered. Fritz, too, after first pronouncing that Rex had a range of "one and a half notes," complimented Harrison on his innate sense of rhythm and promised that he would find a way to write the melodic lines to the contour of his voice.

But the most persuasive argument for Harrison's eventual agreement to play Higgins in the musical came from the actor himself—his desire to spend time in New York. "I felt it might be a very good thing to put some distance between myself and my problems . . . so as to take a long look at my marriage to Lilli, but also to see how really deep my feelings were for Kay."

Before Alan and Fritz returned home to write more of the show, they toured Covent Garden—arriving at four a.m., the time the flower and vegetable merchants set out their wares. Fritz absorbed some of the sounds of the vendors (which would be so authentically incorporated into his score), while Alan expanded his feeling for their special language. He was fascinated with the Cockney way of inserting expletives within words—"unbloodylikely" was eventually transformed into "absobloomin'-lutely" in "Wouldn't It Be Loverly."

Before leaving London to return to Alan's country retreat, the house in Rockland County, to write the play, they met with Cecil Beaton and cast Stanley Holloway in the part of Doolittle. Most important of all, their lawyer Irving Cohen joined them abroad and, sitting with the directors of the British Author's Society, who still controlled the rights to *Pygmalion*, was able to hammer out a contract giving Alan *et al.* complete musical rights to Shaw's play.

Once they had flown back, promising to return with more songs for his approval, Harrison felt he should take voice lessons. After failing miserably to make any headway in creating the smooth legato and pearly tones needed for *bel canto*, although he practiced diligently with a maestro in Wigmore Street, he frantically called Alan and Fritz in New York. Rather than an operatic coach, Alan recommended musical comedy conductor Bill Low, who was leading the London clone of *Guys and Dolls*. Low, who had wide experience with actors, told Harrison not to think about singing the words but to start by just saying them. "There is such a thing as talking on pitch," Low explained, "using only the notes you want to use, plucking them out of the score." This created the style Harrison actually is credited with inventing; so he was the first actor to use Low's method. Of course, this technique becomes transparent and boring unless the performer has enormous vocal declamatory range. Harrison's acting voice, which runs the gamut from falsetto to basso-profundo, adds color to all his songs. He has another asset: an extraordinary sense of rhythm.

Once they arrived back home, not needing another star name on their roster, Alan and Fritz formally signed Julie Andrews, who had indeed not made any commitment after *The Boy Friend*, for Eliza. Then they set about finding the most important member of any musical's creative team—beyond the composer, librettist, and lyricist—the director. He is the axle upon which the musical's wheels will spin.

Everybody wanted Moss Hart, but Levin announced to the crestfallen pair that Moss was busy writing a show. Alan and Fritz, feeling strongly that at last they were on the right track, waited until they had finished "Wouldn't It Be Loverly?" and then two songs for Rex: his opener, "Why Can't the English," and "I'm an Ordinary Man." Now they gave Herman the okay to call, intending to play them for the great Hart, along with "Just You Wait, 'Enry 'Iggins," "The Embassy Waltz," and "The Ascot Gavotte"—that is, most of the songs in the first act.

Kitty Carlisle Hart recently remembered that

they came down to Beach Haven, New Jersey, where we had a summer home. It was a very simple community in those days and we had no piano there. We were living in a rented house, but I had been teaching a class in singing and I had found a piano across the weeds (a lot of cattails waving in the breeze) from the back of our house. There was a little kindergarten that operated in wintertime for the local children, in the basement of which they had a little upright piano with some keys missing. Fritz and Alan and Moss and I trooped over there, and Fritz played the score and Alan sang. It was love at first sight. Irresistible. It swept everything before it.

Moss was doing a play then called *In the Pink* with Jerome Chodorov and Harold Rome, but Alan and Fritz's show was something that he just *had* to do. So he called Rome and Chodorov and told them. They were so gallant and so nice about it . . . the kind of friends you don't find very often. They said, "Mossie, do what you have to do," although it was a bitter blow to them. But in those days, it seems to me that people in the theater behaved better. They were more "friendly-like" as it says in *Pygmalion,* and they helped each other. They went out of town to each other's shows. And they never took credit. Moss and George [S. Kaufman] went out to see *The Women* which Clare Boothe Luce wrote and they never said boo about what they did to help the show. They would never take any credit.

There is a certain smell to an incoming Broadway show by which record people can sense a hit a mile away. *Lady Liza* had that aura about it, and once Goddard Lieberson, the head of Columbia Records' musical division, mentioned the *Pygmalion* adaptation to his boss Bill Paley, who headed the Columbia Broadcasting System, Paley asked Herman Levin to see him. In short order CBS had agreed to finance the entire production to the tune of $400,000—an enormous budget at the time—and of course Columbia Records would make the original cast album.

Only one snag remained, setting the starting date. It all depended on when Rex Harrison could free himself from *Bell, Book and Candle* and could come to New York for rehearsal. Attendance at the play had surprisingly risen, and it looked like Harrison was in for a long run. Herman flew to London and arranged with the British producer, Hugh "Binkie " Beaumont, to close his still profitable show. In return, he gave money, a percentage of the profits, and a guarantee that when *Lady Liza*, or whatever it was to be called, opened abroad, it would be under Beaumont's aegis.

At last in July of 1955 the tentative date for first rehearsals—January 3, 1956—was fixed, with the Broadway opening planned for March 15.

My Fair Lady, arguably the crown of the American musical, showed such intelligence, taste, and style that it deserves a scene and song rundown of libretto and score.

ACT ONE

Overture

The rather brief introduction begins with the theme of "You Did It." After its bustle, we settle down rather commercially on the show's two hit tunes, "On the Street Where You Live" and "I Could Have Danced All Night." But as the curtain rises on a scene outside Covent Garden, we see the commotion, the flower-sellers and costermongers dance, and we are hit over the head with a "London Bridge Is Falling Down," which ends with "my fair lady." Does Loewe mean to show the audience where they are or is he trying to implant a cockney pronunciation of "Mayfair lady" in our ears?" No matter, we are led smoothly into

Scene 1: Outside the Opera House

The opening brings on Freddy, who accidently knocks over Eliza's flowers. Her diatribe and wails are noted by Higgins, which leads him to the song "Why Can't the English (Teach Their Children How To Speak?)." The idea comes from a line in Shaw's preface: "The English have no respect for their language, and will not teach their children to speak it." But it is more than a line in a play. It encapsulates Shaw's personality. Higgins, like Shaw, is imperious and insulting and reserves his passion—and a grand passion it is—for the English tongue. When Higgins tells Eliza to stop her mangling of the language or else to go out into the rain, she shows her spunk by answering that she has as much right to be "here, if I like, same as you." Higgins-Lerner-Shaw answers:

> A woman who utters such depressing and disgusting sounds has no right to be anywhere—no right to live. Remember that you are a human being with a soul and the divine gift of articulate speech: that your native language is the language

of Shakespeare and Milton and The Bible; and don't sit there crooning like a bilious pigeon. [Lerner will paraphrase this and it will become a turning point of the first act.]

Higgins's prowess in identifying regional accents is beautifully lyricized by Lerner in the verse.

Higgins: Hear them down in Soho Square,
Dropping aitches ev'ry where
Speaking English any way they like.
You, sir, did you go to school?
Cockney: What ya tike me fer, a fool?
Higgins: No one taught him "take" instead of "tyke."
Hear a Yorkshire man, or worse,
Hear a Cornish man converse.
I'd rather hear a choir singing flat!

Chickens cackling in a barn,
Just like this one,
Eliza: Garn!
Higgins (spoken): Garn! I ask you, sir, what sort of word is that?
(Sung) It's "Aoooow" and "Garn" that keep her in her place,
Not her wretched clothes and dirty face.
Why can't the English teach their children how to speak?
This verbal class distinction by now should be antique . . .[1]

Higgins plants the idea with Colonel Pickering, a colleague he has met by chance, that with his teaching, in six months he could pass this guttersnipe off as a duchess. The balance of the scene, after an exuberant Higgins has tossed his coins at Eliza and gone off with Pickering to discuss dialects, is totally original, and "Wouldn't It Be Loverly" gives real insight into Eliza. The song with its tender melody lets us peer beyond her ambition and see a side of her that Shaw never exposes. Yes, she hopes to be "a gol in a flow'r shop, who speaks genteel" but she would like "someone's head restin' on my knee, / warm and tender as he can be / Who takes good care of me. . . ." In succeeding refrains of the song, choreographer Hanya Holm had a chance to introduce another element, the "ensemble," and to present what was lacking in *Pygmalion*—the lighthearted touch, especially dancing.

Scene 2: A Tenement Section—Tottenham Court Road

This was one of the scenes that opened up the book. Alan and Fritz discussed what Eliza was doing between the end of Shaw's first act and the beginning of

---

[1] Harrison disapproved of the first version of this lyric, which he said made him feel "like an inferior Noël Coward." It began:

Why can't the English teach their children how to speak?
In Norway there are legions
Of literate Norwegians . . .

Lerner began again with the rhyme scheme as above.

"Ere, 'ow do I know you're tykin' me down right?" Eliza (Julie Andrews) queries Higgins (Rex Harrison) in the opening scene. MOCNY.

the second, when she appears at Higgins's study. "She probably went home to her parents," he first wrote. Realizing she had no parents, he remembered Shaw tells us she lives alone, has no mother, and that her father is a dustman living with another woman. "So that is precisely what we wrote. Eliza goes home, meets her father and two of his cronies. Brief scene between them. Song by father to establish his character, 'With a Little Bit of Luck' and on to the next scene." The song sums up Doolittle's character far better than the interminable philosophical speeches Shaw puts in his mouth.

Scene 3: Higgins's Study

This is the scene where Eliza comes to Higgins for lessons. At first he wants to throw her out, for he has already notated her brand of cockney, but he finds her so "deliciously low" that when Pickering says he would be curious to see if Higgins could prove his boast to turn the "squashed cabbage leaf into a duchess," he picks up the gauntlet of Pickering's challenge. Mrs. Pearce, the house-keeper, stands in for their conscience. She asks, "What's to become of her when you've finished?" Eliza changes her mind, but now a determined Higgins seduces her with chocolates and dreams of taxi rides. As soon as he gets Mrs. Pearce to take Eliza out and clean her up, we get an insight into Higgins's contradictory character.

Shaw's cue concerning the song "An Ordinary Man," better known as "Let a Woman in Your Life," is twofold. First he states, "When you let them in

Eliza is induced to stay in Higgins's house as the subject of an experiment. Higgins seduces the flower girl with "have a chocolate, Eliza?" MOCNY.

your life you find that you're driving at one thing and the woman is driving at another." Then, refuting Mrs. Pearce's assessment of his character, he adds, "Here I am, a shy diffident sort of man . . . yet she's convinced that I'm an arbitrary, overbearing bossing kind of person. I can't account for it."

Lerner reverses the order to far better advantage. He tells us, "I'm an ordinary man," but from the first scene we have seen he is far from ordinary. He (and perhaps Loewe more so) fills the song with violent contrasts, the misogynist, the lover of quiet and solitude comes through when the song climaxes with the dire consequences of letting "a woman in your life."

The song's form, a verse followed by long A B A C sections (done three times), holds the long tour de force together. The second verse and chorus (my favorites) are printed below.

I'm a very gentle man;	Your reply will be concise.
Even-tempered and good-natured,	And she'll listen very nicely
Whom you never hear complain;	Then go out and do precisely what she wants!
Who has the milk of human kindness	You were a man of grace and polish.
By the quart in ev'ry vein.	Who never spoke above a hush
A patient man am I	Now all at once you're using language
Down to my fingertips;	That would make a sailor blush.
The sort who never could, ever would	Oh, let a woman in your life
Let an insulting remark escape his lips.	And you are plunging in a knife!
Just a very gentle man.	Let the others of my sex
	Tie the knot around their necks;
But let a woman in your life	I'd prefer a new edition
And patience hasn't got a chance.	Of the Spanish Inquisition
She will beg you for advice;	Than to ever let a woman in my life!

Scene 4: Tenement Section

A short scene to break up the long prior one in which Doolittle is informed that Eliza has moved in with Higgins. Thinking the worst, the cadger heads straight for Higgins's house. His lines serve double duty as a reprise and plot developer:

A man was made to help support his children,	With a little bit of luck,
Which is the right and proper thing to do.	With a little bit of luck,
A man was made to help support his chil-	They'll go out and start supporting you!
dren—but	

Scene 5: Higgins's Study

Doolittle arrives, and after an amusing scene elucidating the dustman's twisted logic, Higgins buys him off with a five-pound note. Now, after a particularly exhausting session learning her vowels, comes Eliza's diatribe, "Just You Wait!"

The song, done with humor and charm, captures the vernacular in which she imagines a dire fate in store for Higgins.

One day I'll be famous! I'll be proper and prim;
Go to St. James so often I will call it St. Jim!
One ev'ning the King will say, "Oh, Liza, old thing,
I want all of England your praises to sing.
Next week on the twentieth of May
I proclaim Liza Doolitle Day!

All the people will celebrate the glory of you
And whatever you wish and want I gladly will do."
"Thanks a lot, King," says I, in a manner well-bred;
"But all I want is 'enry 'iggins 'ead."
"Done," says the King with a stroke.
"Guard, run and bring in the bloke."

The crucial scene that follows is made all the more powerful by the introduction of a series of vignettes showing Higgins's attempts to get Eliza to master the long "A" of English speech. (The phrase "The Rain in Spain" and most of his other phonetic exercises come from the screenplay.) Shaw's play includes a short scene where Higgins tries to teach Eliza the alphabet, but then *Pygmalion* skips on to her presentation at Higgins's mother's tea party. But Lerner knew that the moment Eliza mastered what Higgins was trying to teach her, an outburst of song had to follow. He opted to build up to it with a cinematic technique using blackouts to indicate the passage of time.

Higgins's servants at first sing of their concern for their employer: "Poor Professor Higgins, on he plods, against all odds. Every night, One a.m., Two a.m. . . ." But soon they realize how sardonic he is. When Eliza makes no progress, they go on to fear for their own sanity and change the lyric to: "Quit, Professor Higgins . . . hear our plea or payday we will quit, Professor Higgins."

Even Pickering has turned against him and asks him to be reasonable.

HIGGINS: *(Rising)* I am always reasonable. Eliza, if I can go on with a blistering headache, you can.
ELIZA: I have a headache, too.
HIGGINS: Here. (*He plops the ice-bag on her head. She takes it off her head and buries her face in her hands, exhausted to the point of tears*) (*With sudden gentleness*). Eliza, I know you're tired. I know your head aches. I know your nerves are as raw as meat in a butcher's window. But think what you're trying to accomplish. (*He sits next to her on the sofa.*) Think what you're dealing with. The majesty and grandeur of the English language. It's the greatest possession we have. The noblest sentiments that ever flowed in the hearts of men are contained in its extraordinary, imaginative and musical mixtures of sounds. That's what you've set yourself to conquer, Eliza. And conquer you will. (*He rises, goes to the chair behind his desk and seats himself heavily.*) Now try it again."
ELIZA: The rain in Spain stays mainly in the plain.

The moment is sheer magic. This clever bit of dramaturgy which never fails to elicit empathetic applause is infinitely moving. It is brought about not by

The exhilarating moment when Eliza masters "the rain in Spain stays mainly in the plain," leading to a wild fandango. Andrews, Harrison, and Robert Coote as Colonel Pickering. MOCNY.

Higgins's slave-driving pigheadedness, but by his kindness. At last Eliza sees he has a heart, and she wants to please him. Lerner's stage direction ("with sudden gentleness") is a give-away. Higgins's and Pickering's response, "I think she's got it," whose inherent rhythm is ♪ ♪ ♪ | ♪ ♪ must have suggested the habañera that later turns into a wild fandango and a *jota*.

Riding on the tide of the excitement, Pickering and Higgins calm down by discussing how they will surprise the latter's mother by introducing Eliza at Ascot. Now that it is time for Eliza to retire, she sings her first solo as a "lady." The song, "I Could Have Danced All Night," is so well known that there is no need to have it reprinted here, but one should not overlook the servants' counterpoint to Eliza's euphoria: "You're tired out, / You must be dead, / Your face is drawn, / Your eyes are red . . ." etc. It turns the second chorus into an ensemble number. And director Hart had the good sense to reprise the song yet again, after the servants leave. He lets Eliza drift off to dreamland in the last quiet refrain.

Now since it is a foregone conclusion that Eliza and Higgins will pull off their triumph, the rest of the act seems to be headed downhill. But danger and empathy for this *new* Eliza piques our interest, and we look behind the light-hearted words. Eliza sings: "I only know when *he* / Began to dance with me . . ." Obviously she has fallen for Higgins, and we remember he will "never let a woman in his life."

Scene 6: Near the Race Track Meeting, Ascot

A short scene, not in Shaw, in which Pickering prepares Mrs. Higgins to meet Eliza.

Scene 7: Inside a Club Tent, Ascot

The civilized sang-froid of the upper class English has never been more caustically portrayed in song. Added to that were a chorus that hardly opened their mouths, Loewe's delicious measured gavotte, and Cecil Beaton's black-and-white creations to clothe the assemblage.[2] The entire number takes its cue from Lerner's stage direction: "The crowd awaiting the Ascot opening race sing with a minimum of movement." All this prepares us for the most stylized tea party in literature since *Alice in Wonderland.* But the entire Ascot scene, even to the lack of color in the costuming, is not gratuitous. It is meant to show Eliza (with Higgins and Freddy to a lesser degree) as the only ones there who are flesh and blood. The stylization returns, and Eliza shocks the assemblage with her line, when in the excitement of the race she exhorts her horse to "move your bloomin' arse!!!"

---

[2] The inspiration for this number came from Vincente Minnelli, who had suggested to Arthur Freed that the colorful *American in Paris* ballet would be more effective if the Beaux Arts Ball which preceded it were to be costumed in shades of black, gray, and white.

Cecil Beaton's magnificent black, ivory, and gray costumes make their own statement about the sang-froid of the English gentry. Only Eliza (kneeling, center) adds a touch of life. MOCNY.

Scene 8: Outside Higgins's House—Wimpole Street

Freddy Eynsford-Hill, who was captivated by Eliza at the tea party, is given this moment in "one."[3] Lerner maintained that the inspiration for this, one of his most popular songs, came from a memory of his childhood. When he was ten he often sat across the street from the house of the prettiest girl in his dancing class. He hoped to be able to dash across and speak to her alone, without the competition of his classmates. "She never came by," he wrote. "I found out later I had the wrong address."

Throughout the musical British audiences hardly could fault Lerner's Englishness, but they did object to the Americanism of "On the Street . . ." when they commonly say, "In the Street . . ." In spite of the error the song was as popular with the British as it was with the Americans.

---

[3] The short scene played near the footlights is necessary to prepare for the big set of the Embassy Ball to come

Scene 9: Higgins's Study

After her failed performance at Ascot, Eliza was understandably reluctant to go to the Embassy Ball, so a short cajoling scene for Higgins was added. Always used to having his own way, aghast at the prospect of losing the bet, Higgins paints a stunning picture of Eliza's conquests at the gala. Set to a charming waltz, the song remained one of Alan's favorites. "Fritz and Moss Hart and I were absolutely positive it would be Rex's *pièce de résistance*," he wrote. The song and scene were dropped after one performance in New Haven. The brilliant lyric, one of Lerner's finest, is printed below.

Oh, come to the ball,
Oh, come to the ball,
It wouldn't be fair
To the men who'll be there
To deny them
All the dreams you'll supply them.
There even may be
A dashing marquis
Who, feature by feature,
Will swear you're the creature
He always prayed for,
Single stayed for.
If you aren't there
His complete despair
Will be painful to see,
So come to the ball,
Come to the ball,
Come to the ball,
With me.

Consider the lord
So frantically bored
He's leaving his kin
And becoming an In-
Dian lancer,
Hoping danger's the answer.
Your innocent glow
Will dazzle him so
That glancing at you
Will restore his illu-
Sions, and as he glances,
Farewell, Lancers.
Should he die, alas
At the Khyber Pass,
What a loss it would be.

So come to the ball,
Come to the ball,
Come to the ball
With me.

I can see you now in a gown by Madame
Worth,
When you enter ev'ry monocle will crash.
I can see you now, like a goddess come to
earth,
I can hear the ladies' teeth begin to gnash.
Little chaps will wish they were Atlas,
A Queen will want you for her son.
Portly men will wish they were fatless,
And the married men will wish they were un.

What a triumph, through and through,
What a moment, what a coup.
And off by a wall,
Unnoticed by all,
A man with a smile
You could see for a mile
Will be standing,
His dimensions expanding.
The pride in his eyes
Will double their size,
As sweetly and neatly
You somehow completely
Electrify them,
Lorelei them.
Search the world around
There could not be found
Someone prouder than he.
So, come to the ball,
Come to the ball,
Come to the ball
With me.

Once "Come to the Ball" was cut the only thing that remained of the scene was Eliza's stunning entrance down the staircase. Originally it had been planned to have her make her entrance in Scene 11, but with all the exquisite costumes Beaton had designed for that scene, the sight of Eliza in gown and train, even as breathtaking as it was, got lost.

Scene 10: The Promenade of the Embassy

A brief scene wherein dialectician Karpathy (called Nepommuck in Shaw), a former student of Higgins's, is detailed to find out who Eliza is.

Scene 11: The Ballroom of the Embassy

We witness Eliza's triumph, for as Higgins predicted in "Come to the Ball," she is the envy of all the women and the desiderata of all the men. As the curtain

A dangerous moment near the end of Act I when Karpathy (second from left), an expert at unmasking frauds, asks to dance with Eliza. MOCNY.

descends on the act, the swirling dancers create the lavish beauty necessary for a first-rate musical while the suspense Lerner creates is far greater than what we find in Shaw.[4] Lerner sends the audience out for intermission without telling them whether or not Karpathy has penetrated Eliza's elegant mask and discovered the cockney underneath.

ACT TWO

Scene 1: Higgins's Study

Now in the song "You Did It" Higgins tells what occurred at the ball—although we saw it happening in dumb show. Both Pickering and Higgins completely ignore Eliza's contribution to their charade. At last, after Pickering retires, Eliza gets to the heart of Higgins's ingratitude. She throws his slippers at him and shouts, "What's to become of me?" When Higgins replies that she might marry and that his mother "could find some chap or other who would do very well," we are treated to one of Shaw's most penetrating barbs concerning marriage:

ELIZA: We were above that in Covent Garden.
HIGGINS: What do you mean?
ELIZA: I sold flowers. I didn't sell myself. Now you've made a lady of me, I'm not fit to sell anything else.

The argument accelerates and the scene ends as Eliza decides to leave. Before she does, she reprises "Just You Wait." When she first sang it, the song was in the realm of pure fantasy. Now she realizes that her desertion will hurt Higgins, and Lerner gives her fresh, stinging words. "You will be the one it's done to, and you'll have no one to run to," she shouts, and her song conveys her scorn (and love) far more vividly than ever before.

Scene 2: Outside Higgins's house—Wimpole Street

Freddy, too, gets his reprise of "On the Street Where You Live" but is interrupted in his reverie by Eliza's actual appearance. When he begins to flatter her with poetry, trying to emulate Higgins's professorial manner, he merely adds fuel to the fury that had begun in "Just You Wait." She interrupts him with "Words! Words! Words! I'm so sick of words! / I get words all day through, / First from him, now from you!" This leads into her own tour de force, "If you're in love, Show Me!"

---

[4] The moment looks backward to Hammerstein's *Show Boat* and forward to Sondheim-Lapine's *Into the Woods*. In all these cases the plot seems neatly tied up with a ribbon. But we dread the upcoming consequences of these happy endings to come in Act Two.

Success. A smug Higgins recounts his victory to his staff as an approving Pickering looks on.
MOCNY.

Scene 3: Flower Market of Covent Garden

Lerner had learned from Hammerstein that the first act must introduce the new songs and that the second, apart from having one or two new numbers, should consist mostly of reprises. Now the reprise pays off. Eliza, not knowing where she belongs, goes to Covent Garden and enters upon a reprise of "Wouldn't It Be Loverly?" Of course, her old cronies don't recognize her.

The mood changes when she runs into her father, now dandified. He is furious at Higgins, he tells his daughter, because he wrote an American millionaire that Doolittle was the most original moralist in England. "It put the lid on me, right enough!" he adds. "The bloke died and left me four thousand pounds a year in his bloomin' will." Now the woman he has been living with for many years wants to get married. The song "Get Me to the Church on Time" has an authentic British music-hall quality and like "A Little Bit of Luck," usually stopped the second act.

Scene 4: Upstairs Hall of the Higgins House

Higgins, discovering Eliza's departure, is frustrated. He can't function without her ordering his morning coffee, overseeing his appointments, etc. He and Pickering are befuddled as to why she left. As Higgins sings "A Hymn to Him," the rather pretentious title for the song generally known as "Why Can't a Woman Be More Like a Man," we realize again how self-centered Higgins is.

The much married Alan, who later in his life often joked that between him and Rex they had supported more women than Playtex, was fond of telling the story of the genesis of "A Hymn to Him." He and Rex were walking down Fifth Avenue, complaining about alimonies and what the bitter women had put them through, when Rex suddenly stopped and said, "Alan, wouldn't we be better off if we were homosexual?" In truth it is a delicate line Lerner treads, never toppling over because he has so well established both Higgins's and Pickering's utter chauvinism. The ending is sheer fun.

Higgins: Why can't a woman behave like a man?
Men are so friendly, good-natured and kind;
A better companion you never will find.
If I were hours late for dinner, would you bellow?
Pickering: Of course not.

Higgins: If I forgot your silly birthday, would you fuss?
Pickering: Nonsense.
Higgins: Would you complain if I took out another fellow?
Pickering: Never.
Higgins: Why can't a woman be like us?

Scene 5: The Conservatory of Mrs. Higgins's House

Eliza has gone to the only person who showed her any compassion, Mrs. Higgins. When Higgins comes to take her back home, she refuses to go. Lerner gives Eliza a wonderful composite speech, actually her only big pronouncement, which seems to gather together Shaw's rambling diatribes:

I should never have known how ladies and gentlemen behave if it hadn't been for Colonel Pickering. He always showed me that he felt and thought about me as if I were something better than a common flower girl. You see, Mrs. Higgins, apart from the things one can pick up, the difference between a lady and a flower girl is not how she behaves but how she is treated. I shall always be a flower girl to Professor Higgins because he always treats me as a flower girl and always will . . .

Higgins asks her how she will make her way in the world. When Eliza threatens to live by tutoring the phonetics he taught her and then insults him further by saying she will marry Freddy, he totally loses control. Eliza is in the driver seat as she sings perhaps the show's only weak song, "Without You," leading into yet another reprise, "You Did It."

Eliza: I shall not feel alone without you. I can
    stand on my own without you.
So go back in your shell,
I can do bloody well
Without . . .

Higgins (Triumphantly): By George, I really did
    it!
I did it! I did it!
I said I'd make a woman
And indeed I did! . . .

With the words "I shall not be seeing you again, Professor Higgins," Eliza walks out. That effectively ends Shaw's play.

It is well known that Shaw did not intend to leave the outcome ambiguous, for he wrote a long postlude explaining that "Eliza was not coquetting; she was announcing a well-considered decision." He supports his premise by saying that Eliza was young and attractive while Higgins was much older. Bringing in his knowledge of actors (and sounding very much like Higgins), he makes light of Higgins's attainment in transforming Eliza, citing, "such transfigurations have been achieved by hundreds of resolutely ambitious young women since Nell Gwynn set them the example by playing queens and fascinating kings in the theatre in which she began by selling oranges." The capstone of his argument comes when he tells us Eliza knew "Freddy had the makings of a married man, while Higgins was a confirmed bachelor."

Lerner does not agree. His romantic concept (he was later in life to become the older teacher of several young wives) is at odds with Shaw's. But he is not 100 percent sure. In his preface to the published version of the show he writes of Shaw's postlude, saying "Eliza ends not with Higgins but with Freddy and—Shaw and Heaven forgive me!—I am not certain he is right." Certain or not, he solves the ending beautifully—in terms of the romantic musical theater of the '50s.

Scene 6: Outside Higgins's House—Wimpole Street

leads directly into

Scene 7: Higgins's Study

Higgins enters in a bellowing rage but makes the terrifying discovery that he's grown accustomed to her face. It is the first extended moment of tenderness we have had from Higgins in the show. He cannot maintain it long and soon fantasizes how Eliza will ruin her life by marrying Freddy and come crawling back to him. "But I shall never take her back, / If she were crawling on her knees./Let her promise to atone! / Let her shiver, let her moan, / I will slam the door and let the hell-cat freeze!" Still, he is too involved to let it go at that. He suddenly abandons the rhetoric he has used in all his songs of the musical and speaking in simple human terms, confesses how vital Eliza is to him: "Her joys, her woes, her highs, her lows, are second nature to me now, / Like breathing out and breathing in . . ."

He goes into his studio, turns on the recording machine, and listens again to her voice and his own. He does not see Eliza walk into the room and stand by the machine, but after a replay of his line, "It's almost irresistible, she's so deliciously low, so horribly dirty," she says aloud, "I washed my face and hands before I come, I did."

Lerner's stage direction is consistent with Higgins's character. "He straightens up. If he could but let himself, his face would radiate unmistakable relief and joy. If he could but let himself, he would run to her. Instead, he leans back with a contented sigh, pushing his hat forward till it almost covers his face."

Higgins: (Softly) Eliza? Where the devil are my slippers? (There are tears in Eliza's eyes. She understands.)[5]

The two scenes become the capstone of the play. They not only include role reversal as Higgins this time fantasizes Eliza's dire fate, but his tenderness in "I've Grown Accustomed to Her Face" becomes a mirror image of hers in "Wouldn't It Be Loverly?" Lerner observed that "Higgins goes through as much of a transformation as Eliza, the only difference being that Shaw would never allow the transformation to run its natural course." In turning Eliza into the sculptor and Higgins into the statue, Lerner has moved this basic Cinderella story onto another, more biting plane.

The song itself, especially in its opening phrases, is an exquisite example of words matching music. Actually, one could hardly find pitches that conform better to the spoken contour of "I've grown accustomed to her face, She almost makes the day begin" (see example below) than the ones Loewe chose. In both cases the last notes, D, the second of the scale, and F, the sub-dominant, are enigmatic, incomplete. The music turns the phrases into interrogations rather than declarations. It is as though Higgins, still true to form, is holding something back, still noncommittal to a woman in his life. We sense it will be many years before Eliza will be able to introduce him to completeness, C, the tonic, the affirmation.

I've grown ac-cus-tomed to her face,——— She al-most makes the day be-gin.

By now the success of *My Fair Lady* has become legendary. As the second Lerner-Loewe opus to win unanimous rave reviews from all the New York critics it ran six-and-a-half years on Broadway and broke all the records then existing for a musical.

[5] Loewe's score at the ending is nothing short miraculous. After Higgins's number in the key of C, when Higgins enters his studio Lowe underscores the scene by modulating to B-flat with a slow, soulful "I Could Have Danced All Night." After Higgins utters his ". . . my slippers," the same theme is repeated "molto maestoso" in the fresh key of E-flat. The trick is the same one Gershwin used (in reverse) to create the exciting ending to his *Rhapsody in Blue.*

Within a few years of its opening it became internationally known, with touring companies visiting Australia, Sweden, Mexico, Holland, Russia, and Israel. Shortly after it opened in New York it began simultaneous long runs in London, Oslo, Stockholm, Melbourne, Copenhagen, and Helsinki. Warner Brothers, who made the film, boasted that after its first year, their revenues from recordings and admissions was eight hundred million dollars. By now it is safe to say the total gross exceeds a billion.

Only a handful of musicals, once the bloom is off the score, seem worthy of revival, but My Fair Lady seems to come around in a major revival at least once every decade. Certifying to My Fair Lady's splendor in 1976, Walter Kerr, a critic not noted for gushing, wrote: "The show is twenty years stronger. So dazzlingly melodic and visually rich in its first act that it scarcely needs a second—and so emotionally blinding in its second that you wonder why you were merely dazzled by the first." Attesting to the show's imperishability he added, "Structurally sound of limb, skipping like springtime and living out its winters on wit, My Fair Lady isn't an entertainment that requires certain performers to bring off the Bernard Shaw based libretto, the leaping Alan Jay Lerner lyrics, the sweeping Frederick Loewe score; it's an entertainment that invites any and all performers to simply lift the lid from its treasure-chest and avail themselves of its glistening baubles."

After My Fair Lady's curtain came down, many of the celebrities at that first performance adjourned to the usual Broadway watering hole, Sardi's. Herman Levin had not thrown the usual opening-night bash there, preferring to reserve a small elegant room at "21" where an intimate group would await the first reviews over supper. Moss and Kitty Hart, the Goddard Liebersons, Rex Harrison, Julie Andrews, and Irene Selznick waited in vain for Fritz, who preferred to mingle with his adoring fans at Sardi's. Alan didn't appear until the party was almost over.

He had gone to his mother Edith's Park Avenue apartment to await the earliest papers. When the first glowing notices appeared, she almost paraphrased Samuel Goldwyn's now classic remark to Richard Rodgers after the opening of Oklahoma! (that Rodgers should shoot himself) when with genuine concern she asked, "What, my son, can you ever do to top this?" The observation was to haunt him for the rest of his life, for although he was to have much success in future projects, My Fair Lady would always represent the very pinnacle of his artistic curve.

After weeks of cogitating the answer to his mother's question Alan came up with the solution. It is an old saying, by now a joke, that circulates among manufacturers of women's dresses in Manhattan's garment district, an area Alan knew well. When they are asked how their next season's line of dresses will look, they usually answer: "The same. Only different." And that is pretty much what their customers want.

After *Brigadoon* and the aborted *Huckleberry Finn*, Lerner owed one more picture to Arthur Freed. When *My Fair Lady* was previewing to ecstatic reviews in Philadelphia in February 1956, Freed flew in from California and suggested he have a look at *Gigi*, Colette's 1942 short story whose setting was turn-of-the-century Paris. It concerned an innocent but not stupid young woman who was being initiated into the arts of courtesanship by a worldly aunt. The idea, which intrigued Alan and might just possibly top his recent success, was pretty much the same as *My Fair Lady*. Only different.

The period and setting were irresistible, and Alan hoped Fritz would join him in creating the score, especially as their relationship throughout the writing of *My Fair Lady* had been unusually free of quarrels. But Loewe, who had never wanted anything to do with movies, was content to spend his days with the young *cocottes* on the Cote d'Azur and his nights at the *chemin de fer* tables in the Cannes casino. One of the Dolly Sisters who saw him nightly described Loewe as sitting "hunched up for hours, gripping a pile of 1000 franc chips; his eyes reduced to smudges of shadow."

Since at that point it was felt only four songs would be needed for the film, after Fritz's refusal Alan agreed to write the scenario and lyrics and complete the score with another composer if Freed would try to get Maurice Chevalier, one of Lerner's idols, to accept the role of Honoré Lachailles, uncle to the hero, Gaston. He made one other demand, that Cecil Beaton be hired to do the sets *and* costumes. It was a foregone conclusion that Vincente Minnelli would direct.

Hoping Chevalier would accept, Alan built up what was a relatively small part of Colette's story, the role of Honoré, into the film's narrator and commentator. Lerner even added a relationship for Chevalier with Gigi's grandmother, Mamita, planning at least three songs for Chevalier. But rather than begin working with Burton Lane in Hollywood, Alan flew to New York to try Fritz once more. "To add a little seasoning," Lerner wrote, "I said I felt it was essential that the score be written in Paris. It would not only be fun . . . but unquestionably it was bound to help . . . to write it in the country in which the story takes place." Loewe read the script and telephoned the next morning to say he was enthusiastic about doing the film.

Shortly after that they flew to Paris and settled in at the Hôtel Georges V. Working only with Alan's title, Loewe had already completed the melody for Chevalier's first song, "Thank Heaven for Little Girls." (Encapsulating the line, "for little girls get bigger ev'ry day," it would become the enigmatic theme of the entire film.) Then they went to Belgium, where Chevalier was appearing in a nightclub. After he agreed to do the film, they asked why at seventy-two he was still performing. His answer: "I'm too old for women, too old for that extra glass of wine, too old for sports. All I have left is the audience." Thereafter Lerner credited those observations as the motif for one of the film's wittiest songs, "I'm Glad I'm Not Young Any More."

Lerner made no secret of the fact that he wanted Audrey Hepburn, who had launched her career as Gigi on Broadway, to star opposite Dirk Bogard as Gaston. But Hepburn did not want to do the role again, and J. Arthur Rank, who had Bogard under contract, refused to release him. Leslie Caron, who since appearing in Lerner's *An American in Paris* had made a specialty of playing sensitive young girls maturing into womanhood (*Lili, Daddy Long Legs*) and who even had the experience of playing Gigi on stage in Paris, accepted. Louis Jourdan, whose career had been stifled by typecasting as the Continental lover, jumped at the chance to play the bored Gaston. When he auditioned, it was discovered he had an excellent singing voice.

With the expense of shooting on location so exorbitant, Freed had planned just a few outdoor shots of Paris, with all the interiors to be shot back in Hollywood. But Alan, who felt Paris was as much a character in the film as Gaston and Gigi, persuaded him that certain key scenes at the Champs-Élysées, the Palais de Glace, and especially the big scene at Maxim's would have more authenticity shot in Paris. Freed had put his career on the line to get the studio heads to agree to the inordinate expense of supporting the entire company abroad; now he warned everyone that "there must be no dawdling."

This was particularly hard on Alan, who often agonized for days over a single line. When he became blocked during the writing of the six songs he had by now slotted for *Gigi*, it became a major catastrophe. Writing the Maxim's number, "She's Not Thinking of Me," for example, Alan held up shooting, to the great consternation of Freed, when he became stuck for a couplet to balance

> Oh, she's shimmering with love!
> Oh, she's simmering with love!

He spent nine days working on it until at last he came up with

> She's so ooh-la-la-la-la,
> So untrue la-la-la-la.

Looking at it later he admitted, "It seems hardly worth the effort."

In *Gigi*, as in everything they wrote together, Alan and Fritz always discussed the dramatic motivation of the songs before they created them. "Lerner always had the idea for the song and would give me the title." Then they would sit together at the piano while Loewe worked out the basic melodic motif. It was essential that they then decide *where the title would come*. This would determine the form. Would the title appear at the outset as in "Gigi" or "I Could Have Danced All Night"? Or would it would come at the end of the A section as in "I'm Glad I'm Not Young Any More" or "Almost Like Being in Love"? Sometimes, as in "They Call the Wind Maria" or "With a Little Bit of Luck," the title comprises most of the short refrain.

Once that title placement had been agreed upon, Loewe would then notate the motif and title. Then he would go to his apartment (in Paris) or when they were on the road, his suite, and return usually the next day, with the completed music. After playing it and adjusting the melody to their mutual satisfaction he would leave the score with Alan (who was an excellent sight reader). Lerner would generally have memorized the tune before he began to work on it. Most times the lyric would come easily, but sometimes the agonizing would begin after a few days. Loewe was far more patient than Lerner's other collaborators were, relishing his partner's dilatory ways especially when they were in Paris, where he could be found nightly at the baccarat tables in private clubs.

Alan, too, enjoyed those evenings, for he began seeing a young French lawyer, Micheline Muselli Pozzo di Borgo. Small, blond, a breathtaking beauty whose English was tinged with a charming accent, she was also very bright. At twenty, having attained the honor of being the youngest *avocat* ever to be called to the French bar, she gained publicity as a defense attorney by handling the cases of rapists, thieves, and murderers. Because of the eloquence of her pleas to the juries, she had been called the French Portia.

Alan, who was very superstitious, always considered thirteen his *lucky* number. Citing *Brigadoon* as his capstone, he once told a reporter that the show was "packed with thirteens. There were thirteen principals, thirteen songs, and thirteen scenes." Now he noted that his relationship with Micheline Muselli would be favored because both her names began with the letter "M," the thirteenth letter of the alphabet.

André Previn, who was the general music director, arranger, and orchestrator for *Gigi*, came to Paris for talks with both Lerner and Loewe. He remembered having dinner with them in the summer of 1957 and meeting Micheline at a restaurant on the Left Bank. When Alan's new heartthrob left the table, Lerner turned to André and said, "Isn't she wonderful?," to which a glib Fritz piped in, "Alan, my boy, *never* fuck lawyers."

By August the screenplay and six of the songs—the score by now expanded to eight numbers—and the principal location photography, which took two months, was completed. Everyone returned to Hollywood, where the newspapers soon announced that the Lerners had separated.

Alan did not worry about the adverse publicity; he was too involved in *Gigi*. There were the two missing key songs, "I'm Glad I'm Not Young Any More," and "The Night They Invented Champagne" yet to be written. Then, after the filming had been completed, the post-production work—orchestration, recording, editing, putting in the titles—had begun, and Alan wanted to supervise it all.

Leslie Caron had recorded her songs in Paris, adding "The Night They Invented Champagne" as soon as it was finished in the Hollywood studio. On listening to the playbacks Alan, André, Fritz, and Arthur decided that although she performed the role delightfully, all of her singing would have to be scrapped. When she was told, Caron was furious, but André Previn made sure Marni Nixon's vocal performance matched Caron's emotional acting, and with Nixon's splendid lip synching technique the result was smoothed out.

When the film was finally "sneak previewed" in January 1958 in Santa Barbara, it was decidedly too long, and although the audience approved heartily of the picture as a whole, Alan and Fritz felt it "a far cry from what they had hoped for. The action was too slow, the music too creamy and ill-defined."

They talked it over the next morning and decided it needed much reworking. Alan felt the scenes in Gigi's house needed to be rewritten and that the charming duet for Chevalier and Hermione Gingold, "I Remember It Well," was muddy. Fritz found the orchestration too lush and "Hollywooden" and wanted a sound that suggested the theater rather than the sound stage. Arthur agreed with them, figuring these considerable changes would cost $350,000.

At first the studio balked at what was then an enormous sum and, citing the audience's favorable reaction at the preview, threatened to release the film as it was and take its chances. When Alan and Fritz in a ploy offered to buy the film outright for three million dollars, returning all the money the studio had invested, MGM had second thoughts. Assuming the team must have inside information on the film's success or at least a guarantee from a corporation for stage and record rights—never once suspecting Alan and Fritz were simply bluffing—they agreed to make the further necessary changes.

In addition to all the scenic changes Alan and Arthur had asked for, the entire score was reorchestrated and rerecorded. The Maxim's sequence, for which the restaurant had been closed for the better part of a week, was reshot. A false "Maxim's" was constructed on the Hollywood lot. The Palais de Glace scene was recut and reedited.

The collaborators' search for perfection was well worth the effort. Gigi won nine Academy Awards, including best picture. Arthur Freed, André Previn, and Cecil Beaton all were awarded Oscars. Chevalier even won a special award. Alan took home two of the golden statuettes, winning the award for Best Screenplay Based on Material from Another Medium, and for Best Song, Gigi. Fritz, who had been recuperating from a heart attack only two months before, decided—against his doctor's advice—to attend the ceremony. He coined a memorable line when he thanked the Academy for honoring his song, "from the bottom of my somewhat damaged heart."

The song, the turning point in the picture, has often been compared to the angry diatribe "I've Grown Accustomed to Her Face" in *My Fair Lady*, sung when Higgins realizes he has fallen in love. In both soliloquies the men thrash like hooked marlins, but Higgins is bloodless, while Gaston is not afraid to show his disappointment at losing someone he carelessly watched grow up or to shout his joy at finding she is now the object of his passion.

To write the song Lerner used a similar technique to one he employed in *My Fair Lady:* he imagined what would have happened in the story. Taking his cue from Collette's story, particularly the lines "she ran to open the door and came back a moment later. Gaston Lachailles, haggard, his eyes bloodshot, followed close behind her." Alan imagined the only thing that would have made a determined man like Gaston drained and bleary: certainly the indecision leading to his realization that he was actually no longer thinking of Gigi as a mistress but as a wife!

Throughout his life, this is the lyric Lerner always cited as his best.

She's a babe! Just a babe!
Still cavorting in her crib;
Eating breakfast with a bib;
With her baby teeth and all her baby curls.
She's a tot! Just a tot!
Good for bouncing on your knee,
I am positive that she
Doesn't even know that boys aren't girls.

She's a snip! Just a snip!
Making dreadful baby noise;
Having fun with all her toys;
Just a chickadee who needs a mother hen.
She's a cub! A papoose!
You could never turn her loose.
She's too infantile to take her from her pen.

Of course, that week-end in Trouville
In spite of all her youthful zeal,
She was exceedingly polite,
And on the whole a sheer delight.
And if it wasn't joy galore,
At least, not once was she a bore
That I recall.
No, not at all.

Hah!

She's a child! A silly child!
Adolescent to her toes,
And good heavens how it shows!
Sticky thumbs are all the fingers she has got.
She's a child! A clumsy child!
She's as swollen as a grape,
And she doesn't have a shape.
Where her figure ought to be it is not.

Just a child! A growing child!
But so backward for her years:
If a boy her age appears
I am certain he will never call again.
She's scamp and a brat!
Doesn't know where she is at.
Unequipped and undesirable to men.

Of course, I must in truth confess
That in that brand new little dress
She looked surprisingly mature
And had a definite allure.
It was a shock, in fact, to me,
A most amazing shock to see
The way it clung
On one so young.

Ah!

She's a girl! A little girl!
Getting older, it is true,
Which is what they always do;
Till that unexpected hour
When they blossom like a flower...!

Oh, no. . . !
Oh, no. . . !
But. . . !
But. . . !

There's sweeter music when she speaks,
Isn't there?
A diff'rent bloom about her cheeks,
Isn't there?
Could I be wrong? Could it be so?
Oh, where, oh, where did Gigi go?

Gigi, am I a fool without a mind
Or have I merely been too blind
To realize?
Oh, Gigi, why you've been growing up before
    my eyes.
Gigi, you're not at all that funny awkward little
    girl I knew!
Oh, no! Overnight there's been a breathless
    change in you.
Oh, Gigi, while you were trembling on the
    brink
Was I out yonder somewhere blinking
At a star?
Oh, Gigi, have I been standing up too close
Or back too far?
When did your sparkle turn to fire?
And your warmth become desire?
Oh, what miracle has made you the way you
    are?

One of the most charming spots in the film was the duet "I Remember It Well" which Lerner took from *Love Life* and rewrote for Chevalier. The French entertainer also had great success with "I'm Glad I'm Not Young Any More," which became a staple of his nightclub act until his death. When the song was recorded for the film Chevalier came to Lerner and asked if his accent was acceptable, to which Lerner replied that he had understood everything. "That's not what I meant," Chevalier retorted. "Was there enough?"

Lerner wrote three refrains for the song. Only two were used in the film, the last would certainly have been cut by the censors. But he did use that third refrain in the stage adaptation of the film which had a brief run a decade later. The unused lyric, a raunchy gem, is printed below:

The worry if you're the best or worst,
And then you find out that you're the first.
I'm glad I'm not young any more.

All others but you she may enchant,
You feel that you should, you do; you can't—
I'm glad I'm not young any more.
No desperation,

No midnight pacing of floors,
What liberation
To know that child on all fours
Couldn't be yours.
Too old for a wedding day in June
But not for a week-end honeymoon,
What better life could heaven have in store?
Oh, I'm so glad that I'm not young any more!

By 1958 when Alan was forty, had two Academy awards to his credit, and had written the most lauded stage musical as well as the film musical that received the most Academy awards ever, it seemed he was at the peak of his career with the world in his pocket.

His lavish ways always gave the Broadway and Hollywood columnists a field day. The story about his buying Fritz a Rolls Royce to match one he just purchased with the quip, "Well, it's only fair, you paid for the lunch," while not actually true was entirely possible. Money and possessions counted, some said, while human relationships had very little meaning for him.

"He fell in love with Micheline and he left, that's basically what happened," his daughter Jennifer Fraser recently told this writer. As for Nancy, their mother, "she had no choice," Jennifer's sister Liza Bibb added.

His wife and daughters were still living in the Rockland county house when Alan brought Micheline over to America and installed her in a lavish suite overlooking Central Park in New York's Stanhope Hotel. Soon they moved to the St. Regis and eventually to a large furnished apartment on Sutton Place. Then, on Christmas Day 1957, only two months after Alan's divorce from Nancy, Micheline Muselli became the fourth Mrs. Lerner. His three other wives had been docile enough as to the separations and divorces. With Micheline it would become quite another matter.

Hammerstein

1956–1958

IN THE MID-'50s, while the number of new musicals opening on Broadway was shrinking every year, a great expansion was occurring in the television industry. The medium had come of age, color had arrived, and now that the TV screen was no longer a minuscule rectangle it tried desperately to compete with cinema and stage in artistic merit. Once it had advanced beyond replays of old movies, jugglers, clowns, and talk shows, it attempted to raise its artistic standards, first by offering intimate dramas. But then television realized it had an advantage over other media. It could bring celebrities into the parlor.

Soon well-known stars appearing in new versions of old musicals become a staple. These productions were known as "spectaculars" (which most of them certainly weren't), and advertisers were able to create enough hoop-la to attract huge audiences. *Annie Get Your Gun, Kiss Me, Kate, Bloomer Girl, High Button Shoes, Anything Goes, Lady in the Dark,* and many others were successful. The networks even tried commissioning their own musicals—generally from stories in the public domain like *The Gift of the Magi, Little Women,* or *Marco Polo.* But none of these seemed to attract the public.

When NBC hit upon the idea of mounting the stage production of *Peter Pan,* "family television," an idea whose time had come, was born. Filming this show under commercial sponsorship and showing it to a much larger audience than could possibly have seen it in the theater rescued the production from its financial doldrums. After only 140 stage performances, the TV airing made it unnecessary for *Peter Pan* to continue its theatrical run. Parents cuddling their pajama-clad children had seen Mary Martin fly in their living rooms. Now only an elite few needed to see her manipulated by wires onstage. Soon an encore airing of this Broadway show was arranged, and it turned out to be pure gravy for all concerned.

Everyone was now aware that fortunes were to be made by writing for the small screen, and since shows were shot in New York, theater people need not leave Broadway to trek to Hollywood. Nor would anyone be at the mercy of the insensitive movie moguls.

No one relished trying something new more than Dick and Oscar, and when Julie Andrews's agent called shortly after *Pipe Dream* closed in the summer of 1956 and asked if the team would be interested in adapting *Cinderella* for his client, he was given an unhesitant "yes."

Not only was *Cinderella* just the kind of "family show" R & H wrote so well, it was exactly the confection they needed as antidote to the sleaze of their previous effort, *Pipe Dream*. Writing for Julie Andrews, who had been instantly crowned the queen of Broadway when she opened in *My Fair Lady*, was like a breath of fresh air. The triple combination of Perrault's beloved tale, Julie Andrews, and R & H had enough clout for CBS to offer R & H a handsome $300,000 for *one* airing and to blithely bump a perennial Sunday night TV fixture, the enormously popular *Ed Sullivan Show*. CBS's archrival NBC threw in the towel and scheduled an old movie.

The telecast was scheduled for March 31, 1957, with unprecedented publicity, and an original cast album to be released simultaneously. This date was, coincidentally, the fourteenth anniversary of the New York opening of *Oklahoma!* After more than six thousand performances in its five years on Broadway and ten years on the road, *Oklahoma!* had been seen by almost twelve million people. After *Cinderella*'s single performance, CBS estimated it had been watched by one hundred seven million.

Work on the show began at once. Oscar had planned to take Dorothy home to Australia for the 1956 Olympics and hurried to get the script well under way before the journey. For the ninety-minute show (seventy-two minutes when commercials and lead-ins were subtracted), ten songs, about two-thirds of what was necessary for a Broadway production, would be needed. Oscar had many of the lyrics completed before his eastward journey, and the polishing was done by mail.

Adapting such a well-known chestnut might have produced an allegory or a fantastic departure from the original tale, but Oscar stated at the outset he had no intention to "trick up" the script. This was not to become anything like "the story of a Macy's shopgirl who eventually lands the department store owner's son." Yet while clinging to the fairy tale, Oscar discarded the traditional ways of *telling* the story.

One departure was to have a young, jocular fairy godmother instead of the usual crone. Another was to make the stepsisters and stepmother more comic than downright evil. The libretto even managed to create some sympathy for the trio at its finale. Compassionate too, and non-cardboard, were the King and Queen.[1] Both became loving and caring parents, intent on mapping the future of their son, trapped between reality and idealism.

But the greatest dramatic departure was to give Cinderella resolve and determination and even ambition. This is a gutsy girl, an atypical fairy-tale heroine, one whose imagination takes her far beyond pretty things and marriage. Using some of Hammerstein's most charming rhymes she tells us

I'm a young Norwegian princess or a milkmaid   I'm an heiress who has always had her silk
I'm the greatest prima donna in Milan   made
                                                                                  By her own flock of silkworms in Japan!

Although when her godmother suggests she leave her stepmother's hearth this cindermaid can't quite muster up the strength, still Cinderella has enough ingenuity to spawn the idea for the coach and its rodent attendants. Going to the ball is entirely *her* idea. Of the pumpkin she says: "The moon was shining on it now. Well, I was wishing that pumpkin would turn into a great big royal golden carriage that would take me to the ball tonight." The godmother, merely suggesting that "impossible things are happening every day," encourages the girl to wish "very hard" and it might happen. It is Nettie in *Carousel* giving faith to Julie all over again. Hammerstein always made his heroines work to achieve the supernatural.

Now Julie Andrews closes her eyes so tightly and wishes so hard that the viewer might suspect all that follows is a dream. In this case the top-notch score helped to solidify the sense of fantasy even more. Oscar's teleplay teetered on the edge of reality with liberal use of his favorite illusion, the dream.

Hammerstein often said he was always looking for fresh ways to write a love lyric, and in the case of "Do I Love You (Because You're Beautiful)" (or are you beautiful because I love you?), which poses the eternal question "which came first, the chicken or the egg?" in Freudian terms, he came up with an oblique concept perfectly suitable to the is-this-a-dream aspect of his libretto.

Do I love you   Do I want you
Because you're beautiful?   Because you're wonderful?
Or are you beautiful   Or are you wonderful
Because I love you?   Because I want you?
Am I making believe I see in you   Are you the sweet invention of a lover's
A girl too lovely to   dream,
Be really true?   Or are you really as beautiful as you seem?

[1] Producer Richard Lewine, himself a musician, suggested expanding and deepening the monarchs' cameos by according them a duet. "Boys and Girls Like You And Me," a nostalgic song with regal references deleted from *Oklahoma!*, fits like a glove.

The creation and polishing of this song elicited a stream of long letters back and forth across the Pacific which Rodgers reprinted *in toto* in his *Musical Stages*. Oscar wanted to use "Am I telling my heart" instead of "Am I *making believe*" because he felt he "cashed in" on that phrase in *Show Boat*. Dick answered that he had no objection to *make believe* but didn't like "'A girl too lovely to be really true,'" "for the simple reason that it sounds like a split infinitive."

After several sallies back and forth they agreed to wait until they could sit down and talk things over. "When Oscar returned," Dick wrote, "we discussed the problem—and we compromised. Oscar agreed that 'Am I telling my heart' was not appreciably stronger than 'Am I making believe.'[2] And I agreed to accept 'too lovely to be really true,' because no matter how it sounded, it wasn't really a split infinitive."

Rodgers apologizes for printing his complete file on the genesis of this song with "this may strike the reader as more of a case of splitting hairs than infinitives," but he need not. By reading through this extended correspondence one can see the painstaking care with which both men searched for the precise phrase. Each man knew a song must convey a particular feeling within a restricted form and that no word was negligible. Here Rodgers masterfully set Oscar's quizzical poem by using the minor mode at the outset (see Example A). At the end (Example B), lightness and certainty are indicated by the high tessitura and the switch to the major mode. The song's ending tells the audience that these lovers are not dreaming.

Example A

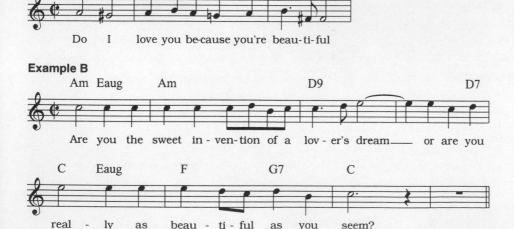

Example B

[2] Actually, "making believe" sings infinitely better than "telling my heart," fits into the fairy-tale genre, and is memorable to boot.

The whole score was imaginative, without a saccharine song in the lot. One has to note that Oscar's libretto and lyrics were the freshest he'd done in years, and it must be added that Dick's score, besides the two aforementioned ballads included bouncy tunes like "A Lovely Night" and "Impossible," all of which received a great deal of airplay. With "Ten Minutes Ago" and "Waltz for a Ball" Rodgers is back in the vein creating some of the verviest waltzes he had penned since *Carousel*.

Unfortunately, when this *Cinderella* was televised in 1957 videotape had not yet been invented, and so that production exists merely in black and white kinescope hoarded by a few collectors and museums. But the score refuses to be confined only to the small screen, and the R & H office began long ago to send it out as a roadshow. It is also—with the usual excursions and interpolations—a favorite of the British pantomime circuit.

Cinderella was remade in 1965, and although it is not the policy of this book to go into sequels or altered versions of original works, one cannot leave any discussion of this truly spectacular "spectacular" without deploring the remake done five years after Oscar's death and produced by Richard Rodgers. It actually abuses this charming work. The original script was scrapped in favor of a dreadful teleplay by Joseph Schrank. Oscar has to be turning over in his grave every time it is aired.[3]

One might forget the whole thing except that *this is the version* that has been transferred to tape, is televised frequently, and is being sold in stores as *the* Rodgers and Hammerstein *Cinderella*.

No longer are we in the timeless elegance of the original setting. Now the tale is set in grim medieval Grimm land. The sisters and stepmother are mean and bitchy. The godmother is a serious cliché. Poor Celeste Holm, looking very much like Billie Burke in *The Wizard of Oz*, is trapped in the role. Everyone else gives cardboard performances, the worst offenders being a blowsy Ginger Rogers as the Queen and Walter Pidgeon looking very much like a windup doll of a King. Even Lesley Ann Warren, a charming actress here at the outset of her career, spends the time mugging and rolling her eyes instead of seeming to "believe."

Yet another version was presented by the New York City Opera in November 1993. Replete with up-to-date though anachronistic jokes as conceived by Steve Allen, this version of the libretto is a travesty. Scheduled to coincide with the holiday season and hopefully another addition to the year-end "family

---

[3] Still on the subject of posthumous remakes, the *State Fair* travesty to which Rodgers added five tepid new songs (words and music) is at the bottom of the barrel of anything bearing the R & H signature.

show" sweepstakes in the style of *The Nutcracker* and *The Sleeping Beauty*, this version serves to perpetuate *Cinderella*, but it is now twice removed from the original charmer of 1957.

Early in 1958, when Oscar was on the West Coast supervising the filming of *South Pacific*, Joe Fields of the famous theatrical family mentioned a book he had optioned that he felt would make a splendid musical. It was called *The Flower Drum Song*. Oscar found the title intriguing and asked for a copy to be sent to his hotel.

C. Y. Lee's sensitive novel was about a college-age Chinese and his efforts to satisfy his own romantic nature while kowtowing to the tradition of his wealthy and implacable father. In larger theme the novel dealt with a young man's search for his identity. The characters appealed mightily to Oscar and Dick, and they decided to make it their next project.

After the recent stage debacle of *Pipe Dream*, Oscar, at sixty-three, did not feel up to adapting the libretto alone, so it was decided that he and Joe Fields (whom he had known since they were both seven years old) would collaborate and Fields would even co-produce. Unfortunately, in transforming Lee's *The Flower Drum Song* into the Hammerstein-Fields *Flower Drum Song*, the collaborators added a musical comedy subplot that only served to trivialize the original compassionate book.

In the novel young Wang Ta, a pre-med student who is deeply dependent on his wealthy but irascible father, becomes involved with three girls: Linda Low, a singer who turns out to be a gold digger; Helen Chao, an unattractive seamstress, originally a friend of Linda's, who falls desperately in love with Wang Ta and who commits suicide when he refuses to marry her; and Mei Li, a fragile young beauty recently arrived with her father from China. After Mei Li, who has been falsely accused of theft, flees his father's house in tears at the end of the novel, Wang Ta goes in search of her in the busy streets of Chinatown. The ending is reminiscent of DuBose Heyward's *Porgy* with an extra twist. We realize that through his love for this naïve girl Wang Ta has become emancipated enough to give up the allowance he receives from his father. At the novel's close, his resolve leads him to abandon the university education his father chose for him and to go into business with another Chinese-American, a friend long ago assimilated into California culture. In sum, he is demanding his piece of "The American Dream."

The collaborators kept all three women in their story, but by adding a smart-talking agent, Sammy Fong, they were able to create a five-pointed plot. Wang Ta and Mei Li became the main romantic story line, while Sammy Fong and Linda Low, now a nightclub singer, became the subplot. The Hammerstein-

Fields libretto now had Mei Li betrothed to Sammy under the terms of a family contract signed when they were children. Getting her unbetrothed and into the arms of Wang Ta becomes the evening's argument.

Helen is brought in as well. In such a musical comedy situation, her suicide would certainly be out of the question. Had *Flower Drum Song* been written pre-R & H, Oscar would surely have paired her off with a beau at the final curtain. As it is we never do find out what happens to her, but Helen, in the person of Arabella Hong, a classically trained soprano with a flawless voice, is given the show's best ballad—the exquisite (and demanding) "Love, Look Away."

If because of its truncated plot *Flower Drum Song* is minor Rodgers and Hammerstein, it need not have been so. Oscar was perfectly capable of dealing with the sensitive characters, emancipation, and yes, even the suicide. Some of his lyrics are exquisite, and Rodgers's score shows that each man was capable of turning out fresh, imaginative work.

The plot of *Flower Drum Song* borrows from *Oklahoma!* and *The King and I*.

Miyoshi Umeki sings her flower-drum song for the older generation. L to r: Keye Luke, Juanita Hall, Umeki, Eileen Nakamura, and Conrad Yama. MOCNY.

Both deal with one of Oscar's favorite themes, the breakdown of the old social order. Sammy is a close cousin to Ali Hakim, *Oklahoma*'s peddler-conniver, while Wang Ta's father and aunt are slaves to tradition not unlike the King of Siam.

Realizing that they were treading on familiar ground, R & H were eager to hire young collaborators to pump fresh blood into a show which they sensed mightily resembled *South Pacific* and *The King and I* in its East-West rapproachement. To this end they surrounded themselves with people they had never worked with before, like co-producer/co-librettist Joseph Fields. They abandoned their habitual designer, Jo Mielziner, engaging Oliver Smith to create freshly imaginative settings. Then they hired Carol Haney, a jazz-oriented choreographer, and arranger-orchestrator Luther Henderson to supervise the dance arrangements. Unfortunately Gene Kelly was chosen to direct, since Jerome Robbins was busy with *Gypsy*.

"Gene Kelly hadn't directed on stage for years," James Hammerstein said, "and was very slow in the book scenes. Directing isn't like riding a bicycle. If Gene had done it five years earlier and they [R & H] had been five years younger, it would have been one of the better ones." Its energy, casting, humor, and especially its score, brash by times, sensitive by turns, all contribute to giving the audience a good time. But unfortunately "good time" musicals, like so many of the Porter, Gershwin, and Kern shows—although their songs are eminently extractable—do not revive very well.

Oscar's lyrics contrasted old and new cultures. Beginning with a serene "You Are Beautiful," we are given a clue to Wang Ta's character. In the verse he recites his philosophy in what purports to be an ancient Chinese poem:

Along the Hwang Ho Valley   Came drifting down the stream.
Where young men walk and dream   I saw the face of only one
A flower boat with singing girls   Come drifting down the stream . . .

"You Are Beautiful" not only shows Wang Ta's love of his classical Oriental heritage, it shows his eagerness for a love relationship. Creating an interesting song that does two things at once was the aim as well in "I Enjoy Being a Girl," which Linda sings in the very next scene. Not only did Oscar pen a hit, but he shifted gears admirably into the contemporary with Linda's recitation of her credo *and*, at the end, a clue to her assimilation. That, unfortunately, is the pattern throughout the show: one song à la classical China followed by one of contemporary San Francisco.

In recent times, because the lyrics have been labeled sexist, the song has come in for its share of brickbats by women's groups, although it does not seem to have been offensive in its time.

I'm a girl, and by me that's only great!
I am proud that my silhouette is curvy,
That I walk with a sweet and girlish gait,
With my hips kind of swively and swervy . . .

. . . I'm strictly a female female
And my future I hope will be
In the home of a brave and free male
Who'll enjoy being a guy
Having a girl like me.

The very next song, "I Am Going To Like It Here," is a tour de force. Here Oscar experimented with a classical rhyming scheme—ABAB; BCBC; and so on. To create a more delicate but unified design than would be possible by having the title recur at the end of each stanza—the expected popular pattern—Hammerstein begins each succeeding verse with the *second line* of the one before (no minor feat if one wants to create an expanding lyric). Then he sets an even more rigorous discipline on himself by using only five rhyme sounds in the entire song. Rodgers, too, joined in the experiment by creating a gently flowing, almost classical Chinese art song. It begins with a singsong motive that gradually wanders into foreign keys for a central peak, but it graciously returns to the original theme at its end.

This is a "book song," sung by Mei Li to let us see her character, but the form, rhyme scheme, choice of language, melody, and harmony paint a portrait not unlike a Chinese brush drawing. Plotwise, the song is a dead giveaway. Accenting Wang Ta's and Mei Li's parallel penchant for the classical Oriental traditions assures us, even this early in the play that they will end up together.

I am going to like it here.	A	All the people are so sincere,	A
There is something about the place,	B	There's especially one I like,	D
An encouraging atmosphere,	A	I am going to like it here.	A
Like a smile on a friendly face.	B	It's the father's first son I like!	D
There is something about the place.	B	There's especially one I like,	D
So caressing and warm it is—	C	There is something about his face.	B
Like a smile on a friendly face,	B	It's the father's first son I like	D
Like a port in a storm it is!	C	He's the reason I love the place.	B
So caressing and warm it is,	C	There is something about his face,	B
All the people are so sincere,	A	I would follow him anywhere . . .	E
Like a port in a storm it is.	C	If he goes to another place. . .	B
I am going to like it here!	A	I am going to like it there!	E

In succeeding songs like "Grant Avenue," "Chop Suey," or "The Other Generation," as the titles would indicate, Hammerstein mixes the cultures. Tom Shepard deemed "Grant Avenue" one of the worst songs they ever wrote. "It was Rodgers trying to be rhythmical with words, which he hadn't been in a long time. Even Hammerstein's rhymes are clumsy and forced. 'Have a new' rhyming with 'avenue' and the like make it a tortured song."

As for "Chop Suey," that is even worse. Oscar ends his lyric by rhyming "chop suey" with "Dewey," "screwy," "St. Louis" (Lewie), "gooey," and, finally, "Drambuie." In "The Other Generation," although he tries to come up with amusing ideas about the lack of communication between the American- and Chinese-born, most of these end up as cliché.

Oscar was, as we have mentioned, better at provoking a smile than a guffaw. His characters were often amusing, but he was no more than adequate with jokes. One feels he did not really care for these characters, and so many of the songs end up as rather tiresome chow mein.

In a way this is true of Richard Rodgers's contribution as well. He was not tired, but because of the failures of his two previous shows he kept his eye on the commercial. Chez Rodgers this always precluded his having done his most distinguished or daring work. He turned out some crowd-pleasing, hummable, and frankly charming songs, the kind of score he had composed with Hart— the kind he *should* have written for *Me and Juliet*.

It was not an easy score to write as the story's emphasis constantly fluctuated because of frequent changes in the book. Weeks before the Boston tryouts, the plot metamorphosed from a story about a sensitive, transplanted Chinese man to a routine musical comedy-farce about two couples. This was because "the Chinese boy was not that good and not that masculine," says James Hammerstein. Anyone would have to struggle not to be upstaged by delicate Miyoshi Umeki, who had performed so brilliantly opposite Marlon Brando in *Sayonara* she had been chosen to play Mei Li. In her quiet way she dominated every scene. Her counterpart in the love triangle was belter Pat Suzuki. "The two girls were marvelous and things started to change—though it was not their intention." At last in Boston, the show-stopping song "Don't Marry Me" was added, and it somewhat redressed the masculine balance, but in sum, *Flower Drum Song* is a distaff show.

Finding an adequate cast of Chinese singing actors (or ones who could be made up to look Asian) was an inordinate task. Both Umeki and Suzuki are of Japanese origin. The two leading men were Hawaiian and Caucasian while many in the singing and dancing choruses[4] were Polynesian, Indian, and Caucasian. The balance of Asians was recruited from the West Coast. But that was in the faraway time before the middle '80s, when Actors' Equity began checking the ethnicity of casts. In truth, *Flower Drum Song* would be easier to cast in the '90s with the easy availability of so many fine singing and dancing Asian actors.

---

[4] *Flower Drum Song* was one of the last shows to hire a *separate* chorus of dancers. This was because the second act included Carol Haney's formulaic dream ballet (à la Agnes de Mille), in which the hero struggles to choose between his American or Chinese girl. By 1960 a big show-stopping ballet had become obsolete, and most chorus casting demanded people who could sing *and* dance *and even* act.

The show came in for more than its share of birth pangs, notably the replace-ment of leading man Larry Storch by Larry Blyden in Boston, and the addition of "Don't Marry Me" for the latter. But even before *Flower Drum Song* left New York for its month of previews in Boston, dance arranger Luther Henderson says he had no music for the crucial nightclub scene.

"Work out a ballet on anything you like," Rodgers told choreographer Haney and orchestrator Henderson, "and I'll write a number that conforms to the same rhythm."

"How about 'Limehouse Blues?' " said Henderson off the top of his head.

"Fine," snapped Rodgers.

"And that's what Carol and I did. Just before we went to Boston, Rodgers handed me 'Fan Tan Fannie,' whose rhythm fitted perfectly into the already choreographed slot."

When it opened at the St. James Theatre on December 1, 1958, *Flower Drum Song* was deemed a hit—but not a hit of the gigantic proportions to which the R & H public had become accustomed. In the final analysis, the show was a concession to popular taste by a team whose credo had always been not to concede.

No critic could fault the musical without giving credit as well: Richard Watts in the *New York Post* noted the "attractive songs, and much that is excellent in Hammerstein's lyrics," but added, "with all its Oriental exoticism it is aston-ishingly lacking distinction." Marya Mannes, writing in *The Reporter*, offered a prescient near eulogy for what would turn out to be R & H's penultimate show. "The Chinese in San Francisco cannot be quite so quaint as Rodgers and Ham-merstein make them," she said, adding, "it is very reminiscent, in a faded way, of Rodgers and Hammerstein.

But in spite of all the so-sos, *Flower Drum Song* managed an eighteen-month run on Broadway and a tour that lasted almost as long. It ran for a full year in the West End and was filmed three years later, with Miyoshi Umeki repeating her original role and featuring Nancy Kwan, whose first stage role, two years earlier, had been in *The World of Susie Wong*.

Kenneth Tynan, drama critic for *The New Yorker*, had been impressed by Miss Kwan's Broadway debut. When *Flower Drum Song* opened in New York in 1958 he wrote what is perhaps the most brilliant and coruscating two-line re-view I have ever read. He said, "Rodgers and Hammerstein have given us what, if I had any self-restraint at all, I would refrain from describing as a world of woozy song."

Hammerstein

1958–1960

MARY MARTIN, who had always been Hammerstein's favorite choice for leading lady, brought his next project, *The Sound of Music*, to him. Oscar and Mary had a relationship that went back to their first meeting in Hollywood in 1937. Laurence Schwab, who had co-produced both *The Desert Song* and *The New Moon*, suggested Mary might be right for a part in the Hammerstein-Harbach-Kern production of *Gentlemen Unafraid*, which was almost complete by that time. Schwab's motives were not totally altruistic, for on the strength of one hearing he had signed Mary to a contract at $150 a week to be the leading lady in his next show.

Now all the projects he was currently involved in had fallen through, and he was stuck with paying her then considerable weekly salary. When Schwab predicted his protégée would be overcome with awe and unable to perform if the neophyte were aware of her famous auditioners, Oscar asked Dorothy simply to invite the young singer to tea.

Miss Martin remembered being ushered into the living room, "where a lady and several gentlemen were having tea. The lady was elegant, attractive, cool and the personification of everything I ever dreamed of in a Noël Coward movie. The gentlemen were gentlemen."

After being offered a refreshment and without further introduction, Mary was asked to sing. She attacked a fast rhythm number at a furious pace, and then she was asked for a ballad. "I know one, but you probably don't know it," she ventured. "It's called 'Indian Love Call.'"

Nobody said a word until she had finished, and then a "tall craggy man who looked like a mountain" walked her to the door. "Young lady, I think you have something," Oscar said. "I would like to work with you on the lines and phrasing if you could come to my house every day."

Mary blurted a thank-you.

"And by the way," Mary swears Oscar added with that well-known twinkle in his eye, "I did know that song. I wrote it."

They were to remain fast friends, even though Hope Manning was given the role Mary auditioned for. It was a stroke of luck, for *Gentlemen Unafraid* closed ignominiously in St. Louis, while Martin went on to fame singing Cole Porter's "My Heart Belongs to Daddy" in *Leave It to Me*.

But Mary Martin's voice had so impressed Oscar at that fateful tea party that when he and Richard Rodgers were writing *Away We Go* she was the first to be offered the part of Laurey.

Having already been planning to star in Vernon Duke's *Dancing in the Streets*, which was being brought in under the aegis of Vinton Freedley, she tossed a coin. "I was deeply torn between loyalty to Vinton [producer of *Leave It to Me*] the man responsible for my then single Broadway success, and Oscar, who had believed in me for so long. So I tossed a quarter," she remembered, and the first Rodgers and Hammerstein collaboration lost.

Ironically, *Dancing in the Streets*, which was previewing in Boston at the same time as *Away We Go*, would fold there, while *Away We Go*, aka *Oklahoma!*, went on to—well, you know . . .

Mary went on to achieve superstardom in *One Touch of Venus* (1943) and *Lute Song* (1946). Her next professional involvement with R & H was when, as producers for *Annie Get Your Gun*, they sent her on the road with the show. They too needed a superstar, and Mary "delivered." She made the role her own, almost eclipsing Ethel Merman's flashy performance as Annie Oakley.

Of course that led to Martin becoming an even greater part of the R & H conglomerate from then on, and it is no secret that they tailored the role of nurse Nellie Forbush in *South Pacific* to her exceptional talent. But that was back in the beginning of the decade, and even superstars are constantly in need of new vehicles. By the middle '50s, when Mary's great success with *Peter Pan* had finally run dry, she began to look for a new musical.

Enter Vincent Donehue, a director with whom Mary was in total sympathy and who had directed her in the television version of *The Skin of Our Teeth*. The Thorton Wilder classic was a misguided effort sponsored by a U.S. State Department aimed at showing off American cultural acumen. Mary, never a straight dramatic actress, was miscast and at sea as the flamboyant Sabina, a role Tallulah Bankhead had wallowed in, but Donehue was a strong enough director to wring an adequate, almost sexy performance out of her when the show was adapted for the small screen. And Mary was eternally grateful.

They were to remain friends long after the government-sponsored turkey was forgotten. When Donehue was offered the directorship of a film based on Baroness von Trapp's life, he immediately thought it might lead Mary, who had only had a brief fling with the movies, to cinema stardom.

After the huge success of the German language version of Maria von Trapp's autobiography, Paramount had optioned the story, intending to star Audrey Hepburn. Donehue sent for a copy of the film and had it screened for Mary and her producer husband Richard Halliday in New York. As the teacher of a large group of children in a household run by an imperious, tyrannical man whom the heroine eventually democratizes and falls in love with, Baroness von Trapp's story parallels the hugely successful *The King and I*. Mary could not have been immune to the idea that this plum role had brought Gertrude Lawrence much acclaim, and certainly must have hoped that the premise would work as well for her.

Now there was no doubt that Mary was keen to appear in the role of the Baroness, leader of the singing family that had crossed the Alps to make a dramatic escape from the Nazis—but she felt it should be a play rather than a movie. Like the German-language film, the story was planned to be done with interspersed folk songs, Austrian ländler (although Mary confessed that even at that time she hoped to have one original quasi folk-song contributed by R & H.)

Audrey Hepburn soon demurred, and Paramount dropped its option. Donehue, Halliday, and Martin picked it up immediately and enlisting the help of producer Leland Hayward, went about the considerable task of securing releases from the seven von Trapp children, who were scattered all over the world. Then the producers hired the highly professional team of Howard Lindsay and Russel Crouse, who had fashioned librettos as varied as *Anything Goes, Red, Hot and Blue*, and *Call Me Madam*, to write the book,[1] and early in 1958 approached Dick and Oscar with a request for one or two songs to give their production a lift.

Both composer and lyricist felt that the mixing of Austrian folk songs with the pieces they might contribute could only create a hodgepodge. R & H offered to contribute an entire score, *if* they could co-produce. They admitted frankly that they were enchanted with the project and liked working with Mary. Leaving the onerous job of turning the treatment into a complete libretto to Russel Crouse, Oscar's friend and neighbor in Doylestown, and his partner Howard Lindsay, respected dramatist and actor, would take some of the burden off sixty-three-year-old Oscar, who dearly preferred working out a lyric to writing a libretto. But in the end, R & H said they were forced to decline because their hands were full now with *Flower Drum Song*.

"We'll wait," came the reply.

And so, a year later, in March 1959, with *Flower Drum Song* safely en-

---

[1] As was their habit, Lindsay and Crouse wrote a "treatment," a full outline of their story. They planned to insert the folk songs later and finally to complete the script with dialogue and stage directions.

sconced at the St. James Theatre embarking on what looked like a long run, and the sixty-page treatment of Baroness von Trapp's odyssey before them, Oscar and Dick began the creative task of "spotting"—finding the places where songs will be inserted. Sometimes a musical segment may make whole pages of the libretto unnecessary. At others a song can elucidate character that creates sympathy or antagonism. At still others, songs are inserted (frankly) because the plot is growing tiresome and music will be wedged in for variety. For example, in this show, the second act trio, "There's No Way to Stop It," is more effective than dialogue could possibly be. It shows the Captain's disdain for the Nazis while Elsa, his fiancée, and Max, her friend, are all for capitulating. Again, in the musical's climactic scene, the seemingly naïve "Edelweiss" (which was incidentally the last lyric Oscar Hammerstein wrote), shows von Trapp's sensitivity and love of homeland.

It was not in Oscar's make-up to keep earlier drafts of his work.[2] He often said no one is interested in seeing how much labor goes into a good lyric, only the fruit of that labor. In this, his last show, perhaps because he was collaborating with two other punctilious wordsmiths, noted for hoarding earlier lines that might come in handy at final rehearsal when nothing satisfies, he maintained an extensive file. These early sketches form a wonderful record of the painstaking metamorphosis his words traveled from an idea to a complete poetic entity.

The show's first song originally was called "Summer Music" and then "The Sound of Summer." It began with "The hillside is sweet / Today the air is sweet with summer music." Eventually "summer music" was replaced by "The Sound of Music." Missing in this early version is Hammerstein's much criticized line about a lark learning to pray. But one must remember that Oscar was writing a poem that would be musicalized—changes would be demanded not only by the composer, but in this case by the librettists.

The A^3 of this song was quite different in early draft, but one can understand that the section needed shortening to fit with Richard Rodgers's 16-bar melody. In the ultimate version the single word "blessed" stresses Maria's religiosity and plays down her confusions. Since this is our heroine's opening song, it instantly defines her character. The two versions are printed below.

ORIGINAL VERSION

The hills give me strength
When my heart is lonely,
And lost in a fog

FINAL VERSION

I go to the hills
When my heart is lonely,
I know I will hear

---

[2] Nor, for that matter did Alan Lerner. The only top-notcher who held onto every scrap, sketch, and doodle seems to have been Cole Porter.

Of a thousand fears,
The hills fill my heart
With the sound of music
And my heart wants to sing . . .
Every song it hears.

What I've heard before.
My heart will be blessed
With the sound of music
And I'll sing once more.

The next song, "Maria," had to carry the plot forward. In this list song, which ends with the decision to send Maria out into the real world, Oscar's notes show his struggles analyzing Maria's character; as he put it, "How do you find a word that means Maria?" The margins of his notes show lists of words like "gamin, hoodlum, bother, calamity, disaster, imp, pest, shrew, brat" before he chose "a flibbertigibbet! a will-o'-the-wisp! a clown!" Later in the song the Sisters interrupt each other in their description of Maria:

Unpredictable as weather,
She's as flighty as a feather,
She's a darling,
She's a demon,
She's a lamb.
She'll outpester any pest,
Drive a hornet from its nest,
She could throw a whirling dervish out of
   whirl.

She is gentle,
She is wild,
She's a riddle,
She's a child,
She's a headache,
She's an angel,
She's a girl!

More than two weeks after he had begun the song he was unable to come up with a last line to answer his opening question, "How do you solve a problem like Maria?" The answer is: *there is no answer.* Realizing that, Oscar penned one more ephemeral and exquisite question: "How do you hold a moonbeam in your hand?"

Although Lindsay and Crouse did come through with some exciting theatrics, especially in the second act, one can notice almost from the outset how frequently they fell back on cliché. (Perhaps that is what inspired Oscar and Dick, using the very simplest means, to outdo themselves, hoping to bring freshness to words and music.) Since the '20s one hasn't seen a more creaky lead-in than the one Lindsay and Crouse wrote for "My Favorite Things." Hack musicals of the distant past would introduce a number by having the ingenue or leading-man say, "Remember that old song we used to dance to? How did it go?" And the song would follow. Here is the way L & C handled the musical's third song:

MOTHER ABBESS: And the other day you were singing in the garden at the top
   of your voice.
MARIA: But Mother, it's that kind of song.
MOTHER ABBESS: I came to the window and when you saw me you stopped.
MARIA: Yes—that's been on my mind ever since it happened.

MOTHER ABBESS: It's been on my mind, too. I wish you hadn't stopped. I used
   to sing that song when I was a child, and I can't quite remember—Please—
MARIA (sings): Raindrops on roses and whiskers on kittens . . .

The number itself, another typical list song, gave Oscar far less trouble than
"Maria," for here the lyricist simply needed to put into rhyme all the homey,
ordinary paraphernalia of life. This song has often been singled as being representative of Oscar's "kitschy" side. Its moral—when bitten by a dog or stung by
a bee, one must think of these favorite things in order not to feel bad—seems
creaky indeed. Yet one must not forget the dramatist in Oscar was trying to
contrast Maria's love of the homespun with von Trapp's mistress's devotion to
the luxurious. One should not overlook Rodgers's imaginative musicalization,
which begins in a doleful minor mode and switches to a bright major for the
moral. So fresh is this number's harmonic and melodic design that "My Favorite
Things" has become a favorite improvising canvas for jazz musicians.

Mary Martin had always played wonderfully opposite children. Now, her first
encounter with the von Trapp brood had to be a tour de force. "Do-Re-Mi" is
a superb singing lesson and because the von Trapps *did* eventually become a
noted singing group, the most important song of the score. In addition to all
that, it is sheer fun.

Starting with the introduction of the pitches of the scale (syllables chosen
since antiquity because they were the initial syllables of the Latin Mass), Maria
comes up with her own mnemonic device by connecting each one to a common
homophone: Do, a deer; Re, sunlight; Mi, myself; Fa, far—as in running; Sol
(pronounced So) brings up the image of sewing. On reaching La, any other
lyricist would have stopped and realized the scheme was not going to work, but
Hammerstein insouciantly brushes the problem aside, defining La merely as "a
note to follow so(l)!"

When the children ask what are they to do with these pitches now that they
have learned them, Maria shows them how to substitute words for the notes to
create a musical expression. Ultimately Maria, or more certainly Dick Rodgers,
shows the children what magic they can create by mixing up these few pitches,
and soon they are singing a counterpoint to their and future generations of
audiences' perpetual delight.

From what we have heard thus far, anyone would assume *The Sound of Music*
to be a one-woman show written to accommodate Mary Martin—and one
would not be far off. What Lindsay and Crouse had created was, of course,
operetta, which so often is built around a strong single character. What they
left out was passion and youth. (Even though we have seven young people with
their attendant flirtations, there is nothing like the passion R & H had produced in musicals from *Oklahoma!* through *Pipe Dream*.) Had Oscar and Dick

Theodore Bikel blows his whistle to convene his seven children. Even Mary Martin, their new governess, stands at attention. MOCNY.

been entrusted with the libretto, they might have created one of their signature pieces—the musical play with true ardor. Technically, we certainly would have been introduced to the Baron and his wealthy Austrian fiancée Elsa earlier.

Lindsay and Crouse's plot will dismiss Elsa early in the second act, and the Baron remains a cipher throughout most of the play, but while they hold center stage these two are in direct contrast to Maria. The lyric of their duet, "How Can Love Survive?," brings out Hammerstein's most masterful use of the irreverent. The premise being the *opposite* of romance, the items mentioned are the

Left alone with the youngsters, Mary Martin relaxes the rules and teaches them to sing "Do Re Mi." MOCNY.

antitheses of Maria's treasured favorite things, the copper kettles and whiskers on kittens, etc. The lyric works as direct contrast. With no starving lovebirds in an attic, Oscar could be as snide as Cole Porter.

No rides for us
On the top of a bus
In the face of the freezing breezes—
You reach your goals
In your comfy old Rolls

Or in one of your Mercedeses! . . .
No little cold water flat have we,
Warmed by the glow of insolvency—
Up to our necks in security,
How can love survive?

But beyond Maria's not belonging in the Baron's economic circle, her realization that she is in love with him is enough to frighten her into retreating to the abbey. Fortunately, the Mother Abbess lets her know that the cloistered walls are no place to hide, and the act ends with "Climb Ev'ry Mountain."

Of the song's genesis, William Hammerstein remembered his father being

moved by his deep feeling at the injustice done the von Trapps which forced
them to leave their homeland:

> One day dad and I were standing in the swimming pool out at Doylestown and
> I asked him to tell me more about *The Sound of Music.* So he told me the story
> and it was very touching. And here I was standing in the middle of the swimming
> pool with my father crying. And then he came to a place where he said, "I have
> to write a song like 'You'll Never Walk Alone.' " And he did—but he hadn't
> written it yet."

The lyrics of "Climb Ev'ry Mountain," relying heavily on Oscar's "dream"
crutch, are certainly not astonishing. But the musical and harmonic design,
being one of Rodgers's most brilliant inventions, deserves some analysis. Here
Rodgers has carried off the "hundred million miracles" that he only talked
about in his previous show. First is range. "Climb Ev'ry Mountain," which
sounds almost operatic, seems to encompass far more than its actual range. We
have to look at the page to assure ourselves it is only an eleventh.

Then there is the remarkable opening motive beginning on the mediant and
immediately stepping up a *whole tone,* out of the scale. This gives an emotional
lift as well as a certain religiosity to the line. The III and VI chords (Schubert
used them liberally in his "Ave Maria") add their unique minor color to Rodgers's theme, but the great tour de force occurs in the release. In this bridge
Rodgers imperceptivly modulates to a higher key.

Although Hammerstein repeated the same lyric for the A² and A³ sections,
the great difference is that Rodgers's modulation has succeeded in bringing his
returning melody back now a perfect fifth higher (see Examples A and B
below). Pulling all the stops out, Rodgers intensifies the final line with one
of the strongest contemporary cliches: the I, I+, IV, V. (See asterisked chords
in Example B below.) Altogether it makes for a breathtaking ending to the
act.

Example A

Example B

If "Climb Ev'ry Mountain" shows Rodgers at his best in composing music for a quasi-religious popular song, it must be added that he faced an awesome challenge before writing the considerable choral music needed for the opening scenes of *The Sound of Music*. Setting the Catholic prayer "Dixit Dominus," the morning hymn "Rex Admirabilis," and the "Alleluia" for the nuns made him apprehensive.

"Given my lack of familiarity with liturgical music, as well as the fact that I was of a different faith, I had to make sure that what I wrote would sound as authentic as possible," Rodgers notes in his autobiography *Musical Stages*. "So for the first time in my life I did a little research."

He contacted Mother Morgan, head of the music department at Manhattanville College, who arranged a concert at which Rodgers might hear many different kinds of religous music.

Mary Martin, too, besides spending time with Baroness von Trapp, contacted members of the church to authenticate her performances. A nun she came to know when touring with *South Pacific* became her consultant. Martin had much advice from Sister Gregory of Rosary College, Chicago. Sister Gregory, a tall authoritative personage, was not beyond counseling Rodgers, Hammerstein, and director Donehue on how nuns should act. It was she who lightened Oscar's approach to the lyrics. "Don't make nuns sanctimonious," she commanded.

The simplicity and accessibility of the score of *The Sound of Music* is typical of Hammerstein's best work. Although this writer does not agree with him, Oscar often said that he had not a large vocabulary. What he meant was that he did not use a fancy word when an ordinary one would do. He most admired Winston Churchill's writing style, clarity, and lack of obfuscation. In his own work, he would spend days searching out exact words to convey the precise meaning of even his smallest thought.

A typical example of his artless simplicity is "An Ordinary Couple," the second-act duet that was added in Boston to try to give the Baron, in the person of Theodore Bikel, a firmer profile. Although Martin had good reason to complain that the melody of this last-minute number "goes downhill," she well appreciated the purity and feeling of the lyric. It is a complacent love song, an old-person's love song, with lines like "in the fading sun" or "our children by our side" in the vein of (but not so good as) "The Folks Who Live on the Hill."

The next number, "Sixteen Going on Seventeen," is a naïve love song that is reprised at the end of the play by a suddenly wiser Maria. In its reappearance, its opening lines,

> A bell is no bell till you ring it,
> A song is no song till you sing it,
> And love in your heart wasn't put there to stay . . .
> Love isn't love till you give it away

fairly well encapsulate Oscar's philosophy, which equates generosity and selflessness as the essence of loving. Because of *Flower Drum Song* and *The Sound of Music*'s overly romantic plot, much of Oscar's output was written off by critics as sentimental. Yet his lyrics expressed true sentiment without going over into bathos. That's the kind of man he was.

An early song, a charming sequence for Maria comforting the children during a summer thunderstorm, is "The Lonely Goatherd," and although the number bears absolutely no relationship to the plot, it is so winning that it invariably works in even the most amateur production of this show. What we need here is a bedtime story or perhaps a number like "(Whenever I feel afraid) I Whistle a Happy Tune." But because of Oscar's technique and Dick's Swiss yodel, the number works here just as well. It's like a Hitchcock movie wherein the flaws in the plot are only noticed when you leave the theater and replay the film in your mind.

By the end of August when rehearsals began, the box office had chalked up a record two-million-dollar advance, and there seemed little to fix in the score and plot. When Oscar went for his routine annual checkup in mid-September, his doctor Ben Kean gave him the usual tests, all of which indicated Hammerstein was in robust health. Just before Oscar left the office, Dr. Kean asked if there was anything else he hadn't reported.

"Oh, yes," Oscar said, "I've been awakening in the middle of the night hungry. I take a glass of milk and then it's fine."

Just to be on the safe side, Dr. Kean asked Oscar to come back the next day so they could take some tests to determine if he had an ulcer. Oscar did indeed return for X-rays, and the following day the doctor gave him the report of the lab's findings. It was not an ulcer, but cancer.

Dr. Kean immediately reserved a room at New York Hospital and scheduled surgery for the following day, September 19th. Dorothy and the children sat tensely in the glassed-in patients' lounge through the long operation. At last the doctors emerged with the sad news: the carcinoma was Grade IV, which meant it was so advanced that they had to remove three-quarters of the patient's stomach. Even with such radical excision they predicted Oscar would be dead in six months.

Since chemotherapy was then just in its infancy and it was not certain to be helpful in treating the disease when it was in such an advanced state, it was Dorothy's decision that they would wait until they had talked to another cancer expert before telling Oscar the whole story.

Ten days later, after he had made a remarkable recovery from the operation itself, he asked Dorothy why the doctors hadn't told him what they did. "They just cut it out. They cut it all out," she replied.

The Sound of Music's rave review after its first New Haven performance on the eve of Oscar's release from the hospital was the best tonic any patient could have. After ten days at home, Hammerstein felt well enough to travel to Boston, where the show was previewing to great acclaim. There was very little tinkering to be done, but Oscar suggested some changes to the librettists that would strengthen the book by deepening the Trapps' grief over leaving their native Austria. This was helped by the addition of "Edelweiss," which he had begun to write a few days earlier. (His earliest sketches for the song's release reflect his preoccupation with his illness. They are: "Look for your lover and hold him tight / While your health you're keeping." Fortunately they were replaced with a more universal, "Blossom of snow may you bloom and grow / Bloom and grow forever.")

The show opened in New York on November 16th to good notices but not unqualified praise. While noticing its "handsomeness" and "substantiality," Walter Kerr in the *New York Herald-Tribune* called it "too sweet for words but almost too sweet for music." Most of the critics for the workaday papers mentioned the professionalism of Mary Martin's performance and the aptness of the score. But the reviewers for the intellectual journals had a field day lambasting the show's saccharinity. Kenneth Tynan, who was usually an R & H champion, wrote, "It is a show for children of all ages, from six to about eleven and a half." Louis Kronenberger called it "as cloying as a lollipop, as trying as a lisp," while Henry Hewes said, "It emerges as an ideal Easter show for Radio City Music Hall." But none of the reviews dampened the public enthusiasm; they came in droves throughout *The Sound of Music*'s 1,443 performances.

The film version did not appear until 1964. It was a huge success—indeed, the most popular film that had been released up until that time. The film benefited from Robert Wise's imaginative direction and the spectacular scenery of Salzburg, but most of all from the enchanting Maria of Julie Andrews.

Oscar, Dorothy, and the family spent the Christmas holiday in Jamaica, and during that time Oscar was feeling well enough to work on a screenplay of a remake of *State Fair* as well as a revision of *Allegro* that was planned for television.

Throughout the spring of 1960, Dr. Kean monitored Oscar's condition, which outwardly seemed not to have changed much, and began giving him liver and vitamin shots to bolster his ebbing energy.

James Hammerstein, who had played tennis with his father since childhood, recalled Oscar's fierce competitiveness as well as his father's frustration and his eventual determination to adjust to his waning powers.

> It was awful playing tennis with him the first time, knowing that he was fatally ill, but we played. He said, "Come on let's play." I went down to the farm in Doylestown and I tried to look like I was playing a bit . . . and he quit in the middle of the set and said, "I'm tired," and he went up to take a shower and didn't mention anything. I came back two weekends later and he'd leveled another part of the farm, sodded it and he had a croquet court put in and was playing a marvelous game of croquet. He had to have his game. The adjustment to his not being able to play tennis was so swift. Here's a guy who played tennis all his life. Boom. He couldn't play tennis. OK. Let's play croquet.

Two months later, in early July, Dr. Kean showed his patient the recent X-rays which indicated the carcinoma was now advancing. He said their only chance would be to administer heavy doses of chemotherapy. Oscar said he had considered the matter carefully and had decided that he would prefer not to spend his last months in the hospital but to die possibly a little earlier on his beloved farm in Doylestown with his head "on Dorothy's pillow."

Once he had made the decision he jotted a few notes in a diary that indicate he was not yet ready to go. They are part of his last journal entry:

> Today is July 12, 1960, my birthday, I am sixty-five. This is the accepted age of retirement. I do not want to retire, am in no mood to retire. This is considered a good time to come to a stop. Perhaps it is, but not for me. . . . Some day I may leave the theatre, but I couldn't walk out suddenly. I would have to linger a while and take a few last looks, I would have to blow a few fond kisses as I edged towards the stage door . . .

Now Oscar began to say his farewells. On that very day, before his birthday dinner he inscribed copies of a handsome photograph of himself which he gave

to each member of his extended family—always including Stephen Sondheim.[3] He asked each of his children to come and talk to him before the pain took away his powers to concentrate. James remembered that he "went to his study. He talked to us all individually and thanked us for keeping his secret. We all knew of the cancer long before he did." When James, his youngest, burst into tears at his father's hopeless situation, he could only offer a typical Hammerstein outburst. "God damn it!" he yelled. "I'm the one who's dying, not you."

But he did not allow the shadow of death to keep him from being driven to New York and his frequent business meetings with Rodgers at the office or in a restaurant. In his autobiography, Rodgers recalled one of their last lunches at the Plaza Oak Room. Oscar, who ate very little now, was pecking listlessly at his food when a man at a nearby table came over to ask them to sign his menu. "You're the most successful team on Broadway," he said. "Tell me, why do you both look so sad?"

By the beginning of August Oscar could no longer go to New York and spent most of his time in his bedroom. Dr. Harold Hayman, his old friend, now retired, came to the farm to be near him. Oscar had become so thin his adopted daughter Susan said his gray, bony hands atop the white coverlet reminded her of an El Greco. Heavily sedated, he sipped sarsaparilla and ate almost nothing now, but he did come down to watch a sporting event when the family gathered for the weekend.

On Monday, August 22nd, Dorothy was with him when Oscar became delirious and began reciting the names of baseball players from the past. She called Harold Hayman, who came and sat with Oscar. Then he told Dorothy to leave the room. Ten minutes later Oscar was dead.

"He died with dignity," James was to say recently. "He was a terrific guy, very funny, and he had great sense of humor. Very warm. He was not much of a father, he had no particular talent for it. Children just perplexed him somehow but he was very close to my mother. They were very much in love."

On the following evening the lights all over Times Square were dimmed for a minute of reverence for this great theatrical spirit.

---

[3] On Sondheim's copy of the photograph Oscar wrote, "For Steve, my friend and teacher, Ockie."

Hammerstein was over six feet tall. Most of the time, instead of getting up and sitting down he preferred to pace the room while writing his lyrics. At other times he used his knees as a desk. Both pictures were taken in his study at his beloved Bucks County farm. (R & H).

Lerner

1958–1963

IN THE SPRING OF 1958, shortly before Alan and Fritz were to leave for London to prepare for the British production of My Fair Lady, they were honored at a testimonial dinner at The Lambs. They were the second most successful team in musical theater, and it seemed a propitious time for many of their colleagues in the organization that had brought them together to "roast" them for having risen to the very pinnacle of success in the relatively short time of sixteen years. But the hosannas and friendly jibes of their peers were little preparation for the adulation that was to greet them abroad.

Gigi's sweep of the Academy Awards had only added fresh fuel to the excitement that surrounded the London opening of My Fair Lady on April 20, 1958. With minds at ease while Bud Widney "minded the store," seeing to the Broadway company, Alan, Fritz, and Moss, accompanied by the three original stars of the New York company who were coming home—Rex Harrison, Julie Andrews, and Stanley Holloway—went abroad to set the show on its feet.

Forgetting that Shaw was Irish, and feeling as though the Yanks had come across to return their native playwright, Londoners greeted this premiere with more excitement and furor than even the New York one had inspired. Now it is commonplace, but it was a first then, that a theater, the prestigious Theatre Royal, Drury Lane, had been closed for three months to allow for installation of the two enormous turntables needed for the revolving scenery. Never before had a British box office offered tickets six months before the first night. The excitement had even spread to Buckingham Palace where, it seemed, even the Royals couldn't wait to see this phenomenon. A command performance of My Fair Lady before Queen Elizabeth, Prince Philip, and their party had been scheduled only five nights after its debut.

Hugh "Binkie" Beaumont, managing director of H. M. Tennent, the producing firm that dominated the West End, had orchestrated such a splendid publicity campaign that the theater was forced to hire a *dozen* extra clerks to cope with the sacks of advance orders that poured in daily. He filled British papers with stories and photos from the Broadway production, whipping up the interest to a fever pitch as the opening night approached.

The press cooperated with Beaumont, calling the first performance "The Opening of the Century." The *Daily Express* had been announcing FIVE MORE DAYS . . . FOUR MORE DAYS . . . THREE MORE DAYS, etc., until April 30, when the headlines simply screamed . . . TONIGHT.

So successful was Binkie's campaign that although the pre-production costs had amounted to a then staggering £85,000, the show was well into profit five weeks before its opening.

Beaumont, who, it will be remembered, had traded Rex Harrison's release from his *Bell, Book and Candle* contract for the British and Continental rights *plus* 1½ percent of the gross *plus* £25,000 (about $60,000 then), had also "persuaded" the Society of British Authors (who controlled the Shaw estate) to prohibit performances of *Pygmalion* everywhere in the UK for a full ten years. In an aim to keep the songs fresh, he had also convinced Columbia Records to withhold their best-selling record from Britain until after the show opened, and although the songs had made their way across the Atlantic, the cast album was only available on the black market (where it was doing a brisk business.)

My Fair Lady was greeted with a hysteria that outdid even the madness surrounding *Oklahoma!* eleven years earlier, one that would not be repeated until the sell-out shows of Andrew Lloyd Webber. Richard Huggett's excellent biography, *Binkie Beaumont*, gives us an idea of the delirium that gripped London on that day.

> The streets from Drury Lane down to the Strand were lined with so many thousands of people waiting to see *the audience* that it seemed more like a coronation than a theatrical first night. Even by comparison with other grand first nights this was something quite exceptional . . . two black market tickets went for 200 pounds, not to a Texan oil millionaire and his wife, as might be supposed, but to a retired grocer from Bury St. Edmunds who had realised his entire stock of war bonds to provide himself with the necessary cash. At one hundred times their face value he considered he had got a bargain.

Typical of the stories that arose surrounding the legendary run of *My Fair Lady* is one that occurred during the first week of the run which was recounted by the house manager of Drury Lane. He reportedly noticed something so amazing he couldn't believe his eyes. An empty seat in the stalls. Sitting next to it was a middle-aged woman. In the interval he approached her and asked if the empty seat belonged to her.

"Yes," she said sadly, "my husband was to have come with me, but he was killed in a car accident and so he couldn't make it."

"But it's a pity to waste a good seat for this show," said the manager. "Couldn't you give it to another member of the family, or perhaps to a friend?"

"Yes, I suppose I could have done that," said the woman, "but you see, they're all at the funeral."

Alan always referred to Bud Widney as his "production manager" but justified the phrase by adding, "It is like calling a bottle of Dom Perignon a cold drink." Lerner was used to talking things out to a sympathetic ear, provided there was someone's he felt was compatible. "I always felt it with Fritz, later with Moss Hart and always with Bud. He has always understood me well enough to know what I have been searching for, and the kind of theater I am trying to create."

Widney recalled that after the success of My Fair Lady and Gigi:

> I saw a review of The Once and Future King in The New York Times and said, "that's Alan's next musical." What does he do to follow My Fair Lady? He's done the same show twice now. It's got to be something very different. He's struggling, also competing with Rodgers & Hammerstein's major, big shows. So that the notion of doing a big epic kind of show seemed like a different tack for him to take. I liked that aspect of it, I also liked that its hero . . . seemed to me to represent the kind of idealism that would be embraceable, meaningful and extraordinarily useful to have in the world. We had been through a bloodying of nationalistic fervor during and after World War II and the times were ripe for an international statement. The hero of the day was Dag Hammarskjöld and the inspirations and hopes of the United Nations were fresh and budding.

Although Widney conceived the project as an epic, serious show, he and Lerner were aware that the story of King Arthur had already been a Broadway success. In 1927, Mark Twain's novel, A Connecticut Yankee in King Arthur's Court, as adapted by Herbert Fields, had featured a winning score by Rodgers and Hart. It had been revived as a final plum for Lerner's old mentor and idol, Lorenz Hart, in 1943. Hart had embellished the show with his masterful final song, "To Keep My Love Alive."

But Widney was thinking of Camelot in far different terms. The story he suggested spurned the joke of anachronism which had fueled so much of the hilarity of A Connecticut Yankee. There would be nothing like "Thou Swell" or "On a Desert Island with Thee." His concept of Camelot omitted none of its tragic overtones.

The Once and Future King, T. H. White's sprawling 650-page book, is written with acerbic wit and a contemporary slant that, as he listened to Bud (and later the same day to Moss), appealed mightily to Lerner. "I've always wanted to do the Arthurian legend," he told Bud, asking him to make a synopsis at once, because "you could never get Fritz to read the whole thing."

The work is actually four novels in one. The first part deals with Arthur's youth and Merlyn his tutor. It climaxes when Arthur wrests the sword from the stone and is crowned King of England. The other three novels that make up the tetralogy deal with Arthur's subsequent life, Guenevere, his marriage, Lancelot, the establishment of an idealistic society, and its eventual downfall.

When Lerner told Loewe that he wanted to option T. H. White's book so that he could tell the entire story of Arthur in contemporary terms, his collaborator disagreed violently.

Fritz's answer was, "You must be crazy," Lerner reported in his autobiography. "That king was a cuckold. Who the hell cares about a cuckold?" Lerner answered that people had been caring about Arthur for over a thousand years. "Well," Loewe rebutted with the hauteur of the Viennese, "that's only because you Americans and English are such children."

Alan's first move was to contact White who lived on Alderney, one of the Channel Islands, to secure the rights. Disappointment set in when White's agent informed them—Alan, Fritz, and Moss—that the rights to the first book had been presold to the Walt Disney organization who planned to use it as the basis of a feature-length animated film.[1] Whether it was sour grapes or honest belief we will never know, but doggedly determined that the latter part of the *Once and Future King* would make the adult musical he craved, Lerner optioned the last three books of the series and convinced Moss Hart that the story should deal with "the creation of civilization." Adding that the relationships to be found in the characters of Arthur, Guenevere, and Lancelot—a provocative triangle—would make an exciting theater piece, he asked Moss to help him adapt the story. This was a great departure for Alan, who always preferred to work out his libretto alone.

Soon he and Moss were able to convince Fritz into believing the tragic story was preferable to the one about a not-so-bright young man's picaresque adventures, and he too signed on for the project. Fritz, now completely recovered from his heart attack of the year before, joined only on condition that he was not intending to "hurl his energies around the way he did formerly." He elicited a promise from Alan that when winter came, Lerner would come to Palm Springs, the Southern California spa, to work with him. The three men, rolling in money and flushed with the success of *My Fair Lady*, rented offices in New York's Plaza Hotel and decided to produce the show themselves.

[1] The film, *The Sword in the Stone*, released in 1963 (three years after the opening of *Camelot*), recounted the adventures of the young boy who eventually became King Arthur. It was one of Disney's weakest efforts.

It is about as impossible for a creator of a Broadway musical to be peripherally involved as it is to be (according to the old one-liner) a little bit pregnant. *Camelot*, because of the tremendous demands it made on the creative people it involved, holds the record for bad luck.[2] Its costume designer, Hollywood's stalwart Adrian, would die before the show went into rehearsal; while the show was previewing Alan would be hospitalized with bleeding ulcers; and upon Alan's recovery Moss would have a heart attack. The husband of the wardrobe mistress was found dead in their apartment. A chorus girl ran a needle through her foot onstage, and the chief electrician would land in the medical center with a bladder problem.

Bad luck aside, some of the misfortunes that plagued *Camelot* might have been averted had they not, as Lerner put it, "opted for greed" and decided to dispense with the producer, but none of that was apparent to the hopeful triumvirate at the time.

The Harts were taking a holiday in Jamaica when a letter arrived from Alan, saying he had changed his mind; he preferred to write the libretto alone. He wanted Moss only to direct it. According to Kitty Hart, "Moss was bitterly disappointed; he had looked forward with so much pleasure to the collaboration. I knew I had to find something to say to soften the blow. To play for time, I suggested we go for a swim. Standing in the warm Caribbean, I told him, 'You're well out of that. There's no way *Camelot* can ever be as good as *Fair Lady* and you'll be the one who gets the blame. Just do what you did on *Fair Lady*, direct it, and help as much as you can.' "

When they were back in Beach Haven, Moss came into Kitty's bedroom and tossed the play on her breakfast tray, asking her to read it at once. Her opinion was that Fritz was right; the first part was indeed lighthearted and charming, but that the play had no second act.

"There's no menace," she said, adding, " 'you'll have to use Mordred as the villain to create the tension.' I proceeded to ad-lib a whole story line for him. . . . No doubt Alan would have come to this by himself, it was so obvious. But it pleased me to think that Moss had liked my idea well enough to pass it on."

Indeed he had passed it on, and it was incorporated into Lerner's second draft. Now, crafty Mordred, Arthur's illegitimate son by his sister, conspiring to keep his father away on an overnight hunting trip, enlists the King's own palace staff to compromise and expose Lancelot and Guenevere. This twist of the plot magnifies their love into adultery (although the audience never sees more than a fleeting embrace) and leads into Guenevere's near-immolation,

---

[2] Perhaps *42nd Street*'s tragedy, the death of its brilliant director Gower Champion on opening day, is its only lamentable comparison.

which will eventually send Arthur's Camelot crashing around his ears. These machinations having several "fatal flaws" are the stuff of high tragedy. As the champion of justice rather than violence, Arthur's dilemma is: "Do we save the Queen or do we save the Law?"

Although she may not have believed in the plot, Kitty Hart's story improvisation puts the whole second act together and gives a raison d'être for much of the first. Now Lerner was able to work out the story of the second act. He created a masterful near-Shakespearian battlefield ending for the act during which Arthur forgives both Lancelot and Guenevere, but still couldn't solve the last scene.

"This is where I made a contribution," Bud Widney recalled.

At the end of the play, the business of pulling these different factions together and creating a round table and the idea of establishing rule by law had totally dissolved in warfare. Alan said, "I cannot end a musical that way." When he called on me for help that's when I remembered . . . the story of the little boy.

I think T. H. White, when he said Tom of Warwick, meant it to be Thomas Mallory, who picked up the legend and carried it on and let the world know about the idealism. The important thing is not that we are now killing each other because of human frailty, the important thing is that an idea has been born. Some day we can get together and somehow solve this—the once and *future* king. So that's the thrilling part of it. When Alan had that, he was off to the races and said, "We'll do it that way. It's the only way to come up out of that pit."

With the plot now craftily worked out, the only elements left to settle were the attributes of the characters. Alan hewed closely to T. H. White's description of Arthur,[3] creating an "all-pervasive innocence in order to make him so ingenuously charming that one would fall in love with him, forgive him anything . . . even over the hot coals of tolerating his wife's infidelity."

Moss Hart thought of Richard Burton for the role from the beginning. He had tailored his screen adaptation of *Prince of Players*, a biography of Edwin Booth, to the British actor with the sensitive manner and the voice which was one of the most distinctive in the theater. Both of them knew how attractive Burton was to women and approved of his singing when they heard him do some of his native Welsh songs at a party in Hollywood. "Even if I had not," Alan added, "we would not have been concerned about his voice . . . At birth, when a Welshman is slapped on the behind, he does not cry, he sings

[3] Arthur is called "Wart," a farfetched nickname White invented. In the novel it was invented by his chum, Sir Kay, and is supposed to have come from "Art" (short for Arthur) and pronounced to rhyme with "cart." When mentioned on stage it rhymed with "fort," the whole point of the jibe being lost.

'Men of Harlech' in perfect pitch." Even before Burton was auditioned, Julie Andrews, now a star, who was then appearing in the London company of *My Fair Lady*, was the model for Queen Guenevere.

Guenevere became a complex, almost modern woman, albeit still dressed in her medieval trappings. When we first find her, she is angered at being forced into a marriage to a man she has not picked herself. She is willful and coquettish, and her opening number, lightheartedly reflecting the everyday violence that pervaded the Dark Ages, asks: "Where are the simple joys of maidenhood? / Where are all those adoring daring boys? / Where's the knight pining so for me / He leaps to death in woe for me? / Oh where are a maiden's simple joys?"

But Lerner's play is spread over thirteen years, and by the final curtain Guenevere has become wise and contrite and even realizes her enduring love for Arthur as she becomes aware of the havoc her infidelity has caused. Her final speech is a moving plea for forgiveness in and out of the theater.

> GUENEVERE: So often in the past, Arthur, I would look up in your eyes, and there I would find forgiveness. Perhaps one day in the future it shall be there

The first Camelot *reading with the cast. Fritz Loewe is at the piano while Alan, in dark glasses, reads the script and sings the songs. At the table are the other creative members of the staff, while the "angels" sit behind them.* MOCNY.

again. But I won't be with you. I won't know it. (*He holds out his arms. She goes into them. As she withdraws, she looks up into his face*) Oh, Arthur, Arthur, I see what I wanted to see.

Of the three leading characters, Lerner's major departure from White's book was his depiction of Lancelot, painted in the novel as "looking like an African ape, with a face as ugly as a monster's." By contrast, Lerner introduces him as "a striking figure of a man with a stern jaw and burning eyes." In *Camelot* the role is only slightly developed, and Lance comes out as a somewhat good-looking buffoon, rather on the order of *My Fair Lady*'s Freddy Eynsford-Hill. The eventual casting of Robert Goulet, a handsome figure with a stentorian voice but a wooden actor, only served to trivialize the role.[4] Introduced with the preposterous "C'est Moi!," his other number is the hit song of the show, "If Ever I Would Leave You," far superior in quality but often compared to Freddy's "On the Street Where You Live."

In *Inventing Champagne*, Gene Lees points out that Guenevere's character is somewhat weakened by her love of such an "incomparably beautiful but egotistical and shallow a prig." He is quite right in saying that "both Arthur and Guenevere are diminished because we are always defined by what (and whom) we love." Yet in Alan's defense it must be added that since he was already flouting convention by creating a musical tragedy, he sought the insurance of the magnetic attraction. The beauteous Queen falling for the handsome Knight is the stock-in-trade of the Broadway musical. Although Hammerstein had walked into dangerous waters with the miscegenation scenes in both *Show Boat* and *South Pacific*, it was still some years before musicals would touch sensitive themes. Ethnic cleansing, apparent in both *Fiddler on the Roof* and *Cabaret*, and homosexuality as seen in *La Cage aux Folles* and *Kiss of the Spider Woman*, as central themes of musicals were years down the road. Adultery was bad enough, but cuckoldry unaccompanied by a strong sexual desire would have mystified a Broadway audience.

Although Lerner had used the theme of two men who care deeply for each other and who are in love with the same woman in *The Day Before Spring*, *An American in Paris*, and *Paint Your Wagon*, he had never before approached it with such seriousness of intent. Over the years since *Camelot*'s debut this ménage à trois has elicited much psychobabble, perhaps because Alan was honest enough to hew to T. H. White's lines, which state, "Lancelot . . . had been thinking of King Arthur with all his might. He was in love with him."

---

[4] Eventually Goulet and *Camelot* became so intertwined that in 1993 he was able to tour with the show. After three decades he had grown into the role of Arthur himself.

Later in the novel, Lancelot becomes more explicit and tells the Queen, "What you want is two husbands."

Alan was safer in his stock comic male bonding. His depiction of Sir Pellinore and his slavish devotion to Arthur resembles Pickering's relationship to Higgins or perhaps more closely, Sancho Panza's tie to Don Quixote. Here one could say Alan was following a low-comic Broadway tradition.

Much of the passion that Lerner wrote into *Camelot* came from his turbulent relationship with Micheline. Having sired three daughters by two of his wives, he now had a son, Michael (Michel to Micheline) with her in 1958, and perhaps for the first time in his life he was not above "letting a woman in his life." Alan was in love. How else can one explain, when he was at home, his putting aside his ever-present clipboard on which he wrote lyrics to spend a weekly hour in the music room. According to his secretary, Doris Shapiro, they listened to Poulenc, Debussy, and Ravel, "to get away from the Broadway music."

With the first act of *Camelot* fairly complete by the summer of 1960, the Lerners took a house at Sands Point, Long Island. Kitty Hart remembered, "Alan came to pick Moss up where we were—somewhere on Long Island. He came by boat across the sound to bring him back to his house to work on the show. It was very difficult. Micheline was making life very hard for him. When they were working, she would buzz in and out of the room and make comments. Poor soul, she wanted to be part of it. She didn't understand what it meant to be married to a writer."

In *We Danced All Night*, Doris Shapiro, who describes Micheline as "looking like a blonde Carmen—and behaving like one," amplifies the tumultuous events of that time: "With nothing to do, one day she packed up her trunks and went off to Europe, taking with her the beloved son and his nurse. When she got to France, she called Alan to say they were never coming back."

"The impact on me was devastating," Alan wrote. "For three days I could not move out of the chair by the telephone. I felt torn, trapped and helpless. I lost all control of my tear ducts and other bodily functions and still could not get out of that chair. I realized that I must be having something akin to a nervous breakdown. In an imagination warped by depression I was reacting as if I were never going to see my son again."

"I think he tried to kill himself at that point," said Kitty Hart, remembering how devastated Alan was. But with the help of Moss, Fritz, medication, and a round-the-clock psychiatrist, Alan was somehow able to bury himself in work. By early September, with *Camelot* scheduled to go into rehearsal in a few days, Micheline changed her mind and with her mother, Michael, and entourage she returned.

When the show premiered at the giant O'Keefe Center in Toronto, it was four and a half hours long. According to Bud Widney, "There was a lot of junky stuff in there. Dancing around in the forest and the quest. We had a twenty minute long ballet about going off and fighting the dragon, The Black Knight and God knows what. Listen, *Pygmalion* was a play licked. Structured. This was a long saga of legendary stuff that went off onto twenty thousand different directions. Alan had yet to pull it together and still make it end in two and a half hours."

If there was drama going on on stage, the backstage turmoil was even more devastating. Doris Shapiro wrote that "ten minutes before curtain time [of the first preview] I went looking for Alan. . . . In a far corner of the chaos I spied Micheline, seated on a high wooden stool, bare-shouldered, swathed in turquoise silks and glitter, blonde hair floating about her head, angrily pointing a finger at Alan who was down on one knee before her, looking up into her eyes imploringly."

She was berating her husband for forgetting to send the car to pick up her mother and bring her to the theater.

Shortly after that Alan's quiescent ulcer hemorrhaged, and he was rushed to the hospital. With the New York opening scheduled for early December and songs yet to be written, much cutting and rewriting to do, he called his secretary and tried to work but was too ill. Moss had refused to rewrite Alan's dialogue. At last Alan took doctor's orders and rested.

After the first night, when the Toronto critics referred to it as "*Gotterdämmerung* without the laughs," word had already reached Broadway that the leviathan *Camelot* was a "disaster area."

Two weeks later, as Alan was leaving the hospital, Moss Hart was being wheeled in. He had had a heart attack. Fritz stood up in front of the company saying, "Don't worry, we'll find a director. The tour is sold out." He wanted to hire another director immediately, but Alan, who had promised Kitty that he would not replace Moss, wanted to sit it out. Actually Alan was able to cut *Camelot* down to manageable size. Not wanting to undo Moss's work, he hired as titular director Richard Burton's coach, Philip Burton,[5] who was able to take the play through the balance of the Toronto performances.

With Alan recovered and Moss recuperating, the final dizzying cutting for New York began. Fortunately, as Alan noted, "the musical was playing to packed houses which is always a great tranquilizer."

Bud Widney was hurriedly dispatched to London to bring back a boy who

---

[5] Philip Burton, a well known London theatrical coach, took over the responsibility of theatrical training for a young man from a large coal mining family in Wales. In exchange for his instruction, the protégé, Richard Jenkins, honored Burton by taking his family name. As Richard Burton, he joined the Old Vic and became the youngest actor ever to play Hamlet in the troupe's history.

could hold the stage with Richard Burton for the last scene. "When I got back the show was still too long," he recalled. "We went on to Boston and there's where we learned that the size of the theater was ruining us. It's a big panorama but the scenes are intimate. The size of the theater, the O'Keefe in Toronto, made a lot of the intimacy lost. Certainly the humor and wit were lost. One-liners work but humor and wit and nuance are tough unless you get the people involved. In Boston the laughter started instantly with the first songs and continued like a dream."

It was not until the show got to previews in New York that it got down to manageable size. It was easy to cut the jousting scenes and some of Guenevere's byplay with the knights. What was difficult was to slash the gigantic ballet. "Tony Duquette, the costume designer who had replaced Adrian, was famous for his wonderful animal costumes—magic forest creatures," Bud recalled.

> Meanwhile in the choreographic department Hanya Holm was a specialist in creating birds and critters who danced in leotards. She created a magnificent ballet for the magic forest. It was totally nixed by the costumes. All the suggestion was gone, all you had was a bunch of these funny things moving around the stage. That was a great embarrassment and one of the things that when we all gathered the day after the opening to figure out why we had gotten such terrible notices— nobody wanted to speak the truth, and it fell on my shoulders to point that particular problem out. We cut the ballet and, of course the costumes.

But it was too late. When the show opened, the reviews were as Alan had feared. One critic, noting the tremendous trappings, changed the name of the work from *Camelot* to *Costalot*. Most of the others noted that the show had no viable second act. Although the box office had two million dollars in the till, with nothing coming in, it was certain to close without repaying its then unheard-of investment of half a million dollars.

By February, when people were walking out of the theater by the dozens, Moss Hart, recuperated, did an unprecedented thing. He asked Alan and Fritz to spend a week rewriting. They cut two songs, changed the dialogue around them, and finally brought the play down to size.

Then Ed Sullivan devoted his entire Sunday evening TV hour to Lerner and Loewe, the last half of which was *Camelot*. Bud Widney noted that this "sold the show to the public at large, and suddenly banged up the box office. People were not walking out either, because the show was trimmed again."

Lerner's sense of poetry and his feeling for language served him handily, for *Camelot* needed a medieval sound touching on contemporaneity. And his rhyme schemes, often varying the number of metrical feet per line, as in the ultimate line of the verses below brought added surprise to his lyrics. Early in the evening Arthur, terrified of his impending marriage, soliloquizes:

You mean that a King who fought a dragon,	A warrior who's so calm in battle,
Whacked him in two and fixed his wagon,	Even his armor doesn't rattle,
Goes to be wed in terror and distress?	Faces a woman petrified with fright?
Yes!	Right!

Loewe's musical score, while not containing as many hit tunes as *My Fair Lady*, certainly was more ambitious—almost operatic in its introduction of *leit-motifs*. In my estimation it will eventually prove to be the more enduring.

Amid the myriad of rich melodic material one notes three musical tricks that recur frequently. They help to give the score a unity Loewe never showed before. First, the composer uses a quasi-Prokofiev martial idea (Example A) which is often adapted into the "Camelot" theme. Second, Loewe's favorite modulation, by thirds (Example B), works better here than in earlier scores, especially

The court assembled for the resplendent wedding of Arthur and Guenevere. The yards of lamé, satin, and ermine added up to the most expensive scene ever produced on Broadway. MOCNY.

when he uses it downwardly again, as in the release of "Camelot." Last, Loewe's whole tone sequence (Example C) calls up a neat Middle Ages—at least to contemporary ears—sound.

Additionally, Loewe introduces touches of modality, that compliment Lerner's lyrics handsomely.

Example A

Lerner was never wholly satisfied with *Camelot*'s score, especially the lyrics he created for that trouble-fraught second act. He deplored having rushed through the creation of Julie Andrews's second-act solo and confessed he had breached the style of the show in "Guenevere," a choral number that told the story of Lancelot's rescue of the Queen. "I could think of no other way to impart that information to the audience." It was not only a stylistic change, but it was breaking a second major theatrical law: it was describing events that took place offstage.

These flaws aside, I believe the score and libretto of *Camelot* taken as a whole is Lerner's most brilliant achievement. One has merely to look at the lyric to "If Ever I Would Leave You," cited above for its modulations, to realize how deeply the song and lyricist were immersed in the vein of the libretto and its period. The medieval quality which the inverted language gives the title, which sings equally well as "If *I Would Ever* Leave You," puts the onus for parting squarely on Lancelot's broad shoulders. Remember, he has been suffering guilt because of his love for the Queen and has tried to leave Arthur's court for eight years. As the song tells us, the urge for honor, occurring every season for thirty-two seasons, has invariably been squelched by his passion. Yet because their love was hidden, this had to be a love song without ever mentioning the word "love."

It is safe to assume that Lerner worked on the lyric in his usual way. First came the seasonal concept, then the title, then the closing line. In this case, the words "summer, winter or fall" would lead him directly into, "never would I leave you at all." Then the real sweat began, getting all the pieces to gibe. An analysis of the rhyme scheme of my personal favorite sections, the A^1 and A^2, excerpted below, shows the precision with which Lerner worked. Words which could hardly be remembered by the ear (marked A and B rhymes) are picked up and rhymed a full sixteen bars later.

If ever I would leave you	But if I'd ever leave you
It wouldn't [A rhyme] be in summer	It couldn't [A rhyme] be in autumn
Seeing you in summer	How I'd leave in autumn,
I never would go. [B rhyme]	I never would know. [B rhyme]
Your hair streaked with sunlight	I've seen how you sparkle
Your lips red as flame [C rhyme]	When fall nips the air [D rhyme]
Your face with a lustre	I know you in autumn
That puts gold to shame. [C rhyme]	And I must be there . . . [D rhyme]

"If Ever I Would Leave You" had great popular success, but the show's most enduring and ultimately moving song remains the title one. "Camelot," a sing-songy, lilting lyric to an almost martial melody, perhaps best expresses the idealism of Arthur. When it is first introduced, its lyric sounds like seductive boasting: a young man who has been touched by magic, cajoling to get a beautiful princess to stay in his country.

The snow may never slush upon the hillside.	A more congenial spot
By nine p.m. the moonlight must appear.	For happ'ly-ever-aftering than here
In short there's simply not	In Camelot.

But by the song's final reprise the lyric's full import, Arthur's dream as passed on to Tom of Warwick, cannot but bring a tear to even the most jaundiced eye.

Arthur: Each ev'ning from December to December
Before you drift to sleep upon your cot,
Think back on all the tales that you remember
Of Camelot.

Ask ev'ry person if he's heard the story;
And tell it strong and clear if he has not;
That once there was a fleeting wisp of glory
Called Camelot.

Camelot! Camelot!
Now say it out with love and joy!
Tom: Camelot! Camelot!
Arthur: Yes, Camelot, my boy.
Where once it never rained till after sundown;
By eight A. M. the morning fog had flown . . .
Don't let it be forgot
That once there was a spot
For one brief shining moment that was known
As Camelot . . .

Adding to the magical poignancy of those lines is the fact that for decades now, the song has been closely associated with John F. Kennedy and his administration. When Kennedy was assassinated, *Camelot* had closed on Broadway and was playing to a packed auditorium at the Opera House in Chicago.

Louis Hayward was playing King Arthur. When he came to the last lines of that song it was reported there was a sudden wail from the audience. Alan described it: "It was not a muffled sob; it was a loud, almost primitive cry of pain. The play stopped, and for almost five minutes everyone in the theater— on stage, in the wings, in the pit, and in the audience—wept without restraint. Then the play continued." Such is the overwhelming power of the musical theater.

A week after that fatal afternoon in Dealy Plaza, Dallas, when the lives of so many on this planet were suddenly altered, Theodore White went to Washington and interviewed Jacqueline Kennedy for *Life.* "At night, before we'd go to sleep, Jack liked to play some records," he quoted her as saying, "and the song he loved most came at the very end of this record. The lines he loved to hear were: 'Don't let it be forgot that once there was a spot, for one brief shining moment that was known as Camelot.' "

Alan was doubly moved when the following day the *Journal-American* screamed this part of his lyric on its front page. When he saw the headlines while walking home he was so dazed that he did not even buy the newspaper. He did not come out of his trance until he was found himself ten blocks beyond his destination.

The association with that idealized political time changed the public perception of the show. From then on the first act became the weak one and the second, idealistic one the strong one. And so it has remained.

Lerner

1963–1965

ONE COULD FEEL IT HAPPENING even during the writing of *Camelot*; the tension between Fritz and Alan that would lead inexorably to their next break-up had been building. Perhaps because he saw the aggravation resulting in hospitalization deep involvement with the show had caused, Fritz pulled back when at every crisis Alan wanted him to become more deeply involved.

It was understandable. Having been touched "on the shoulder" (as he called it) by his heart attack and now rounding the corner of his sixtieth year, Loewe tried to remain as distant from Alan's self-destructive life-style as possible. It was during this time that Lerner started to take what he called "vitamin shots," (actually methedrine in conjunction with vitamins) administered by Dr. Max Jacobson. Dr. Jacobson, familiarly known to the show business community as "Dr. Feelgood," had offices in New York but was known to accompany wealthy clients wherever they needed him. He had treated many celebrities, including Truman Capote, Cecil B. De Mille, Otto Preminger, and Tennessee Williams. Although *The New York Times* exposed Dr. Jacobson's practice in 1972, it was not until 1975 that the New York State Board of Regents revoked his license to practice medicine.

At the time of *Camelot* in 1961, Fritz wanted to be merely "the composer," uninvolved in plots, divorces, or diatribes. He would have no part of pills— uppers and downers—or shots and never drank anything other than fine wines and champagne. He gave up smoking and when he gambled, seemed to do so without stress. It appeared that he longed to spend his twilight years shuttling between his penthouse apartment in Manhattan's chic Dorset Hotel, his house in Palm Springs, and the yacht he rented on the Riviera.

He had never been competitive with Alan and only smiled at his partner's flamboyance, but he could not sit by after Alan's reconciliation with Micheline, when, during the writing of the show, Alan betrayed their longstanding arrangement and asked his wife to sit in on one of their creative sessions. Now Fritz invited Tammy, the twenty-four-year-old woman he was living with, to the session. It was tit for tat, not unlike the time Fritz had bought a duplicate mink coat for Marion Bell's understudy after Alan had purchased one for Bell, his wife-to-be and the star of *Brigadoon*. Neither Micheline nor Tammy had any creative ideas to offer, and of course, both men could produce no songs, only bile, under those circumstances.

Alan's hiring of publicity man Norman Rosemont, over Fritz's objection, further distanced them. Rosemont was engaged to publicize the partnership as well as their current projects. When most of the publicity that came out concerned only Alan, Fritz felt he had been betrayed. With his usual of paranoia, he remarked to friends that Alan was cheating him in financial matters. In any case, by the time *Camelot* opened in Toronto, the collaborators were estranged except for working sessions. By the time the show reached Boston, Alan could usually be found seated in the front row of the theater while Fritz huddled in the last.

When the press reported their divergences and coolness towards each other, Alan did not comment. Many years later he wrote that he did not know why they didn't create a united front as did Hammerstein and Rodgers or as they themselves had in former days when bad reviews or hints of internal strife buffeted their team. He attributed their disagreements on even simple things to too much success.

"Success can be a creative stimulant," Alan was to write.

> But . . . it can weaken as much as it toughens. It can magnify faults and unearth a few new ones and its only virtue is when it is forgotten. Perhaps I was too disdainful of the words of others and Fritz too vulnerable. Perhaps I misinterpreted our differences as lack of support and he misinterpreted mine as heroics. . . . In the end we were a little like the couple being discussed in one of Noël Coward's early plays. "Do they fight?" said one. "Oh no," said the other. "They're much too unhappy to fight."

After the show opened, and certainly after Moss's final adjustments to the musical in early 1961, Fritz reminded Lerner how reluctantly he had been dragged into the *Camelot* project. He told a reporter that all his life he had wanted to live "real good." Now that he had the money he was determined to do so.

"Do you think it's living real good to be in a hotel room in New Haven—have you ever seen a typical New Haven hotel?—eating stale sandwiches and drinking cold coffee, all because there's something wrong with the second act that only God can fix?"

Considering all the bad blood between Loewe and Lerner, adding that Fritz was basically a bon vivant and that it was too late for him to seek out another lyricist, he had no reason to add further explanation for refusing to continue the partnership. Still, Fritz offered one more. He told Alan that was determined to give up composing entirely, for he felt that the passion he put into his music placed an added strain on his already damaged heart.

That left Alan with no partner.

Richard Rodgers, too, had no partner.

With Oscar's death six months earlier, it was only logical that these two preeminent leaders of musical theater should make overtures to each other. Rodgers, only fifty-eight—a year younger than Loewe—could easily have stepped into Fritz's shoes. And from his own viewpoint, the composer would be following Oscar's last wishes to "work with a younger man" were he to collaborate with the forty-three-year-old Lerner.

Both men were driven to create. In essence, Lerner approached his career for reasons quite different from Loewe's. "I write," he said, "not because it is what I do, but because it is what I am; not because it is how I make my living but how I make my life."

Those same words could as easily have been spoken by Richard Rodgers. Writing music was as important to him as the air he breathed. However, he preferred to breathe that air in a collaborative situation. Having outlasted a giant lyricist in Lorenz Hart and then having discovered the added joy of working with another giant, this time a lyric-librettist, he was eager to continue in the same vein.

But he was canny as well, and he knew of Lerner's dilatory reputation. After several preliminary talks about mutual projects, perhaps because he was older, more ambitious, and more organized, he began immediately to think of doing another show alone as insurance against the project's failure.

Enlisting Samuel Taylor, dramatist of *Sabrina*, to create the libretto, he felt strong enough to handle the score. He had great respect for the talents of Diahann Carroll, the elegant singing actress whom he had known since *Flower Drum Song*. Back then, he wanted to cast her for a part in the show but was unable to make her look Oriental enough.

His new idea was to create a play about an interracial love affair. It would be the kind of play Oscar would have approved of—maybe even would have written. "Having a black actress in the starring role would give the play an extra dimension that made it unnecessary for anything in the dialogue or action to call attention to the fact," Rodgers wrote, adding that "rather than shrinking from the issue of race, such an approach would demonstrate our respect for the audience's ability to accept our theme free from . . . sermons."

Eventually *No Strings*, which was produced successfully in 1962, did sermonize without preaching from a pulpit. The couple meet and fall in love in Paris, then because of her expectation of racial prejudice at home in the South, the black heroine magnanimously sends her white lover to his home in Maine while she remains in France.

Rodgers, who allowed that some of Hart's and Hammerstein's lyric ability had finally rubbed off on him, wrote both the music and the lyrics—the latter being surprisingly good—but it was not a happy experience for him. He missed the appointments, the meetings, the story conferences, the back-and-forth discussions; in short, all the *trappings* of writing a show—perhaps more than its very creation.

Once *No Strings* was well on its way, Lerner and Rodgers reopened their negotiations. Lerner was enthusiastic. "We hit it off right away," he told a reporter. "The thought he began, I'd finish, and vice versa. We realized we had the same attitudes toward the theater, the same ideas about casting, management and breaking of theater molds." This last was an obvious reference to Rodgers's *No Strings* as well as his own *Camelot*.

But they could not agree on the starting project. Lerner wanted to do a biomusical about the couturière Coco Chanel, which did not sit well with Rodgers.[1] At last they agreed on a story Bud Widney found about the last days of Erroll Flynn and the teenage country girl the swashbuckling actor, now an alcoholic, induced to live with him. The book, called *The Big Love*, was written by the girl's mother. Alan and Dick got rather far with this unwholesome tale, which they blew up to be a metaphor for the American preoccupation with Hollywood glamour.

Knowing that stars waiting in the wings were an inducement to raise money from backers, Rodgers and Lerner auditioned and cast Joey Heatherton for the role of the girl. When they approached Robert Preston, who had given such a memorable singing-acting performance in *The Music Man*, to star as their Erroll Flynn clone, he cast doubt on the venture by pointing out the sleaziness of the story. "You guys are the most powerful theatrical force on Broadway," he said. "You must be crazy to involve yourselves in such a tasteless idea." Within a week they had dropped the project.

Then Lerner told Rodgers about a story he dreamed up which concerned an uneducated Brooklyn girl's extrasensory perception. It reflected Lerner's consuming interest in the occult and reincarnation. He called it *I Picked a Daisy* after the title character's name, and it concerned Daisy's matter-of-fact clairvoyance: her ability to predict the future. She finds out while under hypnosis

---

[1] With André Previn as composer and Katharine Hepburn as star, Lerner would write *Coco* in 1968.

that she has lived through many incarnations; meanwhile the doctor has fallen in love with the girl in one of her previous lives.

The idea came about when Alan was listening to a late night talk show about parapsychologists and mystics. Bud Widney recalled the phone ringing at one-thirty in the morning It was Alan, saying:

> "I've always been interested in this area, parapsychology. What do you know about it? I'm sure there's a show in there" . . . . I said "Bridey Murphy"—I don't even know where it came from. I think it concerns a girl under hypnosis who discovers she has a previous life. He said, "That's the show! Get me what you can on Bridey Murphy!"
>
> That was the creation of the show and also the fallacy that tripped us up—because on paper it sounds like a humdinger of an idea. A very ordinary psychiatrist puts a very ordinary girl under hypnosis in order to cure her of smoking and she's prosaic and filled with the clichés of contemporary life, really dull. Then she suddenly pops up with a whole personality from another century that is so charming and so delicious that the psychiatrist falls in love with her.

Inconsistency of character was only part of what plagued the plot of what eventually became *On a Clear Day You Can See Forever*. The duality of stories only meant that when one was going on the audience lost interest in the other. With every reincarnation a new life and love story had to be created. Lerner never found the solution.

But it did feel comfortable to him. It was essentially the same story that Alan had written in *My Fair Lady*, *Gigi*, and *An American in Paris*: the older, omnipotent man transforming the younger woman. In addition it was a scenario he had played out in three of his four marriages, one that unfortunately didn't work in his current situation.

Rodgers was intrigued by the excitement with which Lerner discussed the story. He appreciated the contemporary aspect of the plot and overlooked the libretto's basic flaws. Yet because he was more pragmatic than Alan, he took a wait-and-see attitude. When interviewed by the same reporter who had come from a session with Lerner during which Alan called their new collaboration "a punctuation mark in the history of the musical theater," Rodgers was not so optimistic.

"You mean a question mark," he corrected. "You never know how a collaboration goes until the show is on. Alan and I are not at the end of the line, you know. There is a natural qualm, when you embark on a new phase of your life, but it doesn't frighten me. I rather take it as a challenge."

Once they began to work in earnest they learned much about each other's work habits and Alan was forced to admit he had been spoiled by Fritz's understanding and patience. He complained that Rodgers too had been spoiled—"by Larry Hart who could write a lyric in the middle of a cocktail party." He said

Rodgers could not even understand the time it took Hammerstein to complete a lyric. Alan, who sometimes took several months to complete a lyric, soon took to quoting derisively what Rodgers had told him one day in the office: "Do you know what Oscar used to do? He would go to his farm in Bucks County and sometimes it would be three weeks before he would appear with a lyric. I never knew what he was doing down there. You know a lyric couldn't possibly take three weeks."

"When two artists collaborate," Alan told Miles Kreuger, "they sit down and start working on something. And before I could sit down with Rodgers, I had to sign contracts and we had to discuss subsidiary rights and revival rights and the movie version and cast albums. By the time we'd discussed everything there was no more show left to discuss. It was so enervating that I didn't have any more strength left to write the show with him."

Rodgers was too secure in his art to understand Lerner's almost neurotic search for perfection, his almost childish insecurities that would not allow him to part with a lyric until forced to do so. In gigantic understatment he wrote that "perhaps [Alan] felt uncomfortable working with someone he found too rigid." Yet he needled Lerner further when he wrote that he couldn't understand why, "once having made an appointment, Alan would often fail to show up or even offer an explanation—or if he did arrive, why the material that was supposed to be completed was only half finished."

At least outwardly, Rodgers was open and aboveboard and hated dissembling—and Lerner was famous for lying his way out of things. Before the Labor Day holiday in 1962, Alan told Dick he was planning to remain in New York to work undisturbed round the clock. The next morning, before driving to the country, Rodgers called and was told by Alan's maid that "Mr. Lerner 'e is not 'ere. 'E is in Capri."

The dissolution of the partnership was officially announced in *The New York Times* on July 25, 1963. The reason given was that an insufficient amount of Mr. Lerner's lyrics prevented Mr. Rodgers from finishing the score. The article also noted that Mr. Lerner would possibly be continuing the project with Burton Lane. Rodgers's comment, which indicates all was not "hail-and-farewell," was that he did not know whether there would be "legal complications" to prevent Alan from doing so. One feels Rodgers's anger at Lerner more hotly when one realizes that Rodgers knew no other way of life than to write material and to have it produced. In his long career, once a show was commenced it was seen through to completion—nothing he had ever worked on before was stillborn. He felt Father Time keenly on his shoulder, for he told the papers that what he resented most about the time spent with Lerner was that "he wasted a year of my life."

Rodgers never took legal action to stop Lerner from working with Lane on the project they abandoned. He separated his music from Alan's words and handed a sheaf of mostly incomplete lyrics back to him. In a statement unworthy of himself, Rodgers did take a snide swipe at all his former collaborators when he told a *New York Post* interviewer that he had worked with Lorenz Hart and Oscar Hammerstein, both of whom had something in common with Alan Lerner, "the fact that they did not like to work."

When Lane read the *Times* article about the dissolution of the partnership, he wrote Alan that since the paper had mentioned they might be working again together, he was hereby giving his okay. Before his letter arrived, Lerner had telephoned to confirm the resumption of their collaboration, which had ended with the aborted *Huckleberry Finn*. Alan sent the lyric for "What Did I Have That I Don't Have" and the outlined premise of the show.

An intrigued Lane set immediately to work. As it was to turn out, once Lerner transferred the project over to Lane, the show was still a full three years a-borning. It was delayed by the filming of *My Fair Lady* and then a fundraising extravaganza for President Kennedy's last birthday, which Alan was directing and producing at the Waldorf-Astoria as though it were a full-length musical. But perhaps the biggest, most frustrating, and eventually most devastating blow to Alan and his career during that time was the front-page news of his divorce from Micheline.

Lerner and Micheline had been living their turbulent on-again, off-again life style since their reconciliation at the time of *Camelot*'s Toronto premiere. When in New York they shared the magnificent five-story residence at 42 East 71st Street in New York that Alan had bought in 1960. Alan had made his living and writing quarters on the top floor. It included a place where he could seclude himself to write his lyrics, a windowed closet outfitted only with his chair and high desk similar to Hammerstein's. Micheline's opulent quarters on the floor below, complete with huge marble bathroom, rivaled Cleopatra's. The other floors were furnished with mostly French Empire antiques.

Anyone who was an intimate of either party knew of their constant quarrels and separate sleeping arrangements. Servants were aware that their frequent shouting matches would eventually lead to separation, so no one except the public at large was surprised when on May 14, 1964, *The New York Times* noted that Mrs. Lerner had filed suit for separation on the grounds of "insufferable marital cruelty." What surprised everyone was the acrimony and melodramatics of the action.

Micheline had retained Roy Cohn, who was asking for the then enormous settlement of $5,000 per week and $50,000 in legal fees. Cohn had formerly been council for Senator Joseph McCarthy; Sidney Zion, who collaborated with Cohn on his autobiography, had referred to him as "this legal executioner, this notorious bastard who cared nothing for the conventions, who flouted civil decencies." Alan's lawyer was the powerful, almost invariably successful defense lawyer Louis Nizer.

The tabloid papers reported in lurid detail how Mrs. Lerner had barred Mr. Lerner from their house and even changed all the locks when she discovered her husband had cut off her credit. The *Daily News* reported the next day that Lerner, in a swashbuckling exploit of his own, had gained access to the Tunisian Embassy next door, climbed to the roof, crossed over to his own house, and let himself down floor by floor to confront Micheline at 8:15 the following morning. He said he only wanted to have breakfast with his son.

Two days later the same paper reported an upheaval in the court room:

> Nizer, complaining that Micheline has made "scandalous" charges against his client—ranging from "drug addiction to tax evasion"—finally offered to let Micheline have the mansion's third floor if she lets her husband have the fourth floor. "He will not confront her," Nizer promised.
>
> Roy Cohn, Mrs. Lerner's lawyer, protested that even with eight bathrooms it would be difficult for the estranged couple "to be far apart in this house."

On May 20th the presiding Justice of the New York Supreme Court ruled that "living in such close proximity in an atmosphere of hostility would not be conducive to domestic tranquility regardless of his [Lerner's] professed good intentions."

Lerner was denied the rights to breakfast with his son and was obliged to see Michael during the afternoon, away from the house and then only under a governess's supervision. Alan seemed to be losing every battle.

By summer, Micheline had moved out of the house, for Lerner had given his permission for her to take their son to California to escape New York's heat. After a couple of visits by Lerner to see his son and estranged wife on the West Coast, the United Press International reported that the Lerners had reconciled. Neither party denied the allegation. But the truce was short-lived.

On September 29th Lerner sued again to have the child returned to New York so he could enter school, but the California courts ruled that it was in the child's best interests to be educated there. Alimony was now set at $1,500 per week (a record at that time in New York State). So ended the temporary separation litigation; a permanent agreement would be incorporated into their divorce six months later. If such was possible, things were to become even uglier.

Shortly after the trial opened, on March 3, 1965, under a headline that screamed WIFE SINGS A BRUISE SONG, the *New York Daily News* reported that Micheline burst into tears on the stand when she told of how Lerner threatened to get custody of Michel. (In a curious effort by each parent to make the child his or her own, Micheline always called the six-year-old Michel, while Alan invariably called him Michael. As an adult their son calls *himself* Michael.) She added that Lerner told her he had the money, influence and power to do so."

Sobbing, she said she told Lerner: "This child is my whole life. If you take him away I will kill myself—and I would." She denied Lerner's testimony that she also threatened to kill the child.

A great part of Micheline's testimony concerned Lerner's alleged "addiction" to shots given to him by Dr. Max Jacobson. The lyricist has testified that these were merely vitamins, but Micheline said she was induced to take them herself for two months by her husband. She said she quit because she became addicted and "the shots made me feel very bizarre, very high."

"Was there a sex problem in connection with your marriage?" Cohn asked.

"Yes, on my husband's part," Micheline answered. "I had discussions with him. He refused to come near me."

"Did he tell you what the problem was?" Cohn asked.

"He told me that when he was very much in love with a woman his problem occurred," Micheline said. "But he told me it will pass."

"Were there any occasions when he could not have sex?" asked Cohn.

Micheline said: "Since he went to Dr. Jacobson he was getting worse and worse."

Micheline said Lerner became more and more violent, bit his nails, lost 30% of his vision but told her he couldn't stop the shots, that they made him "write quicker."

She said he used to visit Dr. Jacobson at 4 A.M., sometimes he was there all night long . . .

Nizer countered that the lyricist got vitamin shots from Dr. Jacobson whom he described as "a highly regarded doctor." He said the shots have been approved by a government agency.

"In December 1963 in Acapulco," [Micheline] continued, Lerner "Beat me all over—Arms, legs, everything. I said, 'What have I done?' He said, 'You looked at me with disgust this morning at the pool.' He said: 'If you open your mouth I'll kill you.' So I didn't not open my mouth."

The next day of the trial Micheline and Alan continued to present charges and countercharges. *The New York Times* reported that "Mrs. Lerner accused her husband of associations with 'disreputable' persons including homosexuals." Roy Cohn brought out that Micheline had objected to Alan's relationship with two actors. When Lerner took the stand he said he had merely given the men jobs and fired them because of the gossip. But one has to add that it would be impossible for anyone to work in the theater, especially the musical theater, without hiring and firing a great many homosexuals.

It is ironic that Roy Cohn, who, as revealed after his death from AIDS, was himself an active homosexual, tried to smear Lerner. After Cohn's death members of Broadway's gay community quipped: "Roy Cohn is the man who gave AIDS a bad name."

Now Nizer questioned Micheline about her clothing bills amounting to $45,000 in seven months, imputing that she was purposely being spendthrift in order to obtain a higher alimony. Then to counter her allegations that Alan had had an affair with a former airline stewardess, he implied that she too had an illicit relationship with her son's piano teacher.

Two days later the trial ended with a reported out-of-court settlement in "six figures." Citing that by that time the town house on 71st Street had been sold, the *Journal-American* reported, "Micheline is due to get the bulk of the valuable French antique furnishings of the Lerner home."

By late summer Micheline was to establish residence in Lake Tahoe, Nevada, in order to file for divorce. During all this time Lerner toyed with *On a Clear Day*, writing a line here, a lyric there. Lane, not known for patience, became as frustrated as Rodgers had become. But he did not walk away.

Burton Lane recalled that "when Alan came to me with *On a Clear Day* I thought the premise simply wonderful. I thought the story in the past was too heavy handed and not joyful, but it did have such wit and imagination that I finally got excited about it. How that score ever turned out as good as it did is a mystery to me. I've never had anything like this—the worst two years of my life. Alan was on drugs and I don't know how to deal with people who are on liquor or anything else."

Lane's misgivings about the show extended to the title. Beyond its being a take-off on the then hip phrase that California real estate agents used to swindle prospective homeowners when trying to sell beachfront property near smog-ridden Los Angeles—"On a clear day you can see Catalina"—Lane conceded that he didn't know what connection with the paranormal the phrase had. But since Lerner had given him the title *On a Clear Day You Can See Forever*, he felt he needed to write a title song. If he set out the form and wrote it to a "dummy lyric with lots of leeway," then when Lerner had completed it and set it in the show the title would make sense.

> One night I sat down and wrote a melody. I wrote half a melody really. I didn't have the middle section or the ending but I had half a melody and what was going though my mind was that at no point should he give away the title until the very end of the song. I didn't have any words, but I tried to illustrate it by saying, "On a blue day diddle-da-de-dah-dah-dah." On a grey day. On a happy day. Whatever. But on a clear day on that clear day you can see forever. Saving the title for the very end.

In spite of Lane's outline (and perhaps because of his magnificent melody where changing one eighth note might be considered a sacrilege), Lerner swore he spent three hours every morning, seven days a week working on this lyric. It became his bête noire, and his quest for the perfect lyric was to continue for eight months while he wrote ninety-one complete versions—eight of which he showed to Lane. When his British secretary Stef Sheahan was asked if this was possible, she assented, explaining that "if he was really working hard on a lyric, he could lose about six pounds in a day. You could feel him burning it off in the room. He'd just be there with his white gloves on, which he always wore to keep himself from biting his nails. He had a sort of allergy on his fingers anyway and he would have to put a cream on his fingers. They were just shot to pieces. I don't know if it was a fungal disease or whatever. So he'd put the famous white gloves on, and I used to go and buy another ten more pair." Could Beethoven working on the four complete overtures to his only opera *Fidelio* have sweated more?

But in the end, this song intrigues for its melody and title. The meaning of its lyric is still not clear. Yet there is much more brilliance in the score.

Lane recalled:

One night, we came out of a meeting at the Waldorf where Alan had his offices and I said, "Alan I have an idea for 'Come Back to Me.' I told you that I thought it was the best lyric of all those that I saw." Dick Rodgers had written a melody which I still have never heard. Alan heard my tune. The room in which he had a piano must have been forty feet long. He started to pace the room. Tears came out of his eyes, he was so moved by the tune. The ending had to be changed. If I'd been working with Yip Harburg, the lyric would have been changed in two minutes. Four and a half or five months later, I still did not have the lyric to the last two lines. You go out of your mind.

"Come Back to Me," which has turned into something of a standard, is a splendid example of the contemporary "list song," one of the best lyrics Lerner ever wrote.

I

Hear my voice where you are;
Take a train, steal a car;
Hop a freight, grab a star,
Come back to me.
Catch a plane, catch a breeze;
On your hands, on your knees;
Swim or fly, only please
Come back to me!

On a mule; in a jet;
With your hair in a net
In a towel wringing wet—
I don't care, this is where
You should be.
From the hills, from the shore,
Ride the wind to my door.

Turn the highway to dust.
Break the law if you must.
Move the world, only just
Come back to me.

II

Blast your hide, hear me call:
Must I fight City Hall?
Here and now, damn it all!
Come back to me.
What on earth must I do,
Scream and yell 'til I'm blue?
Curse your soul, when will you
Come back to me?

Have you gone to the moon?
Or the corner saloon

And to rack an' to roon?
Mad'moiselle, where in hell can you be?
Leave a sign on your door—
Out to lunch evermore.
In a Rolls or a van;
Wrapped in mink or Saran;
Anyway that you can;
Come back to me.

III

Hear my voice through the din;
Feel the waves on your skin;
Like a call from within;
Come back to me.
Leave behind all you own;

Tell your flow'rs you will phone;
Let your dog walk alone;
Come back to me.

Let your tub overflow,
If a date waits below
Let him wait for Godot,
Ride a rail, come by mail C.O.D.
Come in pain or in joy,
As a girl, as a boy,
In a bag or a trunk,
On a horse or a drunk,
In a Ford or a funk,
Come back to me,
Come back to me,
Come back to me!

On a Clear Day You Can See Forever finally opened on October 17, 1965. It starred Barbara Harris as the wide-eyed Daisy Gamble and John Collum, who replaced *Gigi* star Louis Jourdan, Lerner's original choice. Robert Lewis replaced Bob Fosse as director. The original producers, Feuer and Martin, were replaced by Alan and Norman Rosemont, and as we know, the original composer, Richard Rodgers, had been replaced by Burton Lane. The theater, the resplendent Mark Hellinger on West Fifty-first Street where *My Fair Lady* had debuted less than a decade before, seemed to be the only thing about the show that was *not* replaced.

The splendor of the opening night party perhaps outdid the premiere. It was attended by seven hundred glitterati. The guests included political advisor Arthur Schlesinger, Senator Jacob Javits, Producer David Suskind, actress Arlene Francis, novelist Irwin Shaw, Columbia Records producer Goddard Lieberson, and the late president's two sisters, Patricia Kennedy Lawford and Jean Kennedy Smith (who, it was rumored, was on intimate terms with Lerner).

Seated at a table next to Ethel Kennedy's brother George was Karen Gundersen, a reporter from *Newsweek* who had been sent up to Boston with Mel Gussow to collect material for a planned cover story on what looked like the biggest Broadway opening of the season. Gundersen was dispatched to do the ancillary interviewing while Gussow was to interview Barbara Harris. Karen had intrigued Alan enough for him to ask her if he could accompany her back to New York on the shuttle plane. During the flight they had become good friends, and that was why he had asked her to come to the party.

As at all opening night parties, the tension grows as the time nears for the early morning reviews to come out. This gala was no exception—actually, because it was such a largely theatrical crowd and the show's out-of-town reception had been so mixed, the group seemed on the verge of hysteria.

The premise of On a Clear Day *played fast and loose with time. Here we are transported back to the eighteenth century. Most critics felt the seesawing, however diverting, made one lose interest in the main thrust of the story.* MOCNY.

When the reviews finally came out, they were tepid. The critics agreed unanimously that Lane had written some of his best music in years, and there were kudos for many of Lerner's lyrics. But most of the aisle-sitters felt that Lerner had overreached when he made Daisy able to predict both the future *and* the past. All felt that the show was overlong. Gundersen and Gussow knew then that they would have to pull the *Newsweek* cover story. (A two-page feature on Barbara Harris was buried inside the magazine.)

"He never knew how to leave out material," said Miles Kreuger, referring to the excessive length of the show. "He was so in love with the idea of reincarnation that it was necessary for him emotionally to wedge into a very charming story a whole subplot about an Onassis kind of character paying the psychiatrist money so he could be sure of coming back and thus being assured of immortality. It should absolutely have been removed from the show. It had nothing to do with it. . . . This is the very basic weakness of his writing."

Burton Lane called it "a disaster for me—a show with such promise." He thought that the show was too serious and overproduced. He blamed Lerner for not seeing the faults when they were pointed out. "I have great faith in my judgment. I have the ability to stand back from my work and say, 'What I've done here I don't like.' And I keep working on it until it *is* right . . . I came to Alan with a lot of suggestions, which he rebelled against or didn't agree with. And had he done his work we'd have had a hit show. There was every reason to believe that *On a Clear Day* would be a hit. We turned out a wonderful score. He had a marvelous premise."

As it was, the New York production made it through 280 performances and even began a small road tour. Then it was bought by Paramount as a vehicle for Barbra Streisand.

Alan's rewritten ending for the film, which leaves the story unresolved, was far inferior to the play. Yves Montand as the psychiatrist was uncomfortable with the American vernacular so evident in all the songs. Barbra Streisand couldn't seem to capture the simplicity of spirit needed for a naïve medium. Before the cameras turned, Lane wired to find out if there were new numbers to be written for the movie. Alan wired back that he did not want to work with him on any new songs.

"Then get another lyric writer," was Lane's acid comeback. In the end, Alan went out to Hollywood, and two new songs were written for the film. But that was the last time they were to work together until fourteen years later when they would if not bury the hatchet, at least hang it on the wall while they came together to write *Carmelina*.

By that time Alan would have married and divorced both Karen Gundersen and British actress Sandra Payne.

Lerner

1966–1973

I T WAS ONLY NATURAL THAT LERNER should have been fascinated by the life story of couturière Coco Chanel, for she, like Alan, chewed up life. Beyond that, also like Lerner, she invented herself.

Born illegitimate in a poorhouse in Saumur in 1883, she would go to her grave as Gabrielle Chasnel because legal revision of the misspelling on her birth certificate would have revealed her bastard origin. She would rise from founding a small boutique in 1910 to establishing the House of Chanel, which would alter the world of fashion entirely. Once successful, she paid off her largish family in order to keep her lineage secret.

As the leader of fashion in the '20s she had invented sportswear, costume jewelry, No. 5 perfume ("the only thing Marilyn Monroe wore to bed"), and "the little black dress," but by the early '50s with the rise of younger couturiers, Chanel had become passé. When she introduced a stunning collection in 1954 and made an unexpected comeback, it was front-page news. Rosalind Russell urged her husband, producer Frederick Brisson, to seek out the designer so she could portray her on stage and screen. Brisson, who had a reputation for driving a hard bargain, changed his mind as often as a chameleon changed colors—so much so that he was known on Broadway as "the lizard of Roz."

In 1960 Brisson optioned her life story from Chanel herself, promising that he would dwell on the early days and the love aspects of the young designer. Even then he had to be pulling Chanel's leg, for Rosalind Russell was then in her sixties, far too mature-looking to play the couturière when young. Now with the backing of Paramount to the tune of almost three million dollars for film rights, Brisson approached Alan, who was at the time deeply involved in *Camelot*.

A year after Hammerstein's death Lerner had tried to engage Richard Rodgers to make the Chanel project their first collaboration, but it aroused no interest then. Still, Alan had faith in the theatricality of a show about fashion in the vein of *Irene* or *Roberta* and picked up his notes on the musical from time to time all through the three-year gestation period of *On a Clear Day*. By that time Rosalind Russell was suffering from acute arthritis and patently unable to play the role, so another star had to be found.

Alan went to apologize to Brisson, he told a reporter for *The New York Times*, "to tell him how sorry I was to have tied up his idea for five years. He asked me whether I had any stray ideas on how it should be done and I started to talk. I talked and talked and suddenly we all realized that we had a play here after all."

What he had talked about was how the designer, who had many love affairs but was married only to her career, was a total iconoclast—actually, she was the prototype of the new, autonomous woman of the '60s. For the first time in his career Lerner was planning to write a "star" part, the kind that Ethel Merman, Carol Channing, or Mary Martin generally chose and which spelled "Broadway glitz" to so many theatergoers.

When he was working on *My Fair Lady* he had discovered the unique joy of writing and making something work for somebody like Rex Harrison. A true theater man, Alan knew how important stars are and spoke about it when he was asked to give the eulogy at Yul Brynner's memorial service. Lerner chose for the theme the idea that there are few stars left and how bereft we all are when a genuine one goes out.

Now it was decided: the star and the fashion show would be the draw. Eventually Lerner contrasted the aging Chanel's independence with the story of a young model utterly dependent on love. Her coming under Coco's demanding wing created tension and—voilà—a subplot.

Coco started to take shape. Brisson announced it for the following season with a star of the first magnitude—although he actually didn't know who that would be. Early in 1965 went back to Chanel to enlist her aid and to tell her that Paramount Pictures was providing the financing. He returned to Paris again a few months later, bringing along André Previn, who had been signed as the composer, and Alan. In Chanel's salon on the Rue Cambon, Alan sang two of the songs they had written.

Chanel outlined the story as *she* conceived it. Totally false, she wanted it to open with her father (who was nowhere around at the time of her birth) cooing over her cradle. He would give her the nickname, Coco[1] and would promise her a life of glory and success.

Lerner said he had conceived the musical as the story of a woman who sacri-

---

[1] Actually that untrue story would be used charmingly in a lyric sung when her father invents her sobriquet. "Mimi . . . ? Momo . . . ? Kiki . . . ? Coco . . . !" he announces.

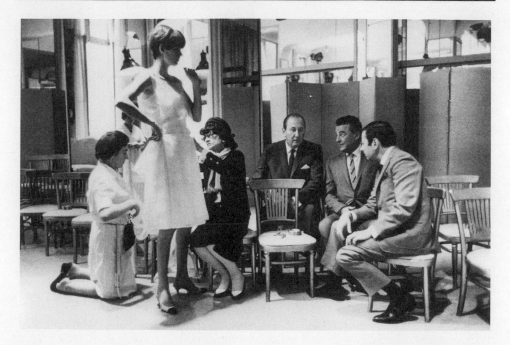

Brisson, Lerner, and Previn watch as Coco Chanel and her assistant (kneeling) pin up a dress on a model. MOCNY.

fices everything for her career. When she attains the peak of success she realizes loneliness is the price she must pay.

"That will be a tragedy," said Hervé Mille, a close friend of Chanel's who was there.

"All right," answered Lerner, "then it will be a *musical tragedy*."

"And who will play me?" asked the eighty-two-year-old couturière suspiciously.

"Hepburn."

Chanel nodded, thinking the story would be about her romantic youth, giving tacit approval to being portrayed by Audrey Hepburn.

"No, Katharine Hepburn," Brisson corrected.

"Katharine Hepburn is too old to play me. Why, she must be close to sixty!" (At that time Hepburn was fifty-eight.)

A year earlier Bud Widney had the idea to bring Katharine Hepburn into the project. "Alan was struggling like hell, Brisson couldn't find anyone. This is a ghoulish story, but I read in the news that Spencer Tracy died. Alan was in California at the time, I was here in the East and I called Alan and said 'call Katharine Hepburn because she hasn't anything to do.' "

He reminded Lerner that Hepburn had been taking care of Tracy in his drunken years and that she was a real theater person. Then Alan called Garson Kanin[2] and told him, "Gar, call her up. Tell her I'm a nice guy." Gar told Hepburn, "You're going to get a call from Alan Lerner. Put your trust in him. Put yourself in his hands and do exactly as he says." This was a real boon to Hepburn, for it offered her the chance to get out of her California life and come back east, which she wanted to do to get away from all that depression. Kate cautioned that she "didn't sing a note," but Alan answered, "You'll learn. We'll give you good teachers."

When Chanel realized that she was to be portrayed on stage as an old woman, she knew she had optioned her life away. Chanel's own vanity aside, Alan had indeed chosen the more interesting part of Coco's life. And in Katharine Hepburn, Brisson had signed up the show's salvation.

Brisson and Previn were eager to get the show into production, but Lerner had to put it aside to write the screenplay for *Camelot*,[3] which finally made it in front of the camera after six years. He had no sooner finished with that then he involved himself in the film version of *Paint Your Wagon*.

Four new songs were needed for the latter film, and although Alan balked at approaching him, it was his new wife, Karen Gundersen, who suggested they drive out to Palm Springs to see if Loewe would be interested. As it was he was not, and Alan collaborated on those songs with Previn. Once they were done, Alan discovered that teaming with André had been stressless—quite different from his recent experience with Lane—and looked forward to completing *Coco* with him.

For his part Previn was even more eager to work with Lerner. Citing how smoothly they had worked together on *An American in Paris* and *Gigi*, Previn said, "If he had suggested we do the telephone book, I would seriously have considered it. That's what I think of Alan."

Now Alan flew to Paris to present "the other Hepburn" to the great Chanel. Hepburn said, "I was scared to death to meet her. I had worn the same clothes for forty years—*literally*—even to the shoes. I thought 'if I don't like her, it will be an *agony*.'" Finally Lerner told her, "You'll like her."

Hepburn agreed that "she was enchanting. The essence of her style was *simplicity*. Exactly what I appreciate most."

---

[2] Garson Kanin (1912–), dramatist and screenwriter, and his wife Ruth Gordon were close friends of both Tracy and Hepburn. Among many other works Kanin had written were *Adam's Rib* and *Pat and Mike*, screwball comedies in which the two stars appeared.

[3] For the film the three stars of the staged *Camelot*, Burton, Andrews, and Goulet, were replaced by Richard Harris, Vanessa Redgrave, and Franco Nero (whose singing was dubbed by Gene Merlino). Their performances were vastly inferior to those of the roles' creators. Perhaps worst of all was Josh Logan's close, almost TV direction, which robbed the musical of all grandeur.

In September 1968 yet another postponement of *Coco* was announced in the papers. This time, Alan had taken a year off to write the screenplay of *On a Clear Day You Can See Forever*. But the show was definitely scheduled to go into rehearsal in the summer of 1969 with Katharine Hepburn as star.

The costume designer for this kind of show would be almost more important than the star. Chanel herself wanted to design the costumes, but Brisson knew taking charge of a huge Broadway production would overwhelm the frail designer, who was then in her mid-eighties. He bowed to Alan's wishes and hired Cecil Beaton, who had worked closely with Lerner on *My Fair Lady* and *On a Clear Day*. Beaton flew to Paris to confer with Chanel. He had admired her work but never liked the woman herself. In his diary he wrote of her venom, her self-centeredness, citing "the terrible strain of listening to Chanel talk about herself without stopping from 1:30 to 5:30." But he eventually said that her talents were very rare, and he forgave her faults, calling her "a genius."

While Beaton was working out the designs, Hepburn was studying voice production. Unfortunately, Roger Edens, the Hollywood coach who had given her the confidence to accept the singing role, had died before the show went into rehearsal. It was a demanding role that required her to be on stage for all except eighteen minutes of the more-than-two-hour show, and she had six big numbers to sing. Hepburn studied and worked hard on her vocal projection, but she never did learn either to come near the pitches Previn's music demanded or to color her voice attractively. Her monotone, heavily amplified, always seemed to cover over the orchestration.

Widney called her singing "the worst sound anybody ever perpetrated on the public in our lives," and attributed it to her lack of musicality. But he added that "she sold a lot of tickets." The public came to hear Katharine Hepburn, but beyond a certain loyalty to her acting ability, they couldn't have been amused when she began to "sing."

Yet her aura attracted them, and singlehandedly she kept *Coco* running for ten months. When her contract had run out and she began to tour with the show, Danielle Darrieux, who was mature, sexy, and really could sing, stepped into the role. Previn said, "It was the first time I had heard my songs sung properly in a year." But without Hepburn the New York company folded its wings and expired two weeks later.

Actually, the show was not very good, and certainly not daring enough. Expensive it was. Besides the fact that Cecil Beaton's costumes smacked more of Victorian London than Art Déco Paris, it was overwhelmed by Beaton's two gigantic basic sets: Coco's apartment and the mirrored salon below, connected with her no-less-famous mirrored staircase. Built on a giant turntable that rose, split, revolved, and descended in whole or in part, the salon décor allowed a rousing fashion show finale with mirrors and platforms turning in different directions.

This mammoth seesaw of a show needed a hummable score. Lacking that it needed to be more avant-garde—even composer Previn said so. Previn and Lerner did not mesh as composer-lyricist, resulting in music and lyrics that were at odds with each other. Lerner's lyrics are straightforward yet sophisticated, although it must be admitted none had the soaring brilliance of which he was capable. Previn mismatched them with his nonmelodic music that tried to be commercial but failed.

Previn said later:

> I had no confidence in those days, and I had no clout. I thought of it as a very French show with *chansons*. I said to Alan very early on, "Maybe we can have a five-piece orchestra in one of the stage boxes." By the time we were through we had turntable stages and Cecil Beaton, and five thousand people on stage, and the whole thing was as un-French as possible. . . . I would say to myself, "This is impossible. We haven't got a plot." But in those days, I thought, "What am I doing? This guy wrote *My Fair Lady*. . . . Leave it alone." And so the show sank.

Even Michael Bennett, the show's choreographer, tried to take *Coco* out of its ordinariness but was thwarted at every turn by Lerner's stock response: "We didn't have to do that on *My Fair Lady*."

All of that left the show with few climaxes, for at the curtain rise, when a despondent Chanel talks to her attorney about trying for a comeback, the audience knows that she will eventually succeed. So the high point of the show—emotionally—is reached in the libretto after Chanel shows her new collection and is devastated because all the Parisian newspapers hate it. Suddenly four buyers from the U.S. representing "Orbach's, Bloomingdale's, Best and Saks"[4] come to the showroom and submit large orders for Chanel to manufacture clothes for the ready-to-wear American market. Because she had given out with her philosophy in her opening number, "The Money Rings Out Like Freedom," we now realize Chanel has achieved her dream again. Financial success. Unfortunately, the message does not endear her to us, and, beyond its title, the song falls flat.

The best song in the show is a throwaway, the kind Lerner does so well—a reverse love song. "When Your Lover Says Goodbye" is a twin brother to the snide "I'm Glad I'm Not Young Any More."

When your lover says goodby,   When your lover flies away
Wise are they who know   Sad are they who learn
When your lover says goodbye . . .   There can be a darker day . . .
Let her go.   Her return.

---

[4] As a sign of the times, both Orbach's and Best have succumbed to recent recessions and have disappeared. Fortunately Saks and Bloomingdale's are still thriving.

Katharine Hepburn as the hard-working couturière prepares for her show, a come-back, while around her all her assistants are exhausted. MOCNY.

She was kind to leave, my friend,
Now you'll have a field day.
You're a fool to grieve, my friend
This is your Bastille Day.

When your lover says goodbye
Wise men all agree,

There can be but one reply . . .
Leave the key.

When your lover says goodbye
You are spared the worst.
Lasting love that doesn't die . . ,
Kills you first.

As opening night approached Lerner exuded nothing but confidence. A month earlier *Vogue* had gushed: "Both Chanel and Hepburn are stubborn charmers, wily craftsmen [note the gender] wise in their work and fortunately they know it." Chanel herself, who had designed and made a sequin dress for the occasion, was planning to fly in for the premiere.[5] Hepburn's name had sold out the house for months, and she was being given standing ovations nightly at the previews.

[5] Two weeks before the opening Chanel suffered a stroke; she died three weeks after *Coco* opened.

Beaton may have been the only naysayer. In a letter to his secretary he confided that the book was "lousy, the music old-fashioned, K.H's singing voice is really pitiful and I can't think the show will get anything but appalling notices."

On opening night when the reviews started coming in he was proved right. They were all bad. The first thing panned was Previn's music. None of the critics cared much for Beaton's costumes, Michael Benthall's flaccid direction, or Michael Bennett's dances. Next came the TV critics, who began whittling away at Alan's libretto and his lyrics.

The show's failure pointed up the difficulty of doing a musical biography, and especially one of a living person. Lerner might have been forewarned had he talked with the various librettists who had struggled and conquered *Fiorello*, *Evita*, and *Funny Girl*. Or had Chanel died while the show was still in preparation, Lerner might have taken a more honest or romantic view of her life. Either way would have added character and been an improvement.

Even if it had a remarkable score *Coco* would not be revivable today because it leaves a bad taste, perhaps because Chanel herself was so disagreeable and just enough of the picture of the woman comes across in the writing to make her unpleasant—without making her a memorable monster. Although *Gypsy*'s Rose was an ogre, she created empathy in the audience by sheer brassiness. Alan's next project would be based on what the public of its day considered an even more horrible monster, Nabokov's Humbert in *Lolita*, *My Love*, and it would be perceived as his worst failure to date. But this time he would get everything right but the timing. Had *Lolita* been presented a generation later it would have been a resounding success.

"It was perfectly all right for me to imagine a twelve-year-old Lolita. She only existed in my head. But to make a real twelve-year-old girl play such a part would be sinful and immoral, and I will never consent to it."

Such was Vladimir Nabokov's indignant reply in 1958 when Hollywood first sought permission to turn his best-selling novel *Lolita* into a film. But he changed his mind two years later, he said, because he had dreamed he saw himself reading his own screenplay. That he was offered $150,000 to write the scenario may have prompted the fanciful dream.

After a search for a sexy prepubescent, Sue Lyon, fourteen and as near as Hollywood could come to a nymphet, made her debut in the title role, while James Mason, who was able to imbue Nabokov's cryptic one-liners with special meaning by raising one eyebrow, portrayed Humbert. Hollywood's master of the macabre Stanley Kubrick oversaw the project, which was released in 1962.

Following the cinema's then moralistic codes, the film emerged as a bowdlerized version of the classic novel. But the public was curious enough about the adventures of the teenager and the mature man as they traveled from motel to motel across the country to allow the film to end up with a three million dollar gross profit, handily repaying its two million cost.

By the spring of 1970, experimentation and exploitation of the offbeat and the sensual were Broadway bywords. Sondheim-Goldman-Prince were readying *Follies*, which concerned ghosts of a theatrical past; Kander-Ebb-Masteroff were writing *70, Girls, 70*, which concerned living superannuated ghosts. Both were headed for the '71 season on the Great White Way. Both were offbeat.

Lolita seemed a perfect subject for a daring Lerner entry into the Broadway sweepstakes, especially after the failure of his attempt at commercialism in *Coco*. Again, Bud Widney came up with the suggestion, "and he loved the idea." Lerner immediately set out to acquire the rights. After a phone call to Nabokov, who had never seen one of his musicals, Alan sent him a selection of original cast albums. Then he arranged a meeting with the novelist, after which Nabokov said he was satisfied. "Mr. Lerner," Nabokov told the press, "is most talented and an excellent classicist. If you have to make a musical version of *Lolita* he is the one to do it."

When Alan asked who might compose this kind of music, Bud brought in some recordings of the music of film composer John Barry. "Alan was sick in bed but he listened to two or three albums and said, 'That's the sound. Let's get hold of him.' "

Although *Lolita, My Love* was Barry's first Broadway effort, it was not his first theatrical score. In 1965 he had written the music for London's *Passion Flower Hotel*. Barry, a Yorkshireman, had the same number of Oscars as Alan, having won two Academy Awards for *Born Free* (one for score and one for song) and another for his score for *A Lion in Winter*. Besides writing the music for most of the James Bond pictures, Barry had also sensitively scored such noted films as *The L-Shaped Room*, *Seance on a Wet Afternoon*, *The Knack*, *The Ipcress File*, *King Rat*, *The Whisperers*, *The Quiller Memorandum*, *Petulia*, *Boom*, and *Midnight Cowboy*.

"After he got his secretary to trace him down," Widney reported, "Alan picked up the phone and said 'my name's Alan Lerner. I want you to write a Broadway show. Would you consider doing it?' Barry said 'yes.' Alan believed in making the phone call to the top. That aspect was why everyone always called him the boss. He made the shows happen. He didn't need a producer, for he was putting all the elements together." Lerner may not have *needed* producer Norman Twain, who had brought *Bajour* and *Henry, Sweet Henry* to Broadway, to help him assemble and supervise his crew, but he certainly could do with assistance to raise the backing, for with the successive non-success of *On a Clear Day* and *Coco*, Lerner was not the sure bet he had been in the days of *My Fair Lady*.

Twain did his job nobly. CBS agreed to make what the producer called "a substantial investment in the show," which was budgeted at $650,000. The timetable was set. Rehearsals would begin in January, followed by a shakedown in Philadelphia, then Boston, and a planned debut on Broadway in late March.

The musical began on a heavy foot with the hiring of opera director Tito Capobianco, whose motto has to have been "shadow and heaviness." Under his direction this was carried over to Ming Cho Lee's normally airy scenery. Unable to move the youngish cast, Capobianco was replaced in Boston with Noel Willman, but the production was stuck with the scenic concepts that he had brought to it, which were very, very dark.

To lighten the important choreographic elements of the show and to balance Capobianco's seriousness Alan hired choreographer Jack Cole, who always seemed at a loss for ideas. He was replaced by Danny Daniels, who put the all important second act trip around America into choreography. But, as Widney recalled, "It required the leading man being involved in the choreography and this leading man took one look at that and said 'Not for me, thank you,' so Danny took a hike." Chorus member Dan Siretta, who had never handled a musical before, was promoted to choreographer.

From the beginning Lerner wanted Richard Burton for Humbert, but Burton said he had had enough of musicals after *Camelot*. Capobianco cast a leading man, John Neville, who was not charismatic or interesting and, worst of all in this role, had no sex appeal. As Widney put it, "James Mason was wonderful in the movie with that suppressed sexual need and vulnerability—but this guy, who was a dear sweet man, was not appealing. Now, after a fiasco in Philadelphia, the new director felt that we shouldn't go with the kind of Lolita who was the sort of obvious sexpot we saw on the screen—that we should have someone more innocent and nubile. Anyway the combination of those two was disaster. Neville as Humbert and the Lolita were both awful."

Receiving some of the worst reviews in his career, Alan closed the show in Philadelphia for a month of repairs. That was when the director was replaced, the choreographer changed, and even the Lolita concept altered. Sexpot Annette Ferra went out and nonchalant young Denise Nickerson came in. But then, when even new director Noel Willman walked out, Lerner put Bud Widney in charge of the direction. Widney felt that the second act needed rewriting and that the sexy concept of Lolita seducing Humbert was the correct one. He urged Alan to return to that.

But before the show could open again, it ran out of money, and it was necessary for Lerner to invest $120,000 to raise the curtain in Boston. He tried making desperate calls to Goddard Lieberson to come up with the additional $300,000 (since CBS were already in for $650,000) needed to continue the Boston production and polish the show prior to bringing it in to New York.

Again, according to Widney, "Goddard said CBS is not about to put money into a failure, for that's what he thought it was. He said, 'I love you Alan, and we made a lot of money on *My Fair Lady* but I can't see throwing money down the tubes.' So Alan was stuck with no way to bring it in unless he put up a big hunk of his own money." Now with Bud and all the critics telling him it didn't work and no money in hand, Lerner had but one choice. He called a halt after nine Boston performances. *Lolita, My Love* closed, never to reopen.

Lolita's dreadful notices began with criticism of Lerner's taking away the novel's suspense by telling the story in flashback—opening with Humbert killing Quilty. But far more offensive to the critics was what was perceived as the musical's immorality. *Boston After Dark* called it a "giant cliché . . . a giant step backward for the American musical." The notices might have been copies of those written after the premiere of *Pipe Dream,* when all the reviewers complained about Hammerstein's "involvement with sleaze." U.S. ethics in the '70s could grudgingly accept Humbert's being seduced by a sexually aware preteen but could not swallow a mature man's seduction of a child-woman—they still can't.

Lerner was so concerned with this question that he passed out audience critique cards in Boston asking, "What do you think of our handling of this girl?" They responded that they didn't like it. None of the critics understood or mentioned the precarious line Lerner was attempting to walk. Of course, the greatness of the novel lies in the eternal enigma Nabokov posed. Who seduced whom?

But it was even more than that. Lolita, Humbert, and Clare Quilty, the experimental poet for whom Lolita deserts Humbert, are all unpleasant characters. The only empathetic person of the four principals is the mother, Charlotte, who Humbert marries just so he can get his hands on her daughter. Performed with an amusing touch of pretension but with great vulnerability by Dorothy Loudon, "Sur Les Quais de Ramsdale, Vermont," in which she tries to turn her rural town into Paris to seduce the worldly Humbert, is a first class showstopper. Unfortunately, Charlotte is killed off in the middle of Act One.

Before she goes, there is a brilliant and imaginative scene during which Humbert fantasizes on ways to kill her. Lerner's stage direction reads

(HE *stops suddenly, his eyes caught by the gun on the wall. Quietly, with complete composure,* HE *walks to it . . . takes it down from the wall . . . walks with measured step to the kitchen door . . . opens it . . . aims . . . shoots.*

(*A thud is heard. A moment later,* CHARLOTTE *enters jauntily carrying a tray of breakfast things* SHE *sets out on a table.*)

During the above, he sings:

I would never have the heart   I'd sicken at the thud.
To shoot her with a gun.   I'd not have the heart, not I.
Farewell, little dream, goodbye.   Farewell, little dream, goodbye.
I could never stand the blood.

He comes up with other methods, all of which he discards. Eventually he enlists the audience as conspirator after Charlotte keeps nattering on about a lost brooch. "She will not stop until my manhood is dangling from her wrist like a charm bracelet," he complains. "What am I to do . . . and how shall I do it." Then he sings:

That torpedoer of hope . . .   Make it slow or make it quick . . .
With my hands or with a rope!   Let her pass on
With a hatchet or a knife.   With the gas on . . .
Blow her up or let her sink . . .   A club . . . !
Line her bathing suit with zinc . . . A tub . . . !
Broken bottle, bat or brick?

(An idea! HE goes to the garden door and just outside it HE finds a small bottle—obviously arsenic. HE pours the contents into Charlotte's glass of juice and tosses the bottle into the garden. Then HE returns to his own chair to read the paper as CHARLOTTE enters, still looking around for her brooch. SHE sits. SHE drinks. Her face becomes hideous with pain. SHE clutches her throat in the traditional fashion, utters a wail that seems to be coming from across the moor and thumps to the floor. A split second later, SHE indicates something is under her back. SHE reaches under and lo! the missing brooch. Overjoyed, she rises and exits into the hall. At a certain point during all this, HUMBERT turns to the AUDIENCE and sings

I would never have the heart   So intense would be the strain
To poison her to death.   To watch her writhe in pain
Farewell, little dream, goodbye.   That I'd be the first to die.
                                      Farewell, little dream, goodbye.

At the end of Humbert's solo, Charlotte reenters. She has found his diary. Horrified and distraught, she is reading from it "her grotesque mother butted in" . . . "that Haze woman," etc. Humbert tries to pretend that she and Lolita are characters in a play he is writing but he can no longer deceive her. In tears, she runs blindly out into the street and we hear a screech of brakes and a thud. A man enters and tells Humbert that a woman, is it his wife?, has been hit by a car. Humbert, still in fantasy, sings:

I would never have the heart
To hit her with a car.
Oh, no! I could not; not
I . . .

At last when a policeman comes in to tell him his wife is dead, he begins to understand that this is reality.

The scene, one of Lerner's best, in its skirting reality is not unlike so much of his work from *The Day Before Spring* and *Brigadoon*. It is also in its ironic way a precursor to the penultimate scene in the Sondheim-Wheeler *Sweeney Todd*.

It is too bad that Lerner could not find a way to deal with the realism that made up the ensuing two-thirds of the evening. The second act becomes one long debauched tour of motels across the states until Lolita finally leaves to run off with Quilty. Humbert spends two years—on stage those years seemed like a decade—searching for her and tracks her down at last. The beginning and end of the musical have a Grand Guignol feeling, for Humbert shoots Quilty. Only afterward does he find out that Quilty too has been deserted by Lolita. Just as the police are about to arrest him for murder, he finally catches up with his Lolita—now married and pregnant by a wholesome young husband. The young couple wants to go off to Alaska to start a new life. Humbert gives her $15,000, which are the proceeds from the sale of her mother's house and a few thousand that were paid him by the University. His final speech, although reminiscent of Dickens's "It is a far, far better thing I do," is most moving.

> "Goodbye Lolita. I hope you will like Alaska. I hope you will have a boy. I hope you will think of me . . . but if the memory does not make you happy, I hope you will not—for I want you to be happy even more than I want you. Do I love you for the first time? I wonder . . ."

As Karen Gundersen Lerner put it in an interview with Gene Lees in 1989, "things were a bit shaky" with Alan's increasing dependence on shots from Dr. Jacobson.

> I became the heavy, saying, "Don't go to him. Where were you last night? You weren't here. You were in Max Jacobson's office." He'd get a shot and then he'd stay there and write. Then he'd get another shot three hours later. He'd stay and scribble. I have some copies of lyrics Alan wrote some of those nights. They were just little dots, all over the place. They weren't coherent. You know rats on amphetamines will eat up their own tails. He was frantic, destroying himself.
>
> When Alan was in the process of writing something, he needed praise from me and others—and he got it.

Karen felt once the work was done and mounted on the stage, and she was sitting around at rehearsals, there was a problem in her expressing some minor opinion. "It did not go down well. I probably made my opinions known too much to him. It wasn't a good idea," she added. "I didn't want to erase my own self . . . I found that I had for a few years, and I didn't want to continue that way."

Lolita's checkered history would have put a strain on any relationship. Alan's fifth marriage had survived happily through the *Coco* period. He was married to Karen for eight years, longer than to any of his previous wives. They actually lived together for six of them. The couple decided to separate when each became interested in someone else during the time *Lolita, My Love* was on the road.

It may have been the recounting of the saga of sixteen-year-old Lolita that put Lerner in mind of the other teenager he had written about, Gigi. She too was deeply desired by an older man. Perhaps Gigi's seduction, since she was being trained as a neophyte courtesan, would not be offensive to American morals. Besides, it all took place in France (where seduction is considered a game) and a long time ago.

In the summer of 1972 the promise of readying *Gigi* for the stage was enough to pull Loewe out of retirement. Alan and Fritz got together again for what would be their final stage collaboration. Alan's excuse was that Edwin Lester, guru of the Los Angeles and San Francisco Light Opera Company, was thinking of mounting a stage production of *Gigi*. It would star a mature Alfred Drake, *Oklahoma!*'s original Curley, in the role of Honoré—the Maurice Chevalier part.

The considerable amount of new material added to *Gigi* on stage proves that neither man had lost his skill and that the collaboration was still viable. Five new songs were written. Among them was a philosophical number reflecting Gigi's juvenile confusion as to where she fits in,"The Earth and Other Minor Things" ("I don't belong where the crowds are / I don't belong where the clouds are / Then where do I belong?"), and a big waltz inserted to build up the part for Drake's Honoré, "Paris Is Paris Again."

The Trouville number which was shown graphically on the screen was expanded into a madcap "I Never Want To Go Home Again," and a scene where Gigi, capitulating to Gaston, utters Colette's beautiful line, "I would rather be miserable with you than without you" climaxes a moving number called "In This Wide, Wide World." But the best of all was "The Contract," a fifteen-minute ensemble piece during which former courtesan Aunt Alicia hammers out a contract for Gigi's maintenance as a courtesan. If her niece is to be a kept woman, she is to be kept in the best of style. An excerpt from this hilarious battle of wills shows Aunt Alicia beating out Lawyer Dufresne as she does throughout the scene.

Dufresne: Madame,
I have been designated by Gaston Albert Philippe Lachailles
To notify you of his passion for your niece.
And further,
Without her permanent companionship,

My client feels his happiness will cease.
So in the customary gentlemanly fashion
I've been authorized to implement his passion.
As this document confirms
Alicia: Yes, yes, of course, let's hear the terms.
I like the horse before the cart:
Which means the coarse before the heart.

Dufresne: Madame,
This is a document of love,
The tender declaration of a man whose aim is
   Spreading sunshine, nothing more.
A contract
With the most generous conditions
He has ever offered anyone before.
Such are his feelings for your niece that he's
   provided
For her as Louis for Dubarry in Versailles did.
A paradise where romance blooms . . .
Alicia: How many baths? How many rooms?
Dufresne: She'll view a lovely world hence-
   forth . . .
Alicia: The view had better not be north.
It's simpler to imagine wonderland
If you have a little floor plan in your hand.
Dufresne: It's all herein

Alicia: Then do begin.
Dufresne: Monsieur will pledge to Mademoi-
   selle
A flat of choice in which to dwell.
A Taj Mahal where love may thrive
The rooms of which will number five.
The flat superbly decorated
In perfect taste and understated.
All this to her Monsieur will give
As long as she or love may live.
Alicia: Indeed! Indeed!
And what if she and love
Live longer than he?
Mamita: Alicia hold your tongue
How crass can you be?
Alicia: There is one thing I am firm in!
What is given her is permanent . . .
Post mortem as well as pre . . .

Gigi boasted a large cast but was still an intimate show. It had a successful many-month run in Los Angeles and on the road but then was overwhelmed by the size of the mammoth Uris Theatre in New York where it opened in November 1973. And even though it had a splendid cast that included, besides Drake, Daniel Massey as Gaston, acerbic Agnes Morehead as Aunt Alicia, and delightful newcomer Karin Wolfe as Gigi, the Broadway public perceived the show as a rehash of the film, the critics ignored the brand new treasures it contained, and it folded in three months.

But the collaboration was reborn, and Alan and Fritz, mellower now, were able to work on the score for another film. Joseph Tandet, who held the rights to Antoine de St.-Exupéry's *The Little Prince,* wanted to make a movie musical of it. Alan, who by now was spending much time in London, inveigled Fritz to come abroad to write the score.

"Eleven years slipped away in a minute," Alan said, "and it was pre-*Camelot* again."

Returning him to a world of fairy-tale fantasy he had not explored since *Brigadoon,* the story of the youngster who finds life in the heavens preferable to life on earth held a mighty attraction for Lerner. He and Fritz quickly wrote a dozen delightful songs. Obviously subtlety had not forsaken Lerner. One look at the Prince's song "Be Happy" illustrates that Lerner's tongue was still in his cheek. He could still write the only happy song of desertion in existence:

If you think you're hurting me
By wickedly deserting me
And leaving me to suffer and to cry . . .
And later die—

You're very wrong,
I'll get along
And so goodbye.

Go ahead, abandon me
Let every horror land on me.
You mustn't let it spoil your holiday.
And by the way,
Before you fly
There's something I
Forgot to say:
I do love you
And want to you to

Be happy. Be happy. Be happy.

Forget you ever knew me
Or loved me at all.

Be happy. Be happy. Be happy.
Forget that I am helpless
And frightened and small.
Be happy. Be happy. Be happy.
Forget that I'll be crying still
And never will
Be happy . . . be happy
Without you.

Lerner had little experience in writing for children, and because of all the subtleties (as in the song lyric above), what was intended to be a juvenile romp into an interplanetary idea came out as a rather confused allegory for adults. The film was a fiasco at the box office, but Alan never blamed his exalted approach. In his autobiography Alan listed his own reasons for *The Little Prince*'s failure.

Fritz wrote the most beautiful score, filled with melody and bubbling with the innocence of youth. Alas, it never was heard on the screen as he had composed it. The director . . . Stanley Donen, took it upon himself to change every tempo, delete musical phrases at will and distort the intention of every song until the entire score was unrecognizable. Unlike the theater where the author is the final authority, in motion pictures it is the director. And if one falls into the hand of some cinematic Bigfoot, the price can be high. In this case it was intolerable, because it undoubtedly was Fritz's last score.

Lerner
1974–1980

I T SEEMED LIKE AN IDEA THAT COULD NOT FAIL. The most prestigious composer of theater and serious music in the United States collaborating with the country's preeminent lyric-librettist; *West Side Story*'s songwriter working together with the man responsible for *My Fair Lady*. In
America's bicentennial year. These two favorite sons, liberals both and roman-
tics, would be writing a musical of tribute, a love letter to their native land. It
would be the two-hundred-year history of how the White House had acted as
a symbol that held the country together, and the concept was so brilliant that
it could not possibly misfire. Called *1600 Pennsylvania Avenue*, planned for a
spring opening so that it would have acquired a patina by that auspicious Fourth
of July 1976, it was one of the most enthusiastically awaited theatrical milestones of the decade.

It was not the first time Lerner and Bernstein had worked together. Although
they mingled in the same circles and knew each other quite well, their styles were
totally disparate. Lerner had his feet firmly implanted in traditional theater with
a large dose of fantasy, while Bernstein's music, even his works for the musical
theater, except for shows written in his youth like *On the Town* and *Wonderful
Town*, were unconventional like *Candide* or messageful like *West Side Story*.

Both alumni of Harvard,[1] they had written two songs which were performed
at their alma mater's Carnegie Hall Benefit Concert in March 1957. They were

---

[1] Bernstein, who was born August 25, 1918, was six days older than Lerner. He had graduated
in 1939 and Lerner in 1940. Bernstein remained close to his alma mater and would cancel all
his engagements from September 1973 to July 1974 while he prepared Harvard's prestigious Elliot
Norton lectures, in which he espoused the superiority of tonal vs. atonal music. Lerner, never
having received his degree formally until after the opening of *My Fair Lady*, had little contact
with the university, although he had donated several valuable manuscripts including the original
My Fair Lady libretto and score to their Widener Library.

"Lonely Men of Harvard," of which the *New York Times* reviewer said, "Mr. Bernstein's tongue is so far up his cheek it is tickling his duodenum," and a serious hymn, "Dedication," whose verses, predicting the gravity with which the pair approached their collaboration, are printed below.

Where a wilderness had stood,
Where no page of time had turned,
Harvard rose that men might know
   The works of God, the works of God.
Where the spirit wandered lost,
Where no torch of truth had burned,

Harvard spoke for men to hear
   The works of God, the works of God.
Still the winds of falsehood blow,
Still through darkness truth must plod.
Harvard, Harvard, onward go!
   So men may know the works of God.

Only three years after they had produced those works that were put into archives and hastily forgotten, Lerner made a statement that seemed to indicate disdain for Bernstein's kind of musical. "Fritz and I don't believe in musical plays with messages," he said, obviously alluding to *West Side Story*, "particularly if the message deals with teen-age rumbles and switch-blade knives. To us the best message a musical can convey is: 'Come back and see me, often.' "

But in spite of that credo, Lerner seemed to eat his words when a momentous musical project actually began to take shape in 1972. Infuriated by the Watergate break-in, Lerner was having dinner with producer Saint Subber when he mentioned he "had been so depressed by the Nixon landslide that he thought he would write a musical about the first hundred years of the White House and other attempts to take it away from us."

Saint Subber went to Bernstein's agent, Robert Lantz, who was only too well aware of his client's feeling about the Republican intrusion into Democratic headquarters which eventually led to the collapse of the Nixon presidency. Lantz believed the pair could work well together. Both giant figures who had often spoken out about their political beliefs, they belonged to the liberal establishment and were both vehemently anti-Nixon.

Bernstein had never forgiven Nixon for ordering the invasion of Cambodia, which had provoked rioting on college campuses across the country. He and other liberals held the President responsible for the confrontations that led to the massacre of students at Kent State. Bernstein's *Mass*, which premiered in 1971, not only attacked Nixon but spoke out against authority, any authority, including the church.

Nixon, who had been invited to come to the premiere, had been warned by the FBI and J. Edgar Hoover to stay away. Although *Mass* was attended by most of Washington's pols, Nixon's absence spoke louder than his presence might have.

The story of the musical Alan first sketched out was not a diatribe, although the collaborators announced they planned to pull no punches. "Alan came to me with an idea for a show that lit up my soul," said Bernstein in 1986. "It was to be a celebration of the White House upstairs *and* downstairs, from its brave, rough beginnings onward through an amazing array of presidencies, warts and all; ups and downs in every sense."

Alan's musical was to encompass a time span that stretched from George Washington's presidency through Theodore Roosevelt's. Its central core would concern the presidents, their wives, and families, fictional servants and *their* attendant families. The principals in the cast would remain the same throughout the generations.

It was a big idea, so big that Bernstein likened its first draft to Wagner's *Das Rheingold.* Then called *Opus One,* in essence it was an idea that came very close to what Alan had tried to do in *Love Life* with little success. That canvas, it will be remembered, the succeeding generations of the Cooper family from the American Revolution to the present, had come out as a revue, or as Alan called it, "a vaudeville." Its impossible story concerned the vicissitudes that beset a family because of changing mores. Throughout the musical there remained one constant—love. In *1600 Pennsylvania Avenue,* as it was now called (the title taken from the White House's street address), the constant was to be another, perhaps less chimeric ideal—that the White House is the symbol of hope and freedom to the world.

Beyond looking at the soap-box oratory that the premise implied, it seems Lerner had again overlooked the theatrical axiom that predicts disaster when you abandon characters. The audience dismisses them too. As the various presidents and their families move out of the White House, the audience loses all concern for or interest in them. Lerner had closed his ears to the loud carping of critics who after *Love Life*'s debacle complained again in print that his story for *On a Clear Day* "brought us close and then snatched us away from the central plot."

Since Bernstein and Lerner had not set a definite time frame for the show's production, each became occupied in his own projects. Lerner, licking his wounds from *Lolita, My Love,* poured his energy into *The Little Prince* and the stage version of *Gigi;* Bernstein busied himself with his conducting and writing works of serious music.

Nothing much was changed about the sprawling project until 1974 when Saint Subber left it. "I loathed the show and I tried desperately to get everyone to abandon it," he is reported to have said. But he was not counting on the determination—and the drawing power—of the combined names—Lerner and Bernstein. Producers Roger L. Stevens and Robert Whitehead, noting that the bicentennial, only two years down the road, was already raising the public's fever, knew the project would sell tickets and stepped in. They hired Frank Corsaro, who had a background in both opera and musicals, to direct.

Producer Robert Whitehead and Alan listen as director Frank Corsaro explains the intricacies of the script to Leonard Bernstein. Co-producer Roger Stevens stands between Corsaro and the composer. MOCNY.

In 1986, with hindsight, Corsaro told Joan Peyser, Bernstein's biographer:

Stevens asked me to listen to the score. I thought it was Bernstein's most original, most interesting, best score since *West Side Story.* It was very long, very vital, and made an enormous impression on me. Then I took the book home to read.

I couldn't believe what was there. It had no foot in any reasonable foundation . . . something was very wrong with it. Yet with these collaborators, how could one say no? I thought I would see what could be done for it. Bernstein was critical of a few things in the book but basically he supported it. When I came into the show, Bernstein was generally behind what was there.

Against his better judgment Corsaro decided to go ahead. He and Whitehead agreed not to proceed with any actual production until the book had been improved and met weekly with Lerner hoping to get him to do the necessary revisions. But the show remained in the same incoherent state in which Corsaro had found it.

"During all of this," Corsaro added, "Bernstein was willing to do whatever he was asked—to change whatever we asked him. He was dashing in and out as usual. Doing a million other things. But the music was never our problem. One minute Alan Jay and I would be on the same wavelength and the next minute I would be talking to a stranger."

There is no doubt that Lerner was heavily dependent on the drugs Dr. Jacobson was supplying him with and was often totally distracted. Besides that, he was in love again. In the autumn of 1974 he flew to Haiti, obtained a Haitian divorce from Karen Gundersen (from whom he had been separated for two years), and on December 10, 1974, the day after his divorce from Karen had become final, he married for the sixth time. His new wife, British actress Sandra Payne, was almost thirty years his junior.

They had met in 1972 when the Lerners attended *Forget Me Not Lane,* a play in which Sandra was starring in London. The relationship began when Alan had gone backstage to congratulate the actress. It was not that Alan, at fifty-six, had grown older, simply that his wives were growing younger. As for the marriage, it was to be brief, less than two years, and was seemingly dissolved almost as gently as it had begun. After the divorce, Sandra Payne went back to England to resume her career.

During much of that time Alan was working with Bernstein and Corsaro trying to whittle *1600* down to playable size. Then in September 1975 *The New York Times* announced the startling news that Coca-Cola would be the sole backer of the show. The article went on to tell how J. Paul Austin, an executive of the company and a friend of Lerner's, had persuaded the soft drink giant to invest $900,000 in the American extravaganza. "Alan Jay went off with these people . . . gave them a résumé and played four or five songs for them," Corsaro complained to Joan Peyser. "There is an affable wonderful way about him and they went ahead without any further investigation."

Corsaro felt caught in a web, knowing that the one who brings in the most money has his way and sensing the producers would go ahead and put the show into rehearsal in the sorry state it now was. At that time Corsaro felt the show projected a "critical, acerbic view of the White House. The black element stood out with aliveness, while the Presidents lay there like some old fossils." Much was yet to be done to get the show to playing size. Bernstein was willing to make changes, but Lerner wanted to leave the book as it was.

Corsaro didn't realize that Lerner was at a loss. Used to dominating his collaborators, he felt intimidated by Bernstein. He had moved into the realm of "high art" and he was content to leave it at that. It was not a musical they were creating but a pageant, an accolade, a cavalcade—and as such Lerner felt continuity or story was superfluous.

Most of the music was in Bernstein's best, angular, concert style. Lerner, who sang his songs almost better than anyone else, always sang at backers' auditions. Of the score for *1600 Pennsylvania Avenue*, he said—without malice—that the music Bernstein set to his lyrics was too difficult for him to sing. "It was the only show I couldn't manage."

The Philadelphia opening on February 23rd was sheer disaster. Many in the audience left after the first act. One angry customer who was refused a refund of his ticket kicked in the box-office glass. Friends of the producers, Jerome Robbins and Arthur Laurents, came down to help, but no one could persuade Lerner either to rewrite the script or to close the show.

The collaborators put on a united front. When Corsaro told Bernstein and Lerner that the basic premise of the piece would never work, Bernstein answered, "We're a couple of rich, old Jews. We're kind of tired. Leave it alone."

And that's what he did. Asking to be fired so he could collect any royalties, Corsaro quit the project ten days before its Washington debut. Choreographer Donald McKayle left with him.

Producers Whitehead and Stevens and angel Austin, representing the Coca-Cola company, wanted the show to come off its pedestal. Feeling the show was patronizing and unfeeling towards the below-stairs staff in the White House, they insisted a new director Gilbert Moses and choreographer George Faison be hired. Both men were black, and over the objections of Bernstein and Lerner they began immediately cutting many references in the show to racial injustice. They felt the Lerner-Bernstein script was token liberalism and inserted to ease the white conscience. Soon the Mark Hellinger Theatre became a battleground. According to Corsaro, Bernstein and Lerner were barred from entering the theater.

André Previn saw the show at this point and said, "I couldn't believe it . . . I had never seen a show by two gigantic people like that so at a loss from the moment one. It looked awful. It sounded awful. It was about *nothing*. For two and a half hours they hammered it in. When at the end of Act One somebody spoke a line which queried the future, from behind the scrim in silhouette came the figure of Abe Lincoln, and the orchestra played *The Battle Hymn of the Republic*, I gripped my seat and said, 'I'm going mad.' "

By the time the show reached New York it was reported that "the entire structure of the story had been razed and the book largely overhauled," but it was to no avail. *1600* barely eked out seven performances.

The show was called "racist" by *Time*, while other critics used words like "embarrassing" and "amateurish." *Newsweek*, trying to make light of this "bicentennial bore," called it "a victim of myasthenia gravis conceptualis, otherwise known as a crummy idea."

Accustomed to failure by this point, Lerner went off with Sandra for a vacation in Bermuda after the show's brief run. Having persuaded his friend to invest nearly a million dollars of Coca-Cola's money in the show, he shrugged his shoulders and said, "Well, these things happen in the theater."

Bernstein was not so cool. He remained in New York and when interviewed told a reporter, "I'm shattered by the whole thing. The score was completely fragmented, not at all as I wrote it." He added that he would not do another musical until the "wounds heal." Apparently they never did for this was his last work for the stage that could be called "a musical." *A Quiet Place*, which premiered in 1983, was a pure opera, the aftermath of his earlier one, *Trouble in Tahiti*.

But he never lost his faith in *1600 Pennsylvania Avenue* and his love for Lerner. "In the three years of writing this work together," he said in 1986, "we grew very close indeed. I came to know and love a most gifted and generous gentleman, a gentleman-genius if you will. I am most grateful for the vast amount of fine fresh material we produced together; and some day, I swear, that material is going to achieve its proper form and become a show that will make us all proud."

Bernstein's esteem is far too lavish, but yet, even though the libretto is a fragmented embarrassment, the composer was right to notice many prideful things in the score.

The opening song, "It's Gonna Be Great" with lines like "Don't let go of the thread / Way up ahead / It's gonna be great. / Stitch by stitch and you'll see / Eventually / It's gonna be great," all set to Bernstein's chugging melody, sets the optimistic mood, opening the show in bright tempo. Later, Bernstein is at his hummable best with another lively song, "The President Jefferson March."

One of the most interesting moments of humor, considered racist but only because it was about domestics and masters, is the lyric Lerner wrote for the servants to sing to bid Abigail Adams welcome.

Welcome home, Welcome home,
Welcome home, Miz Adams.
We're glad that you is here.
De house ain't really finished
But it will be in a year.
Oh, de roof's a little leaky

But only when it rains
Dere's bolts fer all de winders,
All we needs is winder panes.
But if you can stand de wind, you'll love de view
Welcome home, Miz Adams,
Welcome to you.

The emotional melodic line that ran through much of the musical was set to "Take Care of This House," movingly sung by Patricia Routledge as the *first* First Lady to live in the White House. She is instructing a very young black servant lad before the Adamses leave. The song is then reprised by various occupants of the house throughout the evening. As a dedicatory hymn it is perhaps the only song from the show that has had a life beyond the brief Broadway run. It is certainly one of Lerner's finest lyrics.[2]

Here in this shell of a house
This house that is struggling to be,
Hope must have been
The first to move in . . .
And waited to welcome me.
But hope isn't easy to see.
(She turns to little Lud, as if to say goodbye)
(Spoken) Lud, you will stay here after I leave
    and I want you to promise me something.
Little Lud: (sadly) Yes, M'am.
Abigail: (sings)
Take care of this house.
Keep it from harm.

If bandits break in
Sound the alarm.
Care for this house.
Shine it by hand.
And keep it so clean
The glow can be seen
All over the land.

Be careful at night.
Check all the doors.
If someone makes off with a dream
The dream will be yours.
Take care of this house.
Be always on call.
For this house is the hope of us all.

After the debacle of *1600 Pennsylvania Avenue* the critics were merciless with Lerner. They implied that he alone had ruined the project with his fragmented book and that whatever slender talent he possessed should henceforth be applied to lyrics because he had no idea of current libretto construction. Columnists took the opportunity as well to remind the public of Lerner's failures with *Lolita* and *The Little Prince*. Much of their invective was not unlike the brickbats thrown at Hammerstein before he and Rodgers teamed with *Oklahoma!* Oscar, more private, could at that time bury himself in *Carmen Jones* and seems at least outwardly to have had a stronger sense of self-worth. One must contrast this with the certainty that no one in the theater suffered more from self-doubt than Alan. He was notorious for covering his hurts with his urbane insouciance.

But suddenly four major areas into which Alan could plunge his talent, his ego, and his heart were to be offered him simultaneously. And he was to accept them all. The first to come in was an invitation to travel halfway around the world to inaugurate the First International Theatre Forum to be held in January 1977 in Sydney, Australia—that would be his escape. Then there was an invitation to write an autobiography, a contract offered by publisher W. W. Norton—this proved to him at least that readers might still be interested in what literature he could pen. Third to come along that fall of 1976 was an agreement to work again on a lighthearted musical, the fourth collaboration, with his old, sometimes feuding partner, Burton Lane. He took this as an affirmation that

[2] Calling it his own "personal Kaddish," Leonard Bernstein chose to recite these lines at the memorial service for Lerner

LERNER, 1974–1980

367

his talent in the lyric-libretto line was still a viable one. Last, and perhaps most ego-building, was that he had fallen in love with a beautiful young woman less than half his age and she had agreed to become Mrs. Lerner No. 7.

Heavy preparation was necessary for the world-wide conference which was to feature two other towering American contributors to theater—Stephen Sondheim and Hal Prince—as well as representatives from musical theaters all over the world. In Sydney, Sondheim and Lerner conducted master classes on "Lyrics and the Lyric Theater," while Prince held his own forum, "Directing the Broadway Musical." Alan induced his production assistant Bud Widney to come along and stage a splendid closing concert of excerpts from his, Sondheim's, and Prince's productions. Then he inveigled his British secretary, Stef Sheahan, to leave London to travel with the group so she would be available to take the necessary mountains of notes that his multi-ventures would require.

One reason I went along was that the publishers of On the Street Where I Live (as his autobiography was eventually called) wanted a 40 or 60 thousand word synopsis. So he said, "You'll have to come because on all of these massive long flights I can dictate to you and you can type up the notes when we reach our destinations." So we flew to Australia and when we came in to London he delivered the synopsis. I typed in all of these very extravagant hotels everywhere. Alan moved the receptionists out of the way so I could use their typewriters. And then I was there for the beginning of his work with Lane on Carmelina.

With all the arrangements complete, the distinguished trio flew off. Although both of Lerner's companions believed the boss-secretary relationship was a cover-up for a romantic liaison, Sheahan and Lerner's association never went beyond the platonic.

Sheahan told this writer recently:

A lot of people thought there was, including Sondheim and Hal Prince. I haven't met Hal Prince since then, and we were very much all thrown together. But I spoke to him on the phone, and he said, "I did you a great disservice, I really didn't think you were a bona fide secretary—for want of a better word." And I said, "Oh, well, we can all get it wrong at times." It didn't bother me. It only bothered me there in Australia, because I didn't want this to be a cause célèbre . . . Alan didn't like huge gatherings. There was the last, gala night of this symposium. All of them were on stage with the budding lyricists, budding composers, budding theatrical directors.

There were masses of people, and after the show Alan got hold of my arm. He said, "Don't you dare leave my side" and so I went everywhere with him, standing there on the side of him. And I could understand why he said this. These women were just swarming, just engulfing him. "Mr. Lerner it's so wonderful to have you here". . . . and he'd say as he always did, "This is Stephanie Lyall" (as I then was) and they'd just look straight through me and just carry on with "would you like to come out to dinner tonight," or some such remark; and he'd say, "Stephanie and I are going to have dinner afterwards" and it was always this clenched arm around me. Because I could see what was going to happen—he was about to be wafted off into the Australian outbush or something. If I was protective it was because he had a special voice. He could be mean, caustic, elegant, cynical, but he had that gravelly voice. And he had grace, and wonderful manners, and he was *not* pompous. That was the last thing in the world he would ever be.

Besides Sheahan, Lerner had a long association with his indispensable secretary, Judy Insel, who remained in New York looking after his creative and financial concerns. Now that he found staying in London so agreeable, he was often in need of a British amanuensis who might serve as Judy's counterpart. Stef Sheahan was the fastest, most creative secretary, often able to supply just the word he was looking for, and he very often wanted to bring her to New York with him or wherever he was writing a new project. He had never come across anyone like her—the intense working relationship was to last until the end of his life—and his work was far too important to jeopardize it with an amorous dalliance.

"From the first time I worked for him, after I first met him, we sat down, and it just worked," Sheahan recalled of her earliest dictating sessions.

I take shorthand quite quickly, not remarkably quickly, but I don't suppose he had encountered somebody who had taken it that quickly before. So he just romped away, and then if he'd get stuck for a word I'd say "how about this" and then he'd say, "no, not quite right" and I'd say another word, or he'd say something and I'd say, "I think you can do better than that." I was not in awe of him . . . well, I was in one sense, but we were two professional people who were sitting down and each doing his job. . . . I was not treading on Judy's toes. Judy, to my knowledge, had never been involved in the creative process. Judy did a job that was unbelievable to Alan. She's a walking Bible of Lerner information. We hardly ever met because when I did go to work for Alan, which was about 5 months after that initial time, Judy was in another office and Alan and I worked out of his office on 48th and Madison Avenue. And there was my office and Alan's office with a grand piano, and that was when he was doing *Carmelina*. Judy looked after things—everything. She looked after his life and where his children were and all the rest of it.

By that time Alan had already met and fallen in love with young Nina Bushkin, daughter of Joey Bushkin, a well-known jazz pianist who had been a soloist with Tommy Dorsey's band and later a cocktail pianist in smart supper clubs. Nina, a child of the West Coast who lived the well-to-do life in Santa Barbara, was so young that when Alan first introduced her to Burton Lane, he was "stunned." Lane admitted that he "felt like saying, 'Can I carry your books home from school?,' " adding that he thought Alan looked twenty or thirty years older than Nina. Actually Nina, who was only twenty-six, was thirty-two years younger than Alan, who at fifty-eight looked even older—very gaunt and wrinkled probably because of his heavy dependence on amphetamines.

The marriage, which took place in November 1977, was to last hardly three years, during half of which time the couple did not live together. Soon Alan found himself once again in divorce court facing his adversary from the Micheline mayhem, Roy Cohn. This time, although there was less notoriety to do with the case, the deck was heavily stacked against Lerner, for Cohn had allied himself with Marvin Mitchelson, specialist in winning high alimony for Hollywood stars. So although hope sprang eternal in his heart, his personal fortunes at this time were at a very low ebb.

But there was promise. In 1978 Lerner had published *The Street Where I Live* to glowing reviews. The autobiography of the lyricist who seemed to live out his life on the front pages of tabloids was, according to Seymour Peck of *The New York Times*, remarkably reticent, and the critic was astounded that Lerner "avoided all talk of his broken marriages and psychoanalytical soul searching. Mr. Lerner will have none of the confessional approach; he is clearly a civilized, attractively modest, gentlemanly fellow who believes that what the world most wants to know about him is how he came to write his three greatest successes, *My Fair Lady*, *Gigi*, and *Camelot*." The book, witty, charming, and by turns hilarious, is by now in its umpteenth paperback edition and, I am told by its publishers, still selling briskly on both sides of the Atlantic. In this writer's opinion it is the most candid and charming foray into the creation of works for the lyric stage ever penned.

Some surcease from the court battles was available to Alan, who had started on *Carmelina;* and although his relationship with Burton Lane had never been smooth, both men were producing worthwhile material for what was to turn out to be a charming show. They were aiming for a Broadway opening in the spring of 1979.

The renewed collaboration had begun when Alan called Burton to reread the story of an Italian woman in wartime Italy who had brief affairs with a couple of American GIs.

The original story was published in *The Times* of London but had been blown out of proportion by a 1969 film called *Buona Sera, Mrs. Campbell* which had starred Gina Lollobrigida.

Both Lane and Lerner had seen the film and had not cared for it. The film had focused on the men, making the woman little more than an adventuress. Lerner and Lane planned to make her sympathetic and since she was hardly middle-aged, to add a love interest to the story of Mrs. Campbell. Alan said "she has based her life on a lie, and the lie becomes so strong for everyone, and so real, that in the end, the truth is not as happy as the lie. The show really is saying 'honesty is really not the best policy,' kindness is." Can anybody wonder that Lerner would be attracted to his version of the story? It gave him a chance to preach his philosophy of life all over again.

Alan rewrote the plot now so that it concerned a respected young Italian widow whose teenage daughter believes with the rest of the village of San Forino that Signora Campbell is the widow of an American GI killed in a World War II battle in Sicily. Eighteen years after the war, the American battalion that liberated the village decides to organize a reunion. The prospect of all those Americans buying souvenirs galvanizes and excites sleepy, almost bankrupt San Forino—but not Carmelina, who had been involved with three of the soldiers in that long-ago April and has been receiving support checks for "their daughter's maintenance" from each of them for lo, these eighteen years.

After a party to celebrate their arrival, the three soldiers slink off separately to visit Carmelina. Although she tells them the truth, each of the men believes the daughter looks like him and is really his progeny. The daughter, however, is so ashamed by her position and her mother's deception that she attempts to throw herself away and elope with a local fisherman. The three "fathers" interfere and save the day, acknowledging that Carmelina's motives for the deception were unselfish and were arranged to give her daughter some "American creature comforts." By the curtain's fall, Carmelina, not yet forty, who has for the past eighteen years refused to marry her lover Vittorio, finally agrees to their wedding.

Lane, who had his own objections to the film, approved of Alan's revamping of the story. "In the film the guys were very lecherous. I hated it when they left their children to see if they could make it with her . . . it was distasteful to me." The composer recalled that "when we were writing *On a Clear Day*, Alan had written a number called 'Someone in April,' which was originally about loneliness. I never liked it there and it wasn't used, but here I could believe it as the song which the Signora reveals her background. It drew a warm audience response."

It is a key song to the show, and with Lerner's magnificent revised lyric contains not one false narrative step. To avoid sameness, as the suitors enter, Lerner gives each soldier a totally different personality. As a tour de force for the show's star, Georgia Brown, the song actually supplied the "heart" of this one-joke show while stopping the action nightly.

All alone . . . seventeen;
The type that De Sica has in ev'ry scene;
A poor little sparrow in the human storm,
My hands with no other hands to keep them
     warm . . .
I looked and saw a man.

And my life began . . . When

Someone in April—
A stranger in April—
Said could he come in for a while;
Somehow I knew from his smile

That he would be
Gentle with me.
Little by little my heart
Began to fill
Soon we were never apart—
Until . . .

Someone in April
One morning in April,
Before he went out of the door
Said: Thank you for April—
And I was all alone once more.
All alone. Just sixteen.

I was Mimi in the final scene.
I wept—I don't know—till almost four o'clock;
And then very faintly, I heard someone knock.
Come in, I suppose, I must have said
And when I turned my head,
All my sorrow fled . . . for

Someone in April
Was lonesome in April,
As lonesome and helpless as I:
Oh, but how bashful and shy!
Could I . . . ? said he . . .
That is, could we . . . ?
Holding him close for dear life,
I lived again.
Mother and sister and wife . . .
But then—
One day in April
My someone in April
Left roses with love at the door;
That faded in April
And left me all alone once more.

All alone. Blue with cold.
My hands with no other hands for me to hold;

A child with a woman lurking in her breast;
A poor little pigeon in an empty nest.
And then out of nowhere I was blessed . . .
      with

Someone in April
My life had become April
The moment he kneeled at my side.
Something about him implied
He hoped he might
Stay for the night.
Soon all the room in my heart
Was filled again.
Soon we were never apart . . .
But then—
Someone in April
One evening in April
Went out to the neighborhood store—
Leaving the soup to get colder;
Leaving the wine to get older;
Leaving me all alone once more.

Someone in April
It had to be April
That one little month I was with
Braddock, Karzinski and Smith
It had to be
One of the three
All of them came through the door
Like cavaliers.
One of them left me with more
Than tears.

Someone in April—
It happened in April
That one of those generous men
Made certain in April
I'd never be alone again.

Carmelina boasts one other number that has had a life, albeit minor, outside the theater. A favorite of the sophisticated, more mature night-club chanteuse, "One More Walk Around the Garden" is perhaps the most rueful song Lerner ever wrote. This writer finds it not unlike Hammerstein's "When I Grow Too Old to Dream," yet possessing a universal relevance that Oscar's lyric only hinted at. The lyric, intended for the returning soldiers but appropriate to anyone who is closing a chapter of his life, is printed below.

Refrain

For one more walk around the garden,
One more stroll along the shore;
One more mem'ry I can dream upon
Until I dream no more.

For one more time perhaps the dawn will
    wait—
For one more prayer, it's not too late

To gather one more rose
Before I say goodbye and close
The garden gate.

One more rose
Before I close
The garden gate.

In spite of a lovely but sometimes sophomoric score that included "Why Him?" (Why do I dress for him? / Why the pain of eating less for him? / But when I see him coming near me / And my head begins to swim / I keep thinking as I nod / My God! Why him?) and "It's Time for a Love Song." (And I have a love song / A song for the season / That's filled with the fervor of / A man in love / Beyond reason . . .) as well as the songs reprinted above, by the time *Carmelina* was previewing in Washington, D.C., it did not have the smell of success about it. Lerner and Lane squirreled into across-the-hall suites at the by-now infamous Watergate Hotel. They emerged only to pace the Kennedy Center Opera House, a big barn of a theater where *Carmelina* was booked, or to have a work session that usually ended in a squabble.

"I'm a very disciplined writer," Lane rather ungallantly told a reporter for *The New York Times*. "There's no fooling around when I'm working. I get consumed. Alan also works crazy hours, but Alan also takes a lot of trips. It's marvelous to have someone who can write both book and lyrics, but it takes him a long time to get to the point where he really plunges in. I arrive much sooner."

Alan explained that what he thought Burton was talking about was that "I work for three or four months, around the clock and then I stop for two or three weeks. Burton does a certain amount each day." Further explaining his methods he added, "Yesterday I went all day on two lines. I started at 6 A.M. and finished at 2 A.M. I won't tell you what the lines were. You'd think, 'How could that take so much time?' All right, I'll tell you: "Why does he make my feathers fly? / And get my dander up so high?" I was trying to find more typical ways of saying why does he irritate me."[3]

Alan enlisted the aid of his daughter Jennifer as "production adviser," and it was she who told me recently that "a lot was cut, but Dad and Burton were fighting all the time. I think they didn't really like each other. Dad was difficult and so is Burton Lane. Everybody had his own opinion about the way things should be. Besides he really didn't get along with José Ferrer, who directed it. There were things that were being rewritten all the time. You know, the show

---

[3] The couplet was intended for "Why Him?" The lines were deleted from the Broadway version of the score.

actually was pretty well received on the road, it got mixed reviews, but in New York . . . they said the costumes were terrible, the choreography wasn't great. Old-fashioned."

Jennifer went further, wisely describing what happens especially in theatrical previews. She talked about listening to the collaborators argue over a line . . . a phrase . . . a word. "Sometimes in previews you think if you take out one line that whole scene will work. But the audience does not even know that one line was in there before—and they still see a scene that doesn't work. It's so dramatic for the people who are close to it, but it matters little to the audience."

But eventually *Carmelina*, under the direction of José Ferrer, kept its commitment to open at the St. James Theatre on Broadway on April 8, 1979.

Had it played a few more performances there might have been an original cast recording which perhaps might have extended its run, but after tepid reviews it closed seventeen performances later.

Thomas Shepard, who seems to have been in charge of producing and recording every worthwhile show of the '60s and '70s,

wanted desperately for it to work. It's got beautiful things in it. The whole idea of Alan and Burton Lane working together again was so attractive, as were the songs like "One More Walk Around the Garden," that's Alan at his most yearning. Alan is at his best when his concern is for what will be. Unfortunately, the book wasn't consistent . . . old fashioned. The score that was written for it could just as easily have been written in 1945 but nevertheless it felt like a lovely dying gasp from two great writers . . . There were more gasps to come, but in this one you really knew what made them great.

Too bad the greatness was in flashes. The thing that made *Carmelina* so troublesome and so difficult to turn down—and we [at NBC] eventually did turn it down were these flashes of brilliance which made you say, "Oh, my God, if only the whole show could be like that." Because you realize you're in the presence of two great writers, but whether it was the subject matter or the time we live in or the ability of the book to work, the flashes weren't enough to get you through the whole show and I really was sorry about it because the show moved me.

Burton Lane has not given up hope that a revised version can be mounted on Broadway and when this writer talked to him about it in 1992 reported he was working with a lyricist-librettist to redo the score for eventual production.

Alan, who was certainly at a lower financial and critical ebb than he had even been after the closing of *1600 Pennsylvania Avenue*, took off immediately for an extended stay in London. He had been invited to assist in the direction of a *My Fair Lady* company that was forming in the hinterlands. The Eliza of the company was a young actress named Elizabeth Robertson, and although Alan didn't know it at the time, the relationship was to bring him the true happiness and love that had eluded him all his life.

Lerner
1980–1986

"SHE WAS VERY DOWN TO EARTH, very loyal and much younger," said Alan's longtime friend Ben Welles of Liz Robertson. "Alan's fortunes were down, he'd had a lot of flops. He was emotionally insecure and wondering whether his talent had run dry. . . . He went to England, he discovered her and saw that he could really do something with her—and in the process fell in love with her, and vice versa."

Christine Elizabeth Robertson, born in 1954, was thirty-six years younger than Lerner when they met in 1979. She had grown up in Ilford, an eastern suburb of London, the daughter of a policeman and a frustrated actress. Her roots were definitely middle class, although one must add that both her parents were somewhat artistic; her father had played trombone in the British army and her mother had done a little acting.

Liz trained at the Finch Stage School in London, but one could say she was baptized by the fire on stage. As an ingenue she first performed in *Side by Side by Sondheim* and followed this with a role in the same composer's *A Little Night Music*. The unique, almost piercing quality of her singing voice, coupled with her youth, energy, and mastery of the cockney, made her an ideal choice for the role of Eliza in *My Fair Lady*.

The production had been touring when Alan, who had been separated from Nina Bushkin for over a year and was awaiting his divorce, arrived to direct it preparatory to its entrance into London. After her debut in the West End critic Jack O'Brian wrote about Robertson in the *Daily News*. "She is beautiful, young, musically correct, more than a credible dancer, has style, poise and the other requisites for musical comedy stardom."

Since he was also the company's director, Alan maintained a professional attitude towards his Eliza and did not see her socially while they were on tour. But once the troupe opened in London, he told an editor of the *Daily Mail* that the reason he invited Robertson to dinner on the night after the opening was to discuss her performance in a more casual spot. "I would have invited any leading lady out even if I hated her. It was work . . . but I couldn't help thinking how terribly attractive and charming she was.

"I took her to the Savoy which was across the road from the theater, and we talked over the 'Loverly' number. And then we went home for the night— separately. But each time we met we became closer mates." Alan was called frequently to work in Paris on a proposed screenplay for *The Merry Widow* during the next weeks, but whenever he returned he asked Liz out to dinner.

Explaining that being the star of a hit show scares most of the eligible men off, she told Gene Lees, "I wasn't going out with anybody, in fact I was leading a very lonely existence." Lerner admitted he was lonely too. "But I don't think you can put down what happened to us to a negative base. It was all very positive," he insisted.

After his work in Paris "folded," Lerner mentioned that he felt he should go back to New York, but he wanted to stay on simply to be with Liz. "I wanted her to *ask* me to stay, but I wasn't sure she felt as much as I did."

For her part Robertson recalled the special moment when she realized that "New York was much too far away. Paris hadn't been so bad. So I said, 'Oh, please don't go.'"

And he didn't.

"From that moment we have been inseparable. Sometimes it does seem a strain, and difficult for people to believe that two people who have such a disparity of ages and upbringing as we have can be so compatible, but we are. I have never known anything like this in my life."

On August 19, 1981, they were married in a simple Unitarian ceremony in the small town of Billingshurst. Only a dozen intimate friends were invited, and a few days later Liz returned to the Chichester Festival where she was appearing as Nancy Mitford in a pastiche called *The Mitford Girls*. "This is the first time I feel like I'm married," Lerner told a reporter for *The New York Times*, "where I'm not looking at the door ever, and I have no secrets. Bachelorhood is a bore, and very time-consuming—especially if you want to write," he added. "I was never upset when a marriage ended. I would be now."

The relationship seems to have wrought an unbelievable change in Lerner. According to intimates, he stopped using amphetamines (although he still took vitamin shots) and seemed much more controlled and calm. He even *looked* younger; many of the heavy lines that creased his face now seemingly disappeared. Now he rarely went to opening nights and show biz parties, rather naïvely claiming, "That was the trouble in past marriages."

For her part, it is clear that this was a loving relationship, one in which Liz Robertson was never riding on her husband's coattails. The critics and public alike had proclaimed her a star from her first performances as Eliza Doolittle. But it did not stop there. She seemed to shine in every role. Respected British critic Sheridan Morley took special note of her performance in *The Mitford Girls,* and Andrew Lloyd Webber thought so highly of her talent that he entrusted her to take over the solo role in *Song and Dance.*

Lerner, too, put together a one-woman show for her at the Duke of York's Theatre. He wanted the public to see how versatile she was. But she realized how desperately Lerner needed her near him and tried to spend as much time at home as possible. "If I didn't push her," he told a reporter, "she would just stay at home with me all day."

Soon he began thinking of writing a new musical—adapting a play that "he had been in love with ever since he saw it as a teen-ager with the Lunts." (Except for *Pygmalion* he had never adapted a play before.) It was Robert E. Sherwood's 1936 Pulitzer-Prize play *Idiot's Delight.* "I started thinking about doing it as a musical several times over the years, but the time never seemed right. I knew it couldn't be done without a real leading lady and we don't have many of them who can act and sing and have some stature. Then when I thought of Liz I said, 'Here's a real leading lady!' "

The theme of *Idiot's Delight* is such a familiar one in the Lerner canon that Alan felt obliged to defend himself to a reporter from the *Daily News* who attacked him for writing the same story once again. "Several reporters asked me after *My Fair Lady* and *Gigi* opened how I could do two stories in a row about the transformation of young women." Skirting the question, Alan merely smiled and said, "If you are asking me if I would rather not have written *My Fair Lady,* I have to tell you I'm glad I wrote it. I'll admit that the transformation of women is a common theme in my work. It also appears in *On a Clear Day* and in this show where a woman tries to deny her former life."

But Sherwood's *Idiot's Delight* is much more than that, although fundamentally a propaganda play in which, as Brooks Atkinson, wrote "enjoyment keeps breaking in. War," he added, "seemed likely but no one could really believe it." Its message was more about conscientious objectionism to war, isolationism, and the destructive threats of Fascism and Nazism than it was about a woman who pretends to be other than what she is. Had Lerner attacked the real issues as, say, Terrence McNally's recent libretto for *Kiss of the Spider Woman* did, along with his usual musical comedy approach, he might have been able to turn this box-office failure around.

In the original play, Alfred Lunt played a third-rate hoofer named Harry whose act incorporates six bleached blonde strippers. They are booked on a tour of the sleaziest resorts in Europe when they and many other guests are detained by the military in a hotel in the Italian Alps as hostilities leading to World War II escalate. In the cocktail lounge—the unit set of Sherwood's play—Harry encounters a couple of British honeymooners (who will comprise the subplot). But his interest is really aroused when a woman resembling a long-lost love with whom he had a memorable one night stand enters. The woman Lunt fell for, Lynn Fontanne (she calls herself Irene, pronounced Ee-ray-na), formerly a red-haired showgirl, now sports long blonde hair and pretends to be minor Russian nobility whose family fled during the Revolution. She is now the mistress of a wealthy munitions manufacturer.

Throughout the first two acts Harry tries to recall how he knew that face, and it is to Sherwood's credit that he keeps us guessing as to whether the woman is a fraud or not. Sherwood implies that she may be just a pawn in the game of "Idiot's Delight, a solitaire God plays that never ends."

In the last act, just before all the detainees are released to go their various ways, Irene confesses her identity to Harry (still somewhat obliquely.) Having insulted her munitions-maker benefactor partly because of having seen Harry again, but mostly because he is a truly abominable person, she is abandoned by him. The munitions maker joins Harry and the strippers along with the British couple who are already on the train that will cross the border to Switzerland. Irene is left alone as the bombs begin bursting around the hotel. Then, in a surprise romantic twist, Harry pops back to face possible annihilation with Irene, convinced that what he had or might have with her is better than anything else in his mercurial past.

In 1939, the play was filmed with Norma Shearer in a long blonde helmet-like wig and sporting a two-foot-long cigarette holder, while using the phoniest Russian accent ever cinematized. The munitions maker was dourly played by Edward Arnold, but the enigma of the film was Clark Gable, giving his usual surface performance as the hoofer. One didn't know if his bad singing was real or simulated when he sang and tap-danced Irving Berlin's "Puttin' on the Ritz."

With war already on Europe's doorstep, the film lost the play's uncanny prescience. With its unconvincing acting, it seemed more agitprop than Sherwood's sci-fi-fantasy, and with no electricity passing between Gable and Shearer, so dull that it would only delight an idiot.

Lerner's script needed little transformation, but because he finally decided the scene had to be the Austrian Alps and placed the conflict between the Soviets and NATO in the present, he tinkered, perhaps foolishly, with several major details. His original idea was to set the musical in Cuba at the time of Castro's takeover, but then he decided to contemporize the work by moving it to present Afghanistan.

"I wanted to set the show in a day when there might possible be some confrontation between East and West. I started to figure out what would cause it. It was during the days of the hostages so I thought, well, if I were the Russians, I'd march into Afghanistan and that would start the trouble. I started to write it and about six weeks later they did march into Afghanistan—and nothing happened." At last when Lerner read an article in which an official of the Austrian government discussed his country as "living in fear," he began to construct the play.

"Sherwood," Lerner observed, "set his play in the *imminent* future, envisioning World War II." The eternal optimist, Lerner chose to call it the *avoidable* future, or the eve of World War III, and changed the title to *Dance a Little Closer* because, as he explained, "If we do hold each other closer, perhaps confrontation can be avoided."

Having his leading lady Liz portray an American showgirl, now renamed Cynthia Brookfield-Bailey and masquerading as a member of the *British* upper class, was a clever move. To have her traveling with a career politician modeled after Henry Kissinger in the early '80s, when the Soviets and the U.S. seemed embroiled in daily crises that looked like they might lead to the annihilation of the planet, was even more clever. Liz Robertson played the role believably, but partly because leading man Len Cariou didn't have much fire vis-à-vis Robertson, or perhaps because Cariou feared too great intimacy with Mrs. Lerner while Mr. Lerner the director hovered over his shoulder, the relationship never let off any sparks.

In his libretto Lerner committed the same mistake that so grievously flawed *Lolita, My Love,* when he showed, via flashback, the couple in love early in the show. The scene uses a charming love ballad but removes most of the ensuing suspense. Worse than that was what Lerner might have assumed was a contemporary switch. He transposed the subplot honeymoon couple into two gay stewards from the airline, added a bishop detainee, and had the two men plead with the clergyman to be married before the bombs of war exploded.

"I wanted to make a statement about love being essential to all lifestyles," he said. "I have to admit some people have been critical of my including the two boys in the story, but I've found you have to write something because you believe in it."

One can see even in the use of the term "the two boys" how demeaning Lerner's attitude was. Jerry Herman and Harvey Fierstein had treated homosexual love (and marriage) with dignity two years earlier in *La Cage aux Folles.* The switch now was offensive and Lerner's limp lyric that "any love is better than no love" is old hat:

Anyone who loves,
People anywhere;
Anyone who loves,
They deserve a prayer.
We're only living by the hour,

While the sages with the power
Play their game of peace and war
With no shred of pity for
Anyone who lives . . .
Anyone who loves.

For this play, Lerner teamed with one old friend and one new. The former was producer Frederick Brisson, who had served in that capacity for *Coco*,[1] while the latter was Charles Strouse. If it weren't for the former there might not have been the latter.

Early in 1981 Brisson had committed himself to producing the show but a composer had yet to be found. Then the producer threw a big opening night party for his play *Mixed Couples*—which turned out to be a flop. When he walked across the floor and spotted composer Strouse, embarrassment set in because they both knew how hollow the show and the attendant party were.

"I didn't know what to say," Brisson recalled, "except to pat him on the back and tell him how nice it was to see him. Instead, I said, 'Hey, how would you like to do a musical with Alan Jay Lerner?' He said, 'I'd give my right hand.' I said, 'Well, don't do that. If I can put you together with Lerner you're going to need that hand.' "

When Brisson put them together they knew each other slightly but had never collaborated. They had become acquainted when Strouse got on a plane to California and discovered he had forgotten his wallet, money, and credit cards. "Sitting right in front of me was this man I recognized, so I introduced myself and said embarrassedly, 'I forgot all my money.' 'Don't worry,' he retorted, and gave me $40." When Strouse told this writer the story recently he added: "Alan loved this story, and has increased the amount to $90, but it was really $40."

Strouse, who was ten years younger than Alan, got along splendidly with his new collaborator. Lerner appreciated Strouse's facility as well as his lighthearted manner, so different from that of Burton Lane. More than that he relied on Charles's advice in the lyric area, for here was a composer who had produced several shows for which he had been responsible for music *and* lyrics. Collaborating mostly with Lee Adams and Martin Charnin, Strouse had written a string of hits—and a few megahits—since his first smash, *Bye Bye Birdie* (1960). His best known shows before he teamed with Alan were *All American* (1962); *Golden Boy* (1964); *It's a Bird, It's a Plane, It's SUPERMAN* (1966); *Applause* (1970); *Annie* (1977)[2]; *Flowers for Algernon* (1979)

They worked mostly in London. Lerner bought a charming house on

[1] Brisson was one of the few remaining producers who would touch a Lerner show. He had actually made a small profit on *Coco* and had been clever enough to close the show (over Lerner's and Previn's objections) shortly after Katharine Hepburn left.

[2] Although Strouse's follow up to his *Bye Bye Birdie* (called *Bring Back Birdie*) was as resounding a flop as *Yes, Yes, Yvette* was after *No, No, Nanette*, in 1993 Strouse broke the jinx that "the-sequel-is-never-the-equal" with a very good show, *Annie Warbucks*, which brought back many of the characters in his by now classic *Annie*. Unfortunately, the show was never able to make the transfer to Broadway and closed after a few months.

Gramaton Street which he and Liz decorated together. They were planning to make London their home, returning only to Broadway to fulfill their theatrical obligations. But even to his new, warm, and comfortable house Alan brought along his own string of neuroses; especially when Liz might be on tour or away from him, he was still tortured by his silent ghosts and terror of darkness and death. The busier he kept, the less those thoughts intruded. Planning a new project seemed to be the only thing that allayed those fears.

In addition to working with Strouse on the show, Alan loved to dwell on the past. Since his own biography had been so well received, he now decided to write a history of the musical theater. It was a subject that intrigued him greatly and one which he knew intimately, so it was easy for him to obtain a contract with Collins. The attendant advance also came in handy.

His secretary Stef Sheahan took all the notes for the book. When asked about their working arrangement, Sheahan said, "We worked very long hours, I think because he had this terror of being alone. I often said to him 'Look, if you like, I'll leave the office early, I'll switch on all the lights so you won't walk into an unlit house.' It sounds very funny, but it wasn't that funny. He wanted lights on everywhere. He wouldn't leave a house, as most of us do, you know; most of us switch off all the lights and leave a little hall light. But Alan left the house blazing."

By the time he and Charles Strouse had completed their adaptation and were going off to New York for casting and rehearsal, Lerner decided to sell the Gramaton Street house and have all of his and Liz's possessions moved to a large flat he bought on Branagan Street. Alan maintained that it was wasteful to have a whole house just sitting unoccupied in prime real estate area—but the word on the street was that he was broke. He could save face by announcing that he expected his new show to be a hit, to have an extended run, and for them to be away from Britain for many months.

In order to ensure a goodly income so that he would not flounder under rentals and alimonies, his contract called for him to receive a full 8 percent of the show's gross. It is common for the librettist to receive 3 percent, the lyricist 3 percent, and the composer 3 percent (which Strouse did). The extra 2 percent was for directing. If the show was a hit, not even a smash hit, Lerner's share would have amounted to some $20,000 per week. Robertson, too, had an ironclad contract with a guarantee of $5,000 per week.

Producer Brisson, a self-proclaimed "stickler for money," brought the production to Broadway for two and a half million dollars (at a time when *Cats* and *La Cage* were in the four and five million area.) With difficulty he finally convinced theater owners Minskoff and Nederlander to join him in the production and, by scrimping, opened at a bargain thirty-five-dollar top, when other shows were charging forty-five.

Dance a Little Closer looked like a shoe-in; the only other musical on Broadway's horizon that season was *My One and Only,* a revival of an old Gershwin score. In April 1983, while *Dance* was previewing, theater touts were betting on the Lerner-Strouse score while nixing the Gershwin. Unfortunately, all that changed on the night of May 11 when the cheap ticket price became meaningless after *Dance a Little Closer* was roasted by every critic. It joined an elite group of Broadway productions that closed after a single performance. *Dance a Little Closer* became forever known by the Broadway soubriquet of *Close a Little Faster.*

Its score,[3] some of which I have excerpted below, suggests it deserved a better fate. Lerner's lyrics, as in *Carmelina,* often display brilliant flashes of wit while moving the plot along. From the opening number, "It Never Would Have Worked," Lerner clues the audience to Harry's raunchy character:

I met a girl in Tallahassee,	I met this cat in Amarillo.
She was as hot as pompano in a pan.	An acrobat upon the pillow.
I was her man,	She was as wild as any cat in the zoo.
Yeah, but I ran . . .	What did I do?
She was baa-baa-bad—	Man, but I flew . . .
Three bags full.	She liked short and tall,
It never would have worked.	Male and fe-
She was in-	It never would have worked.
Satiable.	Half the town had her key.
It never would have worked.	It never would have worked
Three days—three nights—	Sears and Roebuck
I thought I would die.	Envied her supply.
It never would have worked.	It never would have worked
But it would've been fun to try . . .	But it would've been fun to try . . .

Later, Irene brilliantly limns her relationship with the Kissinger-like politician in "He Always Comes Home to Me."

. . . He postures, he poses,	'N homme fatale,
He dallies, he baits.	Which he does with discrete finesse.
He gathers his roses	But I think he would run
But never pollinates.	Like a deer
And be it because of defeat or ennui	If someone
He always comes home to me . . .	By mistake ever said yes.
It inflates his morale	So he'll woo her, disarm her;
To appear	His wit captivates.

[3] In an unprecedented move, two weeks after closing the company was reassembled to make a cast album. Curiously, the album (perhaps because of the appalling quality of its sound) was not released until 1988. Remastered and transferred to cassette/CD, it was improved and rereleased in 1990 and is now available from TER Records. Listening to the recording, one can hear why, in this writer's opinion, *Dance a Little Closer* deserves to join the pantheon of distinguished flops that include *Mack and Mabel, House of Flowers, Allegro,* and *Merrily We Roll Along.*

A consummate charmer
Who never consummates.
Who when he's exhausted his bright repartee,
He always comes home to me . . .
So I smile at him brightly

And never complain;
And hold on so lightly
He never feels the rein.
And ergo, because he believes he is free,
He always comes home to me.

George Rose, too, playing the politico, had his moment in "A Woman Who Thinks I'm Wonderful" (Docile and sweet she should be with me, / That and nothing more. / Should she by chance disagree with me; / Liebchen, there's the door). Nor were the three girls who played Harry's back-up singers neglected. They were given a perfect comedy number, "Homesick," for this musical about the annihilation of the world.

Bebe: . . . Three Mile Island!
Never thought I'd miss it so.
Daddy says that half the folks
Have moved away;
And the meadow by the fact'ry that
Was green is turnin' grey.
But I sure would wake up smilin'
To be home in Three Mile Islan'
In Pennsylvan-i-ay
In the good old USA . . .

Elaine: . . .Sweet Palm Desert!
On the San Andreas Fault.
Every year we get more land
And never pay,
Cause every year the highway moves

An inch or two away.
What I'd give to be there lyin'
In the blazin sunlight fryin'
In Californ-i-ay
In the good old USA.

Shirley: . . . Love Canal!
Never heard a sweeter name.
Sure the water now and then
Comes out shellac;
And a while ago the birds went south
And none of 'em come back.
But God would I be happy just to see my Ma
    and Pappy
And to hear Niagara fall
In the greatest state of all . . .

Besides the title song, for which Strouse supplied a charming melody, Lerner included his obligatory soliloquy, "Another Life," in which Irene tells Harry why she settles for money rather than love; and the couple's torrid duet, "There's Never Been Anything Like Us" (I've been around / I've had my share / Of tangled arms / And tangled hair / I've touched a cheek / That left me weak / With appetite / But there's never been anything like us! / There's never been anything like us / There never was anything like us—last night).

After the closing Lerner and Robertson returned to London. He told Sheahan that the flat seemed dingy and he was sorry they had sold the house, all of which she ascribed to his discontent with the outcome of *Dance a Little Closer*. But if he was anything, Alan was an eternal optimist, and undismayed he plunged into a new project, an adaptation of the 1936 Universal film *My Man Godfrey*, which starred Carole Lombard and (ex-husband) William Powell. Always the professional, Lerner's first move was to contact Producer Alan Carr and secure the rights to the original story and screenplay.

The zany Depression comedy by Morrie Ryskind and Eric Hatch about the foibles of the very rich was typical Lerner fare—transformation. The story concerned an inane and rather unlikable millionaire family who, in the course of a scavenger hunt—a game of bringing home "disused objects"—pick up a tramp who is looking for a job. Noticing a certain elegance about Godfrey, they return with him and win the prize when they introduce him as "the forgotten man"; they then hire him to stay on to be their butler. Lombard and her sister, Gail Patrick, both fall for him, but what chance has anyone against Lombard?

Later in the scenario, when the family fortunes begin to fail, Godfrey saves the day with his uncanny mastery of the Stock Exchange. All this betrays to this flighty ensemble that before the Great Depression, Godfrey was richer than they and hid out from his family because he was sick of the hypocrisy of the world of the wealthy. He also had far more "class." Although the story opened with recriminations against the mindless and heartless rich, in the end Hollywood (and Lerner) conformed to the happy ending where all the millionaires turn benevolent and Godfrey opens a most successful "City Dump Café," giving jobs to all the "forgotten men."

The story is obviously a kind of male *My Fair Lady*; another *Cinderfella*, reminding us of the Depression and far out of step with '80s prosperity. But unconcerned, Lerner's next move was to his usual outline—in this case, fifty scenes that lay the ground plan for the libretto. Then the usual search for a composer began. Alan auditioned all the fairly well-known composers in England with little success. Finally he went to the office of an old friend, Jim Henney, head of foreign rights at Chappell Music, his publisher, and asked if he knew of anyone on their staff who could be comfortable writing "modern depression type music."

"My friend Jim Henney," thirty-five-year-old pianist-composer Gerard Kenney said recently, "calls and says to me, 'There's a guy over here, a librettist and lyricist and he's been through fifteen or sixteen composers and he's very unhappy. So I gave him your name.' Well, I'd had a few hits over here and was feeling very cocky and said, 'So, what's he done?' He said, 'Other than *Gigi*, *My Fair Lady*, *On a Clear Day*, *Paint Your Wagon*, *Camelot*, and *Brigadoon*, not a lot.' I said, 'You mean Alan Jay Lerner?!' And he said, 'Yeah, you're having lunch with him on Tuesday.'

"So I go over to his house, knees knocking, walk in and there is a beautiful mahogany piano. I played him every song I wrote. My best hits—'New York, New York,'[4] 'The Minder Theme,' 'Son of a Song and Dance Man,' 'Nickles and Dimes'—all the songs I thought he would really be impressed with."

Lerner *was* impressed with Kenney's facility and candor, and by the end of

[4] Not the megahit of the same title by Kander and Ebb.

October 1983 the two began to work together in earnest. But Lerner never mentioned the plum of collaborating on *Godfrey* until weeks later after they had written several songs.

Kenney's music, while having the contemporaneity of a Peter Allen or a Barry Manilow, has much of the verve of Cole Porter and George Gershwin. Perhaps that is because this actual son of a song-and-dance man who grew up with vaudeville around him had been enmeshed in it from the earliest age.

"In my house, if you didn't play, you didn't eat. An Irish-Catholic family with a load of kids. When the war came along, that ruined variety theater for America, but it still stayed very strong in my house. And we had the Cole Porters, the Gershwins, Hoagy Carmichaels all over the piano. And I just went up to the piano and looked at it once and that was the beginning of my love affair with it."

Kenney's training came from playing and singing in gin mills and cafés from St. Tropez to Munich; his jobs included everything from backing stars to honoring customer's requests. After he cut his first record in Britain he became a favorite of the cabaret circuit. Then after the theme for a popular TV series, *Minder,* swept England as well as a few other songs that became popular, his career was launched.

We worked together for about four or five weeks and the first thing we wrote together was a song called "Somebody's Girl." [Thanks for asking me, Joe, / But I gotta say no / I'm somebody's girl. / So sorry, dear Bill / Try Joanna or Jill / I'm somebody's girl /] . . . Liz recorded it on her album. And she did "Some People" and she did a song I wrote in New York with another lyricist called "All Because of Love."

Then one day, we were at his house and he said, "Can I ask you something?" and I said, "Sure, what?" "Do you want to write a musical together?" Well, chills went down my spine. "My God, write a musical with you? You want to write a musical with *me?*" "Did you ever see the movie *My Man Godfrey?* It's one of my favorite movies in the world. You know London Management? We've got a meeting next week to go and sit and watch it. We'll watch it together and then we'll talk about it. We'll take notes and all the rest."

Soon Alan had fleshed out a good part of the libretto, and they had put together a group of truly remarkable songs. Kenney's melodies and sense of traditional harmony mixed with his rhythmic verve seemed to buoy the lyric along. Excerpts from the best, "It Was You Again," a bouncy ballad; "I've Been Married," where much of the humor comes from Lerner's mocking his own marital record; and "Some People," a wise and beautiful song that shows no diminution of Alan's poetic insight, are printed below:

IT WAS YOU AGAIN

It was you again;
It was you again;
Ev'ry face in the crowd,
Shining through ev'ry cloud,
It was you again;
From Deauville to Seville,
Rome to Capri,
Wherever I looked,

Whom did I see?
I saw you again,
It was you again,
Gondoliers sang their song,
But the voice kept belong-
Ing to you again;
Anyone I was with.
No matter who
It was you again,
Always you . . .

I'VE BEEN MARRIED

I have tied the wedding knot
Until the blood began to clot,
For living life connubi'lly
Isn't any jubilee.
I've seen how lovely loving starts
And slowly turns to martial arts—
I've been married.

I've tossed and turned and couldn't sleep
From counting minks instead of sheep—
I've been married.
The bower of love that began so volcanic
In no time at all could have easily sunk the Ti-
   tanic.
The wonder is that I'm alive
Considering the fact that I've
Been married.

When out the window love has flown
Alone together means alone—
I've been married
I practiced writing epitaphs
And read the book of Job for laughs—
I've been married.

The dinosaur could not survive
But little me, I'm here and I've
Been married . . .

If KO'd boxers can revive,
Then so can I and did and I've
Been married,
Hari-karied;
I've been married.
But never again!
Amen.

SOME PEOPLE

Some people, there are some people with a heart of steel;
Some people never feel things that other people feel;
Others may long to be near them;
Some people never know;
If others told them so
They would never hear them.
Some people walk around foolishly with blinkers on;
Some people never care till their chance to care has gone;
One day they'll wake and discover
That other people
Who used to weep'll
No longer shed a tear;
And love sailed away and left them on the pier . . .

By the time he had three scenes sketched out, Alan, Gerard, and Liz did a spot on British television. Alan announced that "he had never had so much fun."

The fun was interrupted, lain aside, as it were, in late 1984, which was some six months after Andrew Lloyd Webber promised Cameron Macintosh that *The*

Phantom of the Opera would be their next project. Michael Walsh, in his biography of the composer, describes how the composer of *Cats* and the librettist of *My Fair Lady* got together.

> It was time to go shopping for a lyricist. Andrew knew that Rice was not interested, for Tim by this time was working with Benny Andersson and Bjorn Ulvaeus of the Swedish rock group ABBA, who were composing the music for his long-planned show, *Chess*. Lloyd Webber wrote to Alan Jay Lerner, and asked the veteran if he would be interested, and he was delighted when Lerner said yes. It was another boyhood dream come true; the chance to work with, as the cliché had it, one of Broadway's living legends. They may have seemed like an unlikely match over an unlikely subject, but Lerner was impressed with Lloyd Webber's obvious passion for *Phantom* and confirmed his judgement about the stageworthiness of the story. "Don't ask why, dear boy," he told Andrew, "it just works."

Lerner had already written the lyric for the big second act song, "Masquerade,"[5] when he began having dizzy spells. Then his memory began to fail him, and he assumed it was writer's block. His daughter Liza feels he was suffering from a brain tumor which was never diagnosed. Reluctantly, although committed to do the show, he wrote to Lloyd Webber telling him he couldn't work with him anymore.

Liza Bibb Lerner, who was close to her father during this period, said recently that Lloyd Webber was more saddened than annoyed when he realized that Lerner would not continue to work with him. "One thing that Andrew never had was a great lyricist. He never got the critics in the States to like him, and he felt that working with Alan Jay Lerner was going to give him the credibility that he needed and the legitimacy that he's never really had. He still wants it," she added.

Alan was truly sorry to give up his work with Britain's most popular composer, but it was impossible to continue. On his better days he continued to see Kenney and move ahead with *Godfrey*. Contrary to his usual working method, he now sometimes asked Kenney to write a particular song for a section of the musical.

" 'Dancin' your Blues Away,' " Kenney remembered, "was written that way. Alan said to me 'Let's make up a dance, when Godfrey gets to the place, called The Dip—a night club song.' They're at the Waldorf in these incredibly beautiful dresses, and I thought what would they do in 1931—how would Frank Capra do it? And I came in with this tune and literally he wrote to that."

Intermittently Alan worked with Sheahan on the history, now at last titled *The Musical Theatre—A Celebration*. But the winter of 1985 was a brutal one in

---

[5] As far as the completed "Masquerade" number goes, Liza Bibb reported that "Webber expanded it and made a different kind of number out of it."

England and Alan, hoping to get over his writer's blocks and dizzy spells, took Liz for a January holiday in the Caribbean. Stef Sheahan remembered that

they got off a boat and he fell very badly in the water and completely wrecked his ankle and it was very badly swollen. They came back to England and we were working on the book and he started this silly cough—I mean it was an uh-uh cough. A nothing cough. And then it sort of got to be a bit stronger—but it was not a grave-digger cough. Alan was a bit of a hypochondriac anyway, and I'm not very sympathetic to that. So I said, "Oh look, do stop worrying, this is not very serious at all." And he went to see his doctor here who said, "No, no, it's fine, don't worry about it."

And then within a period of two or three weeks it started that he was actually being sick at the end of the cough. He would have to leave the room and vomit. Then he went to see another doctor who thoroughly checked him out and said he wanted him to go into Brompton Hospital.

Stef went with him because Liz's appearances in her show conflicted with Alan's appointment. She remembered going along "the Fulham Road and Alan

The last portrait of Alan. Photo courtesy of Roddy McDowell.

turned to me and said, 'Do you think this could be cancer?' and I said, 'Give me a break. You've got to be joking.' And then he went into his room, and I left him."

They did various tests on him and couldn't be sure if it was cancer, but what they knew was that one lung had totally collapsed. Lerner was then moved from Brompton Hospital to the London Clinic.

> They gave him a scan—he practically died having this scan—I mean it was horrendous apparently. I don't know what type of scan it was but he turned completely blue and they had to resuscitate him. He was kept there for a week or so. He knew at that point, I'm sure he knew. I don't know if he knew how bad it was, that it was not just in his lungs, it was in his head too. They said and quite rightly, "Look, there's actually nothing we can do, radiotherapy or chemotherapy—it's not going to do anything except make you feel worse than you do already. Why don't you go home, and we'll always be here when you need us."

This seemed like a logical solution and one with which Stef and Liz agreed, but the moment his children were informed they insisted Alan be transferred to Sloan-Kettering in New York. The change of hospitals notwithstanding, the prognosis after Alan was put through many painful tests (much to Liz's consternation) was the same. Liza, Jennifer, and Michael were all near their father in his final months. Susan, his eldest, was also suffering from cancer, and in an eerily ironic twist, she too was at Sloan-Kettering, being treated for terminal cancer in the same hospital where her father was dying—and where her grandfather had died.

The Lerner children had insisted their father leave London at once, informing no one of his departure. "You know what he did to me?" Gerard asked this writer recently. "Son of a bitch calls me up. I said, 'Where are you and when are you coming back?' He said, 'I'm not.' I said, 'Why not?' 'Because,' he said, 'I'm dying.' And I said 'Why didn't you tell me when we had lunch in London the other day?' and he said, 'I'm afraid it would have spoiled your lunch.' I said, 'I'm [getting] on the plane now. Where are you?' He said, 'Sloan-Kettering.' I said, 'I'll be there.' 'No, NO, NO! Don't bother.' I said, 'What do you mean, don't bother? You son of a bitch, I love you.' And the night before he died I spoke to him and I promised him 'this show will go on. I promise you.' "

On the morning of Saturday, June 14, 1986, a few days after Susan had died, with his three surviving children and the only wife who had brought him lasting happiness, Liz Robertson Lerner, gathered around his bedside, Alan Jay Lerner slipped away. It was 10:15 A.M., and as the word of his death got around the theatrical community, it was decided to lower the lights on Broadway that night in final obeisance.

Just as Hammerstein's works are his memorial, so Lerner's musicals and film scenarios serve to keep his memory burnished, But perhaps Lerner was given an extra bonus in that *The Musical Theatre—A Celebration* would be published later that year.

"I saw [it through] right to the end," Sheahan said. "And he said that if the publishers want this and this out, you know better than anybody whether it should or not [be deleted]. I had typed it and retyped it so I think I knew what he wanted in it." It was certainly a labor of love, and the witty, accurate, and unbiased book, dedicated "To Liz, who is all the music," stands far above any other history of the musical ever written.

Of the obituaries that began to fill the papers—some brickbats and some orchids—perhaps *Time* magazine's was most cruelly honest: "Mr. Lerner's the-ater work never thrived without Mr. Loewe. His later efforts tend to be daring, flawed, and commercially futile."

Critic Clive Barnes, who really understood the kind of musical Lerner was striving to write, ended up comparing Lerner to the two other great twentieth-century lyricists, Sondheim and Porter. Then he added, "He had a way with a song and could make words cling to melodies like ivy to a wall."

Of the masterworks with Loewe he observed:

What they achieved was the particular apogee of one of the Broadway musical's most singular influences—the European operetta. The music was popular yet so-phisticated, and Lerner's plots and lyric fancies had a certain style and flair, a touch of almost patrician class, that marked them out with an off-hand ease.

Times move. Tastes change. Lerner and Loewe. We shall never see their like again. A pity. I had grown accustomed to their grace.

Chronology

Year	Hammerstein	Lerner	US Musicals	UK Musicals	World
1895	Oscar Hammerstein 2nd born July 12 in New York, City.		Exotic musicals: *Aladdin, Jr.; The Viking; The Tzigane; The Sphinx; Kismet* (score by Gustave Kerker).	Gilbert and Sullivan popular throughout English-speaking world	End of Sino-Japanese War. X-ray discovered.
1896			February: Hammerstein (grandfather)'s *Marguerite* (Faust legend) (words & music). September: his *Santa Maria*. Ziegfeld's Broadway debut. Anna Held's US debut in *A Parlor Match*. Also Sousa-Klein: *El Capitan*.	*The Geisha* (Lionel Monckton score) George Edwardes's (Ziegfeld's British counterpart) A *Gaiety Girl* has continuing influence on US musicals	Nobel Prize established. First modern Olympic Games in Athens. Financial Panic (Panic of '96).
1897			Sigmund Romberg born July 29. English musicals supplant French and German imports. Home-grown product Victor Herbert & Harry Smith's *The Serenade*.	American flop, *The Belle of New York*, runs 674 performances in London.	Tate Gallery opens. Kipling publishes *Captains Courageous*. Queen Victoria celebrates her Diamond Jubilee.
1898			*Yankee Doodle Dandy*. Vincent Youmans born September 27.		The Spanish-American War.
1899				*Gaiety Girl. San Toy*, starring Marie Tempest. Noël Coward born December 16.	The Boer War. First popular magnetic recording of sound.

Year				
1900		The Burgomeister, by Frank Pixley and Gustav Luders.		US adopts the Gold Standard.
1901	Frederick Loewe born June 10.	Sweet Marie (probably by O. Hammerstein). Ziegfeld's The Little Duchess, starring his wife, Anna Held.	A Chinese Honeymoon, lavish musical that ran 1,075 performances. Bluebell in Fairyland (pantomime).	Death of Queen Victoria; accession of Edward VII. The Boxer Rebellion in China.
1902	First appearance on stage, P.S. 9. Richard (Charles) Rodgers born June 28.	De Koven's Maid Marian. Raymond Hitchcock in King Dodo.	The Girl from Kay's. Lionel Monckton in A Country Girl.	End of the Boer War. Phonographs spread recordings of popular music throughout the world.
1903	Grandma Nimmo (his favorite) dies.	First tap dances created. The Wizard of Oz (adapted by Frank Baum).		The Panama Canal.
1904	Begins piano lessons.	Kern: Mr. Wix of Wickham. Dorothy Fields born July 15.	Gibson Girls (imported from US) introduced to Britain. The Abbey Theatre founded in Dublin.	Russo-Japanese War. First feature film, The Great Train Robbery.
1905		Broadway's first neon signs. Victor Herbert's Mlle. Modiste.		Einstein's Theory of Relativity.
1906		45 Minutes from Broadway by George M. Cohan.	Belle of Mayfair.	Aga Kahn founds All-India Muslim League. First radio broadcast.
1907		The Merry Widow. The Talk of New York.	The Merry Widow.	China "Open Door" Policy. Cubist show in Paris.
1908	Enters Hamilton Institute, 81st Street & Central Park West. First published piece of writing. Grows to be 6'1½" tall.			Viennese operetta dominates the world. Art Nouveau emerging.

Year	Hammerstein	Lerner	US Musicals	UK Musicals	World
1909			Cole Porter enters Yale. *The Chocolate Soldier*, adapted from Shaw's *Arms and the Man*.	*The Dollar Princess*. Monckton's *The Arcadians*.	Model T Ford is introduced. Diaghilev brings his *Ballets Russes* to Monte Carlo. *Carmen* (French film).
1910	Mother, Alice, dies.		*Naughty Marietta*, Victor Herbert's great success	Gerald du Maurier manages Windham's Theatre.	Telephone is in common use. Accession to the throne by George V.
1911	He and brother Reggie go to Weigart's Institute, a camp in Highmount, NY. Father, William, marries Alice's sister Mousie.		42 new musicals on Broadway. Irving Berlin publishes "Alexander's Ragtime Band." Al Jolson makes his debut at the Winter Garden.	*The Count of Luxembourg*	King George V coronation.
1912	Enters Columbia College.		Mostly Gilbert & Sullian and revues. *The Count of Luxembourg*, *The Firefly* (Friml).		*Poetry Magazine* is founded. Cellophane invented. The *Titanic* sinks. Charles Pathé produces the first newsreel.
1913	Summer: tours Europe		Herbert's *Sweethearts*. Friml's *High Jinks*.	Opening of the Birmingham Repertory Company.	Fox-trot craze sweeps the world. First ships pass through Panama Canal.
1914	Father, William, dies.	Frederick Loewe appears with the Berlin Symphony as piano soloist.	Romberg's first: *Whirl of the World*. Kern's song "They Didn't Believe Me" interpolated in *The Girl from Utah*. Berlin's *Watch Your Step*.	All Germanic operetta swept off the stage in favor of British fare. *Betty*; *The Bing Boys Are Here* (hit song: "If You Were the Only Girl in the World").	August 4: Britain declares war on Germany. US remains neutral.
1915	Makes his stage debut in *On Your Way*, college show, singing and dancing. Good reviews.		*Katinka* (music by Friml). *Very Good Eddie* (music by Kern). Berlin's *Stop, Look and Listen*.		*Lusitania* is sunk. Charles Frohman, the great producer is a passenger and dies. *Birth of a Nation* (film). Dada movement begins.

Year					
1916	Writing debut in *The Peace Pirates* by Herman Mankiewicz. Oscar writes a few routines for "Mank."	Loewe writes "Katrina," which sells 2 million sheet music copies.	*The Passing Show* introduces a Gershwin song. Jazz sweeps the US.	*Chu-Chin-Chow* (runs 2,238 performances). *The Bing Girls.* Bolton-Wodehouse-Kern: *Miss Springtime.*	Rasputin dies. G. B. Shaw publishes *Pygmalion.*
1917	OH2 & Herman Axelrod write libretto of next varsity show, *Home, James.* Oscar tries to enter US Army but is rejected as underweight. Meets Myra Finn. Quits law school to go into the theatre. Hired as assistant stage manager. August 17: Marries Myra. October: Writes lyric "Make Yourself at Home" for *Furs and Frills.*		Victor Herbert is reigning king; *Maytime, Eileen.* London's hit *Chu-Chin-Chow* comes to US. Kern: *Oh, Boy!* and *Leave It to Jane.* Hitchcock's *Hitchy-Koo.*	J. M Barrie's *Dear Brutus. Bubbly*, a revue with George Robey. Bolton-Wodehouse-Kern: *Have a Heart.*	US enters war, April 7. Business off at all operettas as a reaction to the Viennese type show. "Bobbed hair" comes in; worn by women in munitions factories. Tsar Nicholas abdicates.
1918	October: creates concept for *Sometime* (Friml-Young) starring Ed Wynn. October 26: son, William, born. Writes *Ten for Five.*	Alan Jay Lerner born, August 31, New York City. Father, Joseph; Mother, Edith (after the divorce, Mrs. Edith A. Lloyd).	Leonard Bernstein born August 25. *Sinbad*, starring Al Jolson, introduces "Swanee" by Irving Caesar and George Gershwin. Gershwin is famous overnight. *Yip-Yip-Yaphank*, Berlin's wartime revue, is a huge success.	Kern's *Very Good Eddie.* Biggest hit: *The Lilac Domino.*	Armistice between allies and Germany. Original US Dixieland Band tours Europe.
1919	March: *Up Stage and Down* includes two songs, "Weakness" and "Can It," with lyrics by Hammerstein. May 24: *The Light* (serious play, fails after 4 performances in New Haven).		*La, La, Lucille*, first complete Gershwin score. *A Lonely Romeo* introduces first collaboration of Rodgers & Hart: "Any Old Place With You." Enormous growth of broadcasting.	*The Bing Boys on Broadway.* Bolton-Wodehouse-Kern hit *Oh, Joy* (which was *Oh, Boy!* in US).	Bauhaus Design Group is founded. Fritz Lang film, *The.* M. Weill-Brecht: *The Threepenny Opera.* Prohibition in US (1919–33).

Year	Hammerstein	Lerner	US Musicals	UK Musicals	World
1919 (contd)	With Herbert Stothart, writes *Joan of Arkansaw* [sic] (title changed to *Always You*).				
1920	January: *Always You*, with Harbach (Stothart). March: "That Boy of Mine" (lyric in Rodgers & Hart's *You'd Be Surprised*. March: "Always Room For One More," interpolated into Rodgers & Hart's *Fly With Me*. August 17: *Tickle Me*, with Harbach and Mandel (Stothart)(207 performances). November 17: *Jimmie*, with Harbach and Mandel (Stothart) (71 performances).		George White's *Scandals* (Gershwin score). Ed Wynn's *Carnival*. Rodgers & Hart's *Poor Little Ritz Girl*. Kern's *Sally* becomes the most successful musical of the year. Produced by Ziegfeld, sung by Marilyn Miller ("Look for the Silver Lining").	Coward's *I'll Leave It to You*. Irene. Whiteman visits London. The rage for jazz becomes universal.	August: American women get the vote. Drastic change in fashion.—most men are clean-shaven, most women flappers with short hair. First public broadcasting stations open in US and Britain.
1921	April: *You'll Never Know*, Columbia varsity show by Rodgers & Hart, directed by Hammerstein. May: daughter, Alice, born. November 8–19: *Pop*, a comedy by OH2 and A. H. Woods (closed in Atlantic City).		Vincent Youmans's first show: *Two Little Girls in Blue* (lyrics by Ira Gershwin writing as Arthur Francis). *Shuffle Along* by Sissle and Blake. Romberg's *Blossom Time*.	Kern-Harbach's *Sally* conquers London	Ireland to be a free state. President Harding declares war on war. Famine in Russia. Art Deco begins to emerge.
1922	August: *Daffy Dill*, with Guy Bolton (Stothart), lyrics by OH2. *Queen o' Hearts* (Gensler, Wilkinson), lyrics by OH2 (some by Sydney Mitchell).		Worst musical season in 20 years. *Hitchy-Koo* with score mostly by Cole Porter (a flop). Stage debuts of Bobby Clark and Fred Allen.	With the war behind and somewhat forgotten, operetta, UK's chief joy, returns. *Gypsy Princess*; *Lilac Time*.	Last British troops leave Ireland. PEN (organization of writers) is founded.

1923	February 7: New York opening of *Wildflower* with Harbach (Youmans, Stothart) (477 performances). December: *Mary Jane McKane*, book & lyrics by William C. Duncan and OH2 (Youmans, Stothart-Duncan) (151 performances). "Come On and Pet Me" lyric changed to "Sometimes I'm Happy" becomes hit.	Begins to study the piano.	Revival of the revue, with major writers like George S. Kaufman and Marc Connelly. Eddie Cantor stars in *Kid Boots*.	William Walton's *Façade* to poems by Edith Sitwell. Fred and Adele Astaire in *Stop Flirting*. *The Beauty Prize*.	Harding dies; Coolidge is US President. USSR is established.
1924	January: *Gypsy Jim*, a drama with Herbert Gropper (41 performances). February: *New Toys* with Gropper (24 performances); (filmed in 1925). September 2: New York, *Rose-Marie*, book & lyrics by Harbach and OH2 (Friml & Stothart) (557 performances).	Autumn: Enrolled in Columbia Grammar School, New York City.	*Charlot's Revue*, mostly by Noël Coward. The Marx Brothers in *I'll Say She Is!* Porter's *Greenwich Village Follies*.	*Toni*, starring Jack and Jane Buchanan. All the big hits are now coming from America	First Labour Government in Britain. "Fono-Film" talking picture process is developed.
1925	Living in London, supervising British production of *Rose-Marie*. September: New York, *Sunny*, book & lyrics by OH2 & Harbach (Kern) (517 performances). December: New York, *Song of the Flame* (Gershwin-Stothart) (219 performances).		1925 through 1929 is the heyday of the song and dance musical. *Garrick Gaieties*; *Artists and Models*; *Scandals*. Also operettas like *The Vagabond King*.	*No, No, Nanette*. Coward's *On with the Dance*.	Hitler's *Mein Kampf* is published. Scopes trial tests Darwin theory. "Charleston" sweeps the dance world. Chaplin produces *The Gold Rush*.

Year	Hammerstein	Lerner	US Musicals	UK Musicals	World
1926	October: New York, *The Wild Rose*, book & lyrics by OH2 & Harbach (Friml) (61 performances). November: New York, *The Desert Song* (Romberg) (471 performances).		*Charlot's Revue* (second edition). Rodgers & Hart's *The Girl Friend*. Gertrude Lawrence in Gershwins' *Oh, Kay!*	Most British stars are lured to US. *Lady, Be Good!* and *The Student Prince* are major offerings.	Byrd establishes the North Pole. Lindburgh flies non-stop New York to Paris.
1927	March 2: Meets Henry and Dorothy Blanchard Jacobson on SS *Olympia* en route to direct British production of *Desert Song*. November: New York, *Golden Dawn*, book & lyrics by OH2 and Stothart (Kalman-Stothart) (184 performances); Archie Leach (Cary Grant) in cast. *Show Boat* (Kern) (572 performances).	Writes first essay, "The Little House in the Wood" (about composer Edward McDowell's home) for *Columbia News*.	Ziegfeld opens his most lavish new theater with hit show, *Rio Rita*. Other successes in this banner year are *Hit the Deck, My Maryland, A Connecticut Yankee,* and *Funny Face*.	*Hit the Deck* imported from US, and Coward's *This Year of Grace*. Other homegrown products include *Lady Luck* and *The Blue Train*.	Sacco and Vanzetti executed in US. German financial crisis. C. B. DeMille produces *King of Kings* (film). Talking movies arrive—Jolson in *The Jazz Singer*.
1928	September 19: *The New Moon*, book by OH2 and Frank Mandel, Laurence Schwab; lyrics by Hammerstein (Romberg) (509 performances). September 25: *Good Boy*, book by Otto Harbach, Henry Myers; lyrics by Bert Kalmar & OH2 (253 performances). November 21: *Rainbow*, book by OH2 & Lawrence Stallings; lyrics by OH2 and Youmans (You-		*Paris*, by Cole Porter, Rosalie. *Blackbirds of 1928. Hold Everything* (Bert Lahr's debut). *Animal Crackers* with Marx Brothers. *Whoopee* with Cantor.	*Funny Face* with the Astaires. *Show Boat* from America. British musicals include *Lumber Love, Lady Mary,* and *Virginia*.	Amelia Earhart flies the Atlantic. Hoover is elected President of US. Woman suffrage in Britain. First Disney Mickey Mouse. First radio soap opera.

Year				
	mans) (29 performances). (Film version, 1930, called *Song of the West*.)			
1929	May 14: Marries Dorothy Blanchard. September: *Sweet Adeline* (written for Helen Morgan), book & lyrics by OH2 (Kern) (234 performances).	Porter's *Wake Up and Dream* and *50 Million Frenchmen*	Coward's *Bitter Sweet*. Also *Mr. Cinders*.	Stock market crashes, affecting the world. Hardest hit is the theatre. Admiral Byrd flies over South Pole. Toscanini heads New York Philharmonic.
1930	November 11: *Viennese Nights* (film), screenplay and lyrics by OH2; music by Romberg (inc. "You Will Remember Vienna"). December: *Ballyhoo*, book and lyrics by Ruskin & Brill, with additional lyrics by OH2 (68 performances).	Gershwin's *Strike Up the Band* wins Pulitzer Prize. *Girl Crazy* starring Ginger Rogers and Ethel Merman.	A dreary season. Only Coward's *Private Lives*, Rodgers & Hart's *Ever Green*.	Depression. Gandhi and civil disobedience. Photo flash is invented.
1931	February: New York, *The Gang's All Here*, book by OH2 and Morrie Ryskind; lyrics by Murphy & Robert Simon, music by Gensler. March 23: son, James, born. July: *Children of Dreams* (film), screenplay and lyrics by OH2; music by Sigmund Romberg. September: *Free For All*, book by OH2 and Laurence Stallings; lyrics by OH2; music by Richard Whiting (15 performances).	*The Band Wagon, The Ziegfeld Follies, The Cat and the Fiddle, Of Thee I Sing*.	Jack Buchanan and Elsie Randolph rival the Astaires in *Stand Up and Sing*. Gracie Fields in *Walk This Way*.	Worldwide depression continues. Britain suspends gold payments.

Year	Hammerstein	Lerner	US Musicals	UK Musicals	World
1931 (contd)	October: New York, *East Wind*, book by OH2 and Frank Mandel; lyrics by OH2; music by Romberg (23 performances).				
1932	November: New York, *Music in the Air*, book and lyrics by OH2; music by Kern (342 performances). Filmed 1934; revived 1951 (56 performances).	Summer: Sent to study at the Bedales School in Petersfield, Hampshire (England).	*Walk a Little Faster*, Vernon Duke. *Gay Divorce*, Porter. *Through the Years*, with music by Vincent Youmans. *Face the Music*, with score by Berlin.	Evelyn Laye in *Helen*. Coward's *Words and Music*.	Roosevelt is elected US President. Lindburgh kidnapping. Death of Ziegfeld. Debut of Shirley Temple.
1933	September: London, *Ball at the Savoy*, libretto by Grunwald & Löhner-Bada; English adaptation and lyrics by OH2; music by Abraham (148 performances).	Autumn: Enrolled in Choate School, Wallingford, Connecticut. Writes the school victory song, "Blue and Gold Victorious." Co-edits with John F. Kennedy the school yearbook, "The Brief."	*As Thousands Cheer*, with a Berlin score.	*Nymph Errant* and *Gay Divorce*—Porter sweeps the West End.	Hitler becomes Chancellor of Germany.
1934	April: London, *Three Sisters*, book and lyrics by OH2; music by Kern (45 performances).		Spectacle rules the day: *Revenge With Music* (Schwartz and Dietz); *The Great Waltz. New Faces* introduces Henry Fonda and Imogene Coca. *Life Begins at 8:40* features dancer Ray Bolger. *Anything Goes*, a Porter smash.	Vivian Ellis, *Jill Darling* and *Mr. Whittington*. Coward's *Conversation Piece*.	Hitler becomes Fuhrer. Debut of the *Thin Man* Series in films.
1935	January: *The Night is Young* (film), screenplay by Woolf & Schultz; lyrics by OH2; music by		Gershwin's *Porgy and Bess*.	Ivor Novello's *Glamourous Night. Tonight at 8:30* with Coward and Lawrence.	Saar restored to Germany. Swing Era begins. Jazz of Negro or Jewish origin banned from German

	Romberg (songs include "When I Grow Too Old to Dream"). December: *May Wine*, book by Frank Mandel; lyrics by OH2; music by Romberg (213 perf.).				airwaves. Garbo in *Anna Karenina*.
1936	April: *Give Us This Night* (film), lyrics by OH2; music by Korngold.	Summer: Studies music at Juilliard. Autumn: Enters Harvard (majoring in French and Italian literature). Writes lyrics and skits for the Hasty Pudding shows.	*On Your Toes* (direction by George Abbott). Kurt Weill's *Johnny Johnson*.	*Careless Rapture*, Novello. *This'll Make You Whistle* with Jack Buchanan. *Anything Goes* with Jeanne Aubert.	German troops enter the Rhine; the Rome-Berlin Axis. Abdication of Edward VIII. Frank Lloyd Wright architecture is in demand. BBC begins TV service.
1937	March: *Swing High, Swing Low* (film), screenplay by OH2 and Virginia Van Upp (based on *Burlesque*); songs by various composers, lyricists. July: *High, Wide and Handsome* (film), screenplay and lyrics by OH2; music by Kern.	Summer: Again studies music at Juilliard.	George Gershwin dies in July. *Pins and Needles*, labor organization revue. *I'd Rather Be Right*, a topical revue. *Hooray for What* by Arlen and Harburg.	*Crest of the Wave*, by Ivor Novello. *Me and My Girl*, by Noel Gay. *On Your Toes*, by Rodgers & Hart, is the hit of London.	German aggression worsens. British continue policy of appeasement. Spanish Civil War. Paris World's Fair. *Lost Horizon* (film).
1938	June 3–12: *Gentlemen Unafraid* closes on the road in St. Louis (8 performances). Book and lyrics by OH2 and Harbach; music by Kern. October: *The Lady Objects* (film), lyrics by OH2 and Milton Drake (noncollaborative); music by Ben Oakland. November: *The Great Waltz* (film), lyrics by OH2; music by Johann Strauss II.	Writes both words and music to "Living the Life," "By Chance to Dream," "Man About Town" for Harvard's Hasty Pudding show, *So Proudly We Hail*.	Rodgers & Hart's *I Married an Angel*. *Hellzapoppin*; *Knickerbocker Holiday*.	Coward's *Operette* is only fairly successful.	Germany annexes Austria. The New York World's Fair. The Lambeth Walk latest dance craze. Walt Disney's *Snow White and the Seven Dwarfs*.

Year	Hammerstein	Lerner	US Musicals	UK Musicals	World
1939	March: *The Story of Vernon and Irene Castle* (film), adaptation of book by OH2 and Dorothy Yost. November: *Very Warm for May*, book and lyrics by OH2; music by Kern (New York 59 performances); (adapted in 1944 into *Broadway Rhythm* (film).	With Stanley Miller, writes "The Little Dog Laughed" for Hasty Pudding show, *Fair Enough*. While boxing he loses retina of eye. Graduates from Harvard.	*Too Many Girls*, Rodgers & Hart.	*The Dancing Years* by Ivor Novello has great success and will run for 5 years. Pop song "Roll Out the Barrel" is an enormous hit.	Germany invades Poland. *Gone with the Wind* (film).
1940	OH2 writes poem, "The Last Time I Saw Paris." May–October: Pageant for New York World's Fair: *American Jubilee*, book and lyrics by OH2; music by Arthur Schwartz. "Your Dream is the Same as My Dream" by OH2 & Harbach with music by Kern put into film *One Night in the Tropics*.	Writing for radio. Marries Ruth Boyd.	*Louisiana Purchase*, *Panama Hattie*, *Pal Joey*.	Mostly revivals, including *Chu-Chin-Chow*. *New Faces of 1940* features "A Nightingale Sang in Berkley Square."	Churchill is Prime Minister. The fall of France. Films: *Fantasia* and Chaplin's *The Great Dictator*.
1941	"The Last Time I Saw Paris" set to music by Kern. December: New York, *Sunny River*, book and lyrics by OH2; music by Romberg (36 performances).	Rejected from US Air Corps because of detached retina. Moves into the Lambs Club. Friendship with Lorenz Hart.	Broadway mounts escape or army shows. *Lady in the Dark* is the only solid hit. Cole Porter's *Let's Face It.*	Revivals continue.	Lend-lease. December: US enters the war after Pearl Harbor. Both US and UK declare war on Japan. Film: *Citizen Kane*.
1942	August: Meets Frederick Loewe. They write 14 songs for *The Patsy*, retitled *Life of the Party*, which opens in October and runs 9 weeks.		*This Is the Army*, Irving Berlin's great score.	*Waltz Without End*, Ivor Novello.	The atomic age begins. Germans reach Stalingrad. Magnetic tape recording. Film: *Mrs. Miniver*.

1943	December 2: New York, Carmen Jones, book and lyrics by OH2; music by Georges Bizet. Filmed in 1954. March 31: New York, Oklahoma!, book and lyrics OH2; music by Rodgers (2,212 performances).	What's Up (written at request of Mark Warnow). Opens November 11, New York. Arthur Pierson assists on book; music by Loewe. Jimmy Savo; George Balanchine (64 performances). Daughter, Susan, born.	One Touch of Venus.	The Lisbon Story. A touring company of Irving Berlin's This Is the Army.	Germans surrender North Africa; Mussolini falls. Films: Stage Door Canteen; For Whom the Bell Tolls. Sinatra is the idol of teenagers.
1944			Porter's Mexican Hayride. Bloomer Girl, with a rich Arlen score. On the Town, the first Broadway work of Comden & Green, with music by Leonard Bernstein.	Very little of lasting value.	Normandy landing; FDR's third term. Olivier's Henry V.
1945	April 19: New York, Carousel, book and lyrics OH2; music by Rodgers (890 performances). August: State Fair (film) (remade in 1962 with additional songs by R. Rodgers).	November 22: The Day Before Spring (Loewe); produced by John C. Wilson (167 performances). Anthony Tudor ballet. Stars Irene Manning and Bill Johnson.	Nothing of lasting value.	Operettas: Three Waltzes, Gay Rosalinda. Also Under Your Hat and Coward's Sigh No More.	Germany surrenders; FDR dies; US drops atomic bomb; Japan surrenders. Bebop becomes a worldwide craze.
1946			Call Me Mister, a mustering out revue. Annie Get Your Gun, Irving Berlin's most acclaimed score. Night and Day, film of Cole Porter's life.	Big Ben by Vivian Ellis. Pacific 1860 by Coward, with Mary Martin.	First meeting of the UN. Xerography developed. British Arts Council is inaugurated. BBC's Third Programme begins transmitting.
1947	October 10: New York, Allegro, book and lyrics by OH2; music by Rodgers (315 performances).	March 13: New York, Brigadoon, book and lyrics by AJL; music by Loewe (581 performances). Moves in with Marion Bell, star of Brigadoon, and soon marries her (wife #2). AJL and Loewe have a break-up.	US theater finding its voice with Allegro, Finian's Rainbow, Brigadoon, High Button Shoes, and Street Scene.	American musicals take London: Oklahoma! and Annie Get Your Gun are big hits.	Truman Doctrine; Marshall Plan, Independence for India. A Streetcar Named Desire. First Edinburgh Festival.

Year	Hammerstein	Lerner	US Musicals	UK Musicals	World
1948		October 7: New York, *Love Life*, book and lyrics by AJL; music by Kurt Weill (252 performances).	Porter's *Kiss Me, Kate*. Musicals aiming for charm: *Lend an Ear* and *Where's Charley?*	The American invasion continues with *Brigadoon* and *Lute Song*. British product: *Cage Me a Peacock*.	End of the British mandate in Palestine. In art, it is the era of Jackson Pollock and Henry Moore. Film: Olivier's *Hamlet*.
1949	April 7: New York, *South Pacific*, book by OH2 and Joshua Logan; lyrics by OH2; music by Rodgers (1,925 performances).	Courting Nancy Olson and divorcing Marion Bell.	The big shows take over: *Miss Liberty*, *Regina*, *Gentlemen Prefer Blondes*.	Ivor Novello's *King's Rhapsody*; also *Tough at the Top*. Coward's *Ace of Clubs*.	NATO established. China becomes Communist. Arthur Miller's *Death of a Salesman*.
1950		March 10: Marries Nancy Olson (wife #3).	Berlin's hit, *Call Me Madam*. Frank Loesser's *Guys and Dolls*.	American hit: *Carousel*. British musical: *Take It from Here*. Revues: *Dear Miss Phoebe* and *Blue For a Boy*.	The Korean War. UN Building completed. Menotti's opera, *The Consul*, runs on Broadway.
1951	March 29: New York, *The King and I*, book and lyrics by OH2; music by Rodgers (1,246 performances).	November 12: New York, *Paint Your Wagon*, book and lyrics by AJL; music by Loewe (289 performances). October 12: Daughter, Liza, born. Film: *An American in Paris*, screenplay by AJL; Gershwin score. Film: *Royal Wedding*, screenplay and lyrics by AJL; music by Burton Lane. Academy Award nomination for song, "Too Late Now."	*Top Banana* (burlesque).	Porter's *Kiss Me, Kate*. Novello's *Gay's the Word*. Vivian Ellis's *And So to Bed*.	Electric power from atomic energy. Cinerama developed. Britain begins rating system for films. *The Archers*, sitcom. Johnny Ray and Elvis Presley are the sensations.
1952		Wins the Academy Award for screenplay of *An American in Paris*.	*New Faces of 1952* introduces Eartha Kitt, Alice Ghostley, Carol Lawrence, Ronny Graham.	*Call Me Madam*, *Zip Goes a Million*, and British hit *Bet Your Life* (adapted from *Brewster's Millions*).	McCarthyism. Contraceptive pill is developed. Elizabeth II's coronation; *The Mousetrap* opens.

Year					
			Wish You Were Here constructs a swimming pool on stage.		Chaplin's film Limelight released.
1953	May 28: New York, Me and Juliet, book and lyrics by OH2; music by Rodgers. (358 performances).	July: Father, Joseph, dies. August 26: Daughter, Jennifer, born. Works on Huckleberry Finn with Burton Lane, but the film is never completed.	A dreadful season, the only sure-fire successes being Kismet and Porter's Can-Can.	From the US: Guys and Dolls, Paint Your Wagon, The King and I, and holdovers Porgy and Bess and Carousel. Two superb British shows: The Boy Friend and Salad Days.	Death of Stalin. Korean armistice. Rosenbergs executed.
1954			Small shows like The Threepenny Opera, The Boy Friend, and The Golden Apple vie with extravaganzas like Pajama Game and House of Flowers.	Can-Can and Wedding in Paris.	Salk Polio vaccine developed. The H-Bomb. High court bans segregation. End of Vietnam War. McCarthy censured.
1955	November 30: New York, Pipe Dream, book and lyrics by OH2; music by Rodgers, (246 performances).		Wholesome, homespun shows like Plain and Fancy vie with sexy ones like Damn Yankees and Silk Stockings.	The Pajama Game, The Water Gypsies.	The Kruschev era. Churchill steps down. US launches satellite. Péron overthrown.
1956		March 15: New York, My Fair Lady, book and lyrics by AJL; music by Loewe (2,717 performances).	The almost through-sung musical arrives: The Most Happy Fella and Candide. Also a snappy revue from London, Cranks.	Fanny and Grab Me a Gondola.	Eisenhower wins second term. The Hungarian revolt. Maria Callas takes Milan and New York by storm.
1957	March 31: TV Special Cinderella, book and lyrics by OH2; music by Rodgers.	Spring: Gigi (film), screenplay and lyrics by AJL; music by Loewe. September: Separated from Nancy Olson. Meets Micheline Muselli Pozzo di Borgo; brings her to US; marries her December 25 (wife #4).	West Side Story and The Music Man (which wins the Tony).	Damn Yankees and Bells Are Ringing.	Shake-up in the Kremlin. Sputnik. Eisenhower Doctrine. Little Rock. Civil rights legislation passes in US.

Year	Hammerstein	Lerner	US Musicals	UK Musicals	World
1958	December 1: New York, *Flower Drum Song*, book and lyrics by OH2; music by Rodgers (600 performances).	*Gigi* wins Academy Award as Best Screenplay, Picture; title song also wins award as Best Song.	*The Body Beautiful* (the debut of Bock & Harnick). Other shows are *Goldilocks* and *La Plume de ma Tante*.	British inventions: *Valmouth*, *Irma la Douce*, and *Expresso Bongo*.	DeGaulle is Premier of France. Alaska is US 49th state. Folk music becomes the vogue.
1959	November 16: New York, *The Sound of Music*, book by Lindsay and Crouse; lyrics by OH2; music by Rodgers (1,443 performances).	Son, Michael, born.	A banner year: besides *Sound of Music*, There are *Redhead*, *Destry Rides Again*, *Once Upon a Mattress*, and *Gypsy*.	*The World of Paul Slickey*.	Hawaii becomes the 50th state. Castro takes Havana. Soviet Lunik III strikes the moon.
1960	August 23: Oscar Hammerstein 2nd dies.	December 3: New York, *Camelot*, book and lyrics by AJL; music by Loewe (873 performances).	Another banner year: *Bye Bye Birdie*, *The Fantasticks*, *The Unsinkable Molly Brown*, *Wildcat*.	Lionel Bart's *Oliver*; also *Lilly White Boys*.	Princess Margaret marries Antony Armstrong-Jones. J. F. Kennedy is elected President of US.
1961			*Carnival*, Jerry Herman's first hit. *Milk and Honey*; also *How to Succeed in Business Without Really Trying*.	*Stop The World—I Want To Get Off* (Anthony Newley).	Soviet Cosmonaut Gagarin orbits earth. Berlin Wall erected. Elvis recordings top best-seller lists.
1962			*No Strings*; Rodgers writes words and music.	*Blitz* (Lionel Bart); Coward's *Sail Away*.	John Glenn orbits earth. US stock market plunges in largest drop since 1929. Cuban missile crisis.
1963			Few good shows, but *Oliver!*, a London import, and *She Loves Me*, Steinbock-Harnick's gem, are outstanding.	*Half a Sixpence*; *Mr. Pickwick*; *Oh, What A Lovely War*.	Pop Art as represented by Andy Warhol. John Kennedy is assassinated. Lyndon Johnson takes over presidency.
1964		May 14: Micheline files separation suit for divorce.	*Hello, Dolly*, *Funny Girl*, and *Fiddler on the Roof* are smash hits; Sondheim's *Anyone Can Whistle* and	*Lock Up Your Daughters*; *Maggie May*. *Robert and Elizabeth*, based on *The Barretts of Wimpole Street*,	Harold Wilson is elected British prime minister. Beatles in US debut. US planes bomb North Viet-

	the Coward-Martin-Gray *High Spirits*, although flops, are interesting.	is a great success and runs 2 years.	nam. Johnson elected US president.	
1965	August 31: Divorce becomes final. October 17: *On a Clear Day You Can See Forever*, book and lyrics by AJL; music by Burton Lane (280 NY performances).	*Do I Hear a Waltz?* (Rodgers-Sondheim). *Man of La Mancha*.	*Charlie Girl* is a big hit (2,202 performances).	First combat troops land in Vietnam. Medicare bill passes. Presley, Beatles, Rolling Stones top recording charts.
1966	November 15: Marries Karen Gundersen (wife #5).	Not many musicals, but every one produced this year becomes a classic: *Sweet Charity*, *Mame*, *Cabaret*, *I Do, I Do*.	American imports.	Floods damage Florence. Mao's Cultural Revolution. Johnson's "Great Society." Jimmi Hendrix helps popularize electric guitar.
1967		*Hallelujah, Baby* (Styne, Comden & Green) and other mediocrities.	*The Four Musketeers*.	Stalin's daughter defects to West. Nasser closes Aquaba Gulf to Israeli ships. First heart transplant. Flower children generation.
1968		Spoofs: *Your Own Thing* and *Dames At Sea* vie with second-rate shows like *Zorba* and *Promises, Promises*. But *Hair* and *Jacques Brel* are outstanding.	Mostly American shows except for *Canterbury Tales* (2082 performances).	Martin Luther King and Robert F. Kennedy are assassinated in US. November: Nixon elected US president. Hit records by Johnny Cash, 5th Dimension, Herb Alpert, Simon & Garfunkel, and Glen Campbell.
1969	December 18: *Coco*, book and lyrics by AJL; music by André Previn (332 NY performances).	*1776*, a hit, and *Dear World*, an interesting failure, among unsuccessful versions of old novels, old movies, old plays: *Georgy*, *Gantry*, *La Strada*, *Billy* (Budd), *Canterbury Tales*.	Lavish flops: *Anne of Green Gables* and *A Tale of Two Cities*.	Astronauts walk on the moon. De Gaulle resigns. Film: *Midnight Cowboy*. Hit albums: *Hair*, *Abbey Road* (Beatles), *Blood, Sweat and Tears*.

Year	Hammerstein	Lerner	US Musicals	UK Musicals	World
1970			*Purlie, Applause, Company, Oh, Calcutta.* Rodgers and Charnin's *Two by Two* stars Danny Kaye.	*Oh, Calcutta.*	Kent State student riots. Airline hijacking. Nasser dies, Sadat takes over. Burt Bacharach and soft rock pop. Religious trend enters pop recordings.
1971		March: *Lolita*, book and lyrics by AJL; music by John Barry. (16 Philadelphia performances; 6 Boston performances; 0 New York performances).	*Follies* (Sondheim-Goldman) runs over 500 performances but still loses money. *Jesus Christ Superstar* (Rice-Webber) and *Godspell* (Stephen Schwartz) are successful examples of the religious musical that is now in vogue.	*Godspell. His Monkey Wife* (Sandy Wilson).	Communist China admitted to the UN. Britain to enter Common Market. Voting age in US changed to 18; Top US charts: George Harrison, Janis Joplin, Carole King, Santana.
1972			*Grease* and *Sugar*. Very slim pickings indeed.	*Tom Brown's School Days. Gone With The Wind* (music by Harold Rome). *Jesus Christ, Superstar.*	European Common Market. Break-in at Democratic headquarters. Nixon wins landslide victory. Films: *The Godfather, Cabaret, Last Tango in Paris, Deep Throat.* Top single: Roberta Flack's "The First Time Ever I Saw Your Face."
1973		*The Little Prince* (film), screenplay and lyrics by AJL; music by Loewe. November 13: *Gigi* (stageplay) (104 New York performances).	*A Little Night Music* (Sondheim). *Seesaw. Raisin.*	*Cowardly Custard* (Noël Coward revue). *The Rocky Horror Show, Joseph and the Amazing Technicolor Dreamcoat.*	End of the Vietnam War. Watergate hearings begin. Syria and Egypt attack Israel. Vice President Spiro Agnew resigns. Album: *Goodbye Yellow Brick Road* (Elton John).

1974	December 9: Divorces Karen Gundersen Lerner. December 10: Marries Sandra Payne, British stage & TV actress (wife #6).	*Over Here* (return of the Andrews Sisters). *Candide* (rewritten and restaged). *The Magic Show.* One of Jerry Herman's best, *Mack and Mabel*, is a failure.	Except for *Billy Liar* (Barry-Black), mostly US imports.	Solzhenitsyn expelled from the USSR. Nixon resigns. Haile Selassie is deposed in Ethiopia. All *in the Family* top TV show. "Streaking" becomes a fad.
1975		*The Wiz* (soul-tinged score). *Shenandoah* (folklike). *Chicago* (Kander-Ebb). *A Chorus Line* (Bennett- Hamlisch-Kleban). *Tremonisha* (Scott Joplin).	*Jeeves.*	US and Soviet spacecrafts link up in space.
1976	May 2: *1600 Pennsylvania Avenue*, book and lyrics by AJL; music by Leonard Bernstein (7 New York performances). September 12: Divorces Sandra Payne Lerner.	*Bubbling Brown Sugar. Rex* (Rodgers-Harnick). *Your Arms Too Short To Box With God.*	*Side By Side By Sondheim. Mardi Gras* (Melvyn Bragg).	Mao-Tse-Tung dies. Satellite lands on Mars. Concorde begins regular trans-Atlantic service. Jimmy Carter wins US Presidency. US and Iran sign $10 billion arms sale. Film: *Rocky.*
1977	January: International Music Theatre Forum in Sydney (Australia)— Sondheim, Prince, and Lerner are featured authorities. November: Marries Nina Bushkin, daughter of well-known jazz pianist (wife #7).	*Side By Side By Sondheim. Annie* (Strouse-Charnin). *The Act* (Kander & Ebb, written for Liza Minnelli).	*Privates on Parade.*	Carter grants amnesty to Vietnam draft evaders. US and Panama sign treaties. *Roots*, adapted from Alex Haley's novel, draws largest TV audience (130 million) in history.
1978	November: Publishes *On the Street Where I Live.*	*On the 20th Century; Ain't Misbehavin'.*	*Evita.*	March on Washington for ERA. 900 cult members commit suicide in Guyana. Top album: *Saturday Night Fever.*

Year	Hammerstein	Lerner	US Musicals	UK Musicals	World
1979		April 8: *Carmelina*, book and lyrics by AJL; music by Burton Lane (closes after two weeks).	*Sweeney Todd*.	*Charly and Algernon* (Strouse). *A Day in Hollywood/A Night in the Ukraine*.	Margaret Thatcher is elected British prime minister. Mountbatten assassinated. Russia invades Afghanistan. Height of disco craze. Film: *Kramer vs. Kramer*. Album: *Bad Girls*.
1980			*Evita, Sugar Babies, 42nd Street, Woman of the Year*.		US breaks off diplomatic ties with Iran. Ronald Reagan elected US President. John Lennon shot dead in New York City.
1981		August 12: Marries Liz Robertson (wife #8), whom he is directing in London revival of *My Fair Lady*.	*Merrily We Roll Along*.	Cats.	US-Iran agreement frees hostages. Reagan, Pope John Paul II are wounded by gunmen. First woman appointed to US Supreme Court.
1982			*Cats, Dreamgirls, Nine, A Doll's Life*.		British overcome Argentina in Falklands war. Princess Grace dies in auto crash.
1983		*Dance a Little Closer* (an adaptation of *Idiot's Delight*), book and lyrics by AJL; music by Charles Strouse (1 New York performance).	*Baby, La Cage aux Folles, My One and Only* (Gershwin's *Funny Face* fitted out with a new libretto).	*Blood Brothers*.	New Roman Catholic code. US invades Grenada.
1984		*My Man Godfrey* (incomplete, no music). Begins writing *The Musical Theatre: A Celebration* (book).	*Sunday in the Park With George, The Rink, The Tap Dance Kid*.	*The Hired Man* and *Starlight Express*—both by Andrew Lloyd Webber. One flop, one hit.	Reagan and Bush reelected. Indira Gandhi assassinated. Toxic gas leaks from Union Carbide plant in Bhopal, India, killing 2,000 and injuring 150,000.

1985	Begins working with Gerard Kenney, composer-singer, setting music to lyrics of *My Man Godfrey*.	A fallow year on Broadway. *Big River*, although showing little merit, wins the Tony.	*Mutiny on the Bounty* (spectacular flop). *Les Misérables*.	Gorbachev elected to take helm of USSR. Italian government toppled by political crisis.
1986	June 14, 10:15 A.M.: Alan Jay Lerner dies. *The Musical Theatre* is published posthumously. (Frederick Loewe dies 1:51 P.M. February 14, 1988).	*Me and My Girl* (another British import).	*Chess* (Tim Rice). *The Phantom of the Opera* (Lloyd Webber).	Spain and Portugal join Common Market. Britain and France plan Channel Tunnel. Space shuttle *Challenger* explodes after launch, killing all aboard.

Glossary of Musical Terms

A ABA A common song form in which the first theme (usually eight bars in length) is repeated and followed by a contrasting theme. This second statement is followed by a return of the first theme. All the As must be essentially the same; however, their endings may vary.

A1,A2,release, A3 A preferred way of referring to the AABA.

Alberti bass Alberti, a contemporary of Mozart, is remembered mostly for the manner in which he broke chords in the left hand (low-high-middle-high) to make them sustain on the harpsichords of his time.

alla breve 4/4 time played rather quickly, so that there are only two counts to the bar. Popular musicians call this "cut time."

appoggiatura From the Italian meaning "to lean." A decorative pitch not belonging to the indicated chord, usually approaching the target (chord) note from above. Appoggiaturas create mild dissonance and can add great intensity to a melodic line.

arpeggio From the Italian word for harp, *arpa*. An arpeggio is a chord whose members are sounded individually.

augmented 1) Raising the pitch a half-tone. 2) A triad with its fifth raised a half tone.

beguine A sensual languid dance said to be of Tahitian origin. Its accent is on the second eighth note (quaver) in the bar.

belting Using the chest voice rather than letting air pass over the diaphragm, to create what is known as a soprano or "head sound."

blue note The flattened 3rd, 5th, 6th, or 7th of the scale when used in conjunction with a major harmony, creates a distinctly

biting (some consider it melancholy) sound typical of the blues.

bridge The B section of an AABA song. A contrasting section, more appropriately called "release" or "channel."

cadence A series of chords leading to a conclusion. The typical interminable, and sometime laughable V-I cadence can be found at the end of most Rossini overtures.

chromaticism The use of accidental tones falling outside the prevailing key signature.

circle of chords The natural progression of dominant sevenths to tonics. The series of chords is usually written in circular fashion:

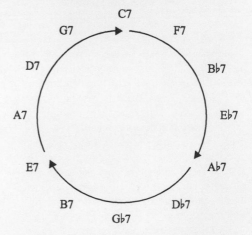

contrary motion (see motion, contrary)

crotchet The British term for quarter note.

cut time (see alla breve)

cycle of chords (see circle of chords)

diminished cliché My own term for a series of four chords comprising the tonic (I), a diminished seventh (usually built on the tonic or the lowered supertonic (I dim or #I dim), the minor supertonic (II), and the dominant (V). These four chords are as common in popular music as the I, VI, II, V series.

fox-trot A dance that spawned many songs, popular from 1914 to 1925. The dance, which uses a box-step, is always in 4/4 time, often with a hesitation on the third beat of the bar. This dance developed during a period of animal-named popular dances, such as the turkey-trot, monkey-slide, and bunny-hug, and is still a favorite of the older generation.

hemiola The deceptive change of meter achieved by false emphasis. Ex: *One* two *three* one *two* three.

leading tone The seventh tone of the scale, so called because it leads smoothly back to the tonic.

legit A shortening of the term **legitimate,** implying proper or operatic singing using resonating head tones rather than chest or "belt" singing.

list song A song whose lyrics present a list of related items.

measure 1) a musical unit divided by a barline. 2) A synonym for *bar.*

melisma The use of several pitches on a single literary syllable. Melismatic passages are most often associated with the baroque but are also frequently found in Middle Eastern, Hebraic, and "soul" music (see example of Cole Porter's "Solomon" below).

motion The movement of melodic as against harmonic lines. Usually referred to as the treble or soprano line against the bass line. There are three kinds: **parallel,** which applies to bass and soprano moving in the same direction; **contrary,** as the word would indicate, is movement in opposite directions; and **oblique,** when one voice stays on the same pitch while the other moves.

motive (or *motif*) The basic germ of a musical idea. A series of notes set in a rhythm that will be used again and again in various ways throughout the song.

oblique motion (see motion: parallel, contrary, oblique)

one-six-two-five A harmonic pattern usually expressed as I, VI, II, V, indicating a series of chords (tonic, submediant, supertonic, and dominant triads) upon which countless melodies have been constructed. As the underpinning of songs from "Heart and Soul" to many of the rock and roll songs of the '60s and beyond, this cliché is the granddaddy of them all.

parallel motion (see motion, parallel)

prosody The blending of words and music. Good prosody coupled with an artist's clear diction will make a lyric understandable.

punch line song One whose lyric contains a surprise at its conclusion, preferable in the very last syllable.

quarter note The basic pulse of most popular songs. The British term is "crotchet."

quaver The British term for eighth note.

ragtime A rhythmic style popular between 1890 and 1914, famous for syncopation and anticipation.

range The vocal palette. In popular songs before the '60s this was limited to a tenth (an octave and a third). Show tunes were permitted a somewhat wider latitude, but it was not until works by Bernstein, Bacharach, Sondheim, and Lloyd Webber, who employed well-trained singers, that a singer's full range was utilized.

refrain The main body of the song, often interchangeable with "chorus." Jerome Kern always called his refrain "burthen."

reprise A custom of repeating a song, often

So   o-lo-mon   had a thou-sand wives—

in several parts of a show, in order to make it indelible.

segue Italian for "follow." A musical direction meaning "to proceed to the next section without a pause or break."

semitone A half tone, the next nearest pitch. In the western musical system, the octave is divided into twelve semitones.

situation song A song in which the protagonist describes a self-involving situation.

skip A musical interval that is more than a whole step.

story song A song whose narrative is its most important feature.

subdominant The fourth degree of the diatonic scale. The subdominant is one of the three principal triads.

suspension A non-chord member that formerly had to resolve to a chord member. Contemporary popular music uses suspensions more liberally than most harmony books permit and does not oblige them to resolve. The most frequently used suspension is the fourth.

syncopation Misplacing accents that are normally felt on the first and third beats of the bar.

tag An extension to a song, sometimes called the coda.

tessitura The general range of a composition. Songs that remain largely around the top of their range are said to have a high tessitura, those that keep punching out the middle or bottom notes are said to have a low tessitura.

transposition Changing the key of a song or composition. Music is generally transposed to place it within the best possible vocal range of the singer.

tremolo A piano technique popular during the Victorian period. The right hand trembles, breaking the intervals of an octave or a sixth. This technique later became the mainstay of tear-jerking scenes in the old Nickelodeon movies.

triad A chord of three notes; two superimposed thirds.

trio The middle section of a song, originally performed in three sections. In contemporary language, this is called an interlude.

tritone The interval of the diminished fifth (or augmented fourth). The tritone was known as the sound of the devil's violin (achieved by retuning) and is assiduously avoided in all exercises in harmony or composition.

upbeat A lead-in or pick-up. The beat before the barline and down beat. Upbeat is so named for the position of the conductor's arm.

vamp A repeated chord pattern, usually ad-libbed until the entrance of the solo performer.

verse Before 1960, the mood-setting, expendable introductory section preceding an ABAC or AABA refrain.

waltz Although usually written in 3/4 time, the waltz is generally performed with one beat to the bar.

whole tone A full step; two half steps.

whole-tone scale A scale made up of full steps. Only two whole-tone scales are possible in our musical system, one beginning on B and the other beginning on C. All others, no matter where they begin, are repetitions of these.

Bibliography

Abbott, George. *Mister Abbott,* New York: Random House, 1963.

Agate, James. *Egos, Vols. 1–9.* London: Hamish Hamilton, Gollancz, Harrap, 1932–48.

Aldrich, Richard. *Gertrude Lawrence as Mrs. A.* New York: Greystone Press, 1954.

Atkinson, Brooks. *Broadway.* New York: Macmillan, 1970.

Bankhead, Tallulah. *Tallulah.* London: Gollancz, 1952.

Beaton, Cecil. *Persona Grata.* New York: Putnam, 1952.

Bordman, Gerald. *American Musical Theatre.* New York: Oxford, 1978.

Bowen, Ezra, ed. *This Fabulous Century, 1920–1930.* New York: Time-Life, 1969.

Brahms, Caryl, and Ned Sherrin. *Song By Song.* Bolton: Ross Anderson, 1984

Burton, Humphrey. *Leonard Bernstein.* New York: Doubleday, 1994.

Castle, Irene. *My Memories of Vernon Castle.* Boston: Little Brown, 1935.

Chevalier, Maurice. *Bravo, Maurice.* London: Allen & Unwin, 1975.

Citron, Stephen. *The Musical From the Inside Out.* Chicago: Ivan Dee, 1992.

Colette. *Seven Stories.* New York: Doubleday, 1933.

Damase, Jacques. *Les Folies du Music-Hall.* London: Spring Books, 1960.

de Mille, Agnes. *And Promenade Home.* Boston: Little Brown, 1952.

———. *Speak to Me, Dance With Me.* Boston: Little Brown, 1973.

Edwards, Anne. *The De Milles.* New York: Abrams, 1989.

———. *A Remarkable Woman: Katharine Hepburn, A Biography.* New York: William Morrow, 1985.

Engel, Lehman. *The American Musical Theater.* New York: Macmillan, 1975.

Everett, Susan. *London: The Glamour Years 1919–39.* London: Bison, 1985.

Ferber, Edna. *Show Boat.* New York: Doubleday, 1925.

Fields, Amond, and L. Marc Fields. *From the Bowery to Broadway.* New York: Oxford, 1993.

Fordin, Hugh. *Getting to Know Him.* New York: Random House, 1977.

Furia, Philip. *The Poets of Tin Pan Alley.* New York: Oxford, 1990.

Gassner, John. *The Theater in Our Times.* New York: Crown, 1954.

Gottfried, Martin. *Jed Harris.* Boston: Little Brown, 1984.

Green, Benny. *A Hymn to Him.* London: Michael Joseph, 1987.

———. *Let's Face the Music.* London: Michael Joseph, 1989.

Green, Stanley. *Encyclopedia of the Musical Theatre.* New York: Da Capo, 1984.

———. *Rodgers & Hammerstein Fact Book.* New York: Lynn Farnol Group, 1980.

Hammerstein, Oscar 2nd. *Lyrics,* rev. ed. Milwaukee: Hal Leonard Books, 1985.

Harrison, Rex. *Autobiography,* New York: Morrow, 1975.

———. *A Damned Serious Business.* New York: Bantam, 1991.

Hart, Kitty Carlisle. *Kitty: An Autobiography.* New York: Doubleday, 1988.

Hart, Moss. *Act One.* New York: Random House, 1959.

Helburn, Theresa. *A Wayward Quest.* Boston: Little Brown, 1960.

Huggett, Richard. *Binkie Beaumont.* London: Hodder & Stoughton, 1989.

Kerr, Walter. *Thirty Plays Hath November.* New York: Simon & Schuster, 1969.

Landon, Margaret. *Anna and the King of Siam.* New York: Macmillan, 1944.

Langner, Lawrence. *The Magic Curtain.* New York: Dutton, 1951.

Lawrence, Gertrude. *A Star Danced.* New York: Doubleday, 1945.

Lee, C. Y. *The Flower Drum Song.* New York: Farrar, Strauss, 1957.

Lees, Gene. *Inventing Champagne.* New York: St. Martins Press, 1990.

———. *The Modern Rhyming Dictionary.* Greenwich, Conn.: Cherry Lane Books, 1981.

Lerner, Alan Jay. *The Musical Theatre.* London: Collins, 1986.

———. *The Street Where I Live,* New York: Norton, 1980.

Logan, Joshua. *Josh.* New York: Dell, 1976.

Loos, Anita. *Gigi,* dramatization. New York: Samuel French, 1952.

Madsen, Axel. *Chanel.* New York: Henry Holt, 1990.

Martin, Mary. *My Heart Belongs to.* New York: Morrow, 1976.

Maxwell, Gilbert. *Helen Morgan.* New York: Hawthorn, 1974.

Michener, James. *Tales of the South Pacific.* New York: Macmillan, 1947.

Minnelli, Vincente. *I Remember It Well.* New York: Doubleday, 1974.

Molnár, Ferenc. *Liliom,* trans. Benjamin Glaser. New York: Samuel French, 1935.

Mordden, Ethan. *Rodgers & Hammerstein.* New York: Abrams, 1992.

Morley, Sheridan. *Gertrude Lawrence.* New York: McGraw-Hill, 1981.

———. *Shooting Stars.* London: Quartet, 1983.

———. *Spread a Little Happiness.* Thames & Hudson, London, 1987.

Nichols, Beverley. *The Sweet and Twenties.* London: Weidenfeld & Nicolson, 1958.

Noble, Peter. *Ivor Novello.* London: Falcon Press, 1951.

Nolan, Frederick. *The Sound of Their Music.* New York: Walker & Company, 1978.

Peyser, Joan. *Bernstein, A Biography.* New York: Morrow, 1987.

———. *The Memory of All That (The Life of George Gershwin).* New York: Simon & Schuster, 1993.

Pinza, Ezio. *Autobiography.* New York: Holt Rinhart and Winston, 1958.

Riggs, Lynn. *Green Grow the Lilacs.* New York: Samuel French, 1931.

Rodgers, Richard. *Musical Stages.* New York: Random House, 1975.

Sheean, Vincent. *Oscar Hammerstein I.* New York: Simon & Schuster, 1956.

Stagg, Jerry. *The Brothers Shubert.* New York: Random House, 1968.

Steinbeck, John. *Sweet Thursday.* New York: Viking, 1954.

Strong, Phil. *State Fair.* New York: Grosset & Dunlap, 1932.

Suskin, Steven. *Opening Night On Broadway.* New York: Schirmers, 1990.

———. *Show Tunes.* New York: Limelight, 1992.

Swain, Joseph P. *The Broadway Musical.* New York: Oxford, 1990.

Taylor, Deems. *Some Enchanted Evenings.* London: Macdonald, 1955.

Teichman, Howard. *George S. Kaufman.* New York: Atheneum, 1972.

Trewin, J. C. *Theatre Since 1900.* London: Dakers, 1951.

Walsh, Michael. *Andrew Lloyd Webber.* New York: Abrams, 1989.

Wilder, Alec. *American Popular Song: The Great Innovators, 1900–1950.* New York: Oxford, 1972.

Wilk, Max. *O K ! The Story of Oklahoma!* New York: Grove Press, 1993.

Credits

OSCAR HAMMERSTEIN 2ND

Lyrics and Music reproduced by permission of Williamson Music Co. (and International Music Publications Limited—asterisked titles only)

From *Allegro*
Lyrics by Oscar Hammerstein II, Music by Richard Rodgers
Copyright © 1947 by Richard Rodgers and Oscar Hammerstein II. Copyright Renewed.
"Allegro" (p. 184), "Come Home" (p. 184), "What a Lovely Day for a Wedding" (p. 183), "Yatata" (p. 183), "The Gentleman Is a Dope" (p. 184)

From *Always You*
Lyrics by Oscar Hammerstein II, Music by Herbert Stothart
Copyright © 1920 by Bambalina Music Publishing Co. (Administered in U.S. by Williamson Music) and Warner Chappell Music, Ltd., London W1Y 3FA (Administered by International Music Publications Limited). Copyright Renewed.
*"Always You" (p. 33), *"The Tired Business Man" (p. 34)

From *Carmen Jones*
Lyrics by Oscar Hammerstein II, Music by Georges Bizet
Copyright © 1943 by Williamson Music Co. Copyright Renewed.
"Dat's Love (Habanera)" (p. 130), "Stan' Up and Fight (Toreador Song" (p. 130)
Copyright © 1944 by Williamson Music Co. Copyright Renewed.
"Dis Flower" (p. 130)

From *Carousel*
Lyrics by Oscar Hammerstein II, Music by Richard Rodgers
Copyright © 1945 Williamson Music Co. Copyright Renewed.
"If I Loved You" (p. 176), "June Is Bustin' Out All Over" (p. 176), "Mister Snow" (p. 176), "A Real Nice Clambake" (p. 177), "Soliloquy" (p. 170), "What's the Use of Wond'rin' " (p. 174), "When the Children Are Asleep" (p. 173)

From *Cinderella*
Lyrics by Oscar Hammerstein II, Music by Richard Rodgers
Copyright © 1957 by Richard Rodgers and Oscar Hammerstein II. Copyright Renewed.
"Do I Love You Because You're Beautiful" (p. 290), "In My Own Little Corner" (p. 290)

From *The Desert Song*
Lyrics by Oscar Hammerstein II and Otto Harbach, Music by Sigmund Romberg
Copyright © 1928 by Bambalina Music Publishing Co. (Administered in the U.S. by Williamson Music) and Warner Chappell Music, Ltd., London W1Y 3FA (Administered by International Music Publications Limited). Copyright Renewed.

*"It" (p. 56), *"The Desert Song"
(p. 56)

From *Flower Drum Song*
Lyrics by Oscar Hammerstein II, Music by
Richard Rodgers
Copyright © 1958 by Richard Rodgers
and Oscar Hammerstein II. Copyright Renewed.
"I Am Going To Like It Here"
(p. 296), "I Enjoy Being a Girl"
(p. 296), "You Are Beautiful" (p. 295)

From *Golden Dawn*
Lyrics by Oscar Hammerstein II and Otto
Harbach, Music by Emmerich Kalman
and Herbert Stothart
Copyright © 1927 by Bambalina Music
Publishing Co. (Administered in the U.S.
by Williamson Music) and Warner
Chappell Music, Ltd., London W1Y 3FA
(Administered by International Music
Publications Limited). Copyright Re-
newed.
*"Jungle Shadows" (p. 59), *"When I
Crack My Whip" (p. 60)

From *I'll Take Romance*
Lyrics by Oscar Hammerstein II, Music by
Ben Oakland
Copyright © 1937 by Williamson Music
and Bourne Music. Copyright Renewed.
"I'll Take Romance" (p. 119)

From *Jimmie*
Lyrics by Oscar Hammerstein II and Otto
Harbach, Music by Herbert Stothart
Copyright © 1920 by Bambalina Music
Publishing Co. (Administered in the U.S.
by Williamson Music) and Warner
Chappell Music, Ltd., London W1Y 3FA
(Administered by International Music
Publications Limited). Copyright Re-
newed.
*"Below the Macy-Gimbel Line"
(p. 37), *"I Wisht I Was a Queen"
(p. 37), *"Jimmie" (p. 37), *"Some Peo-
ple Make Me Sick" (p. 38)

From *The King and I*
Lyrics by Oscar Hammerstein II, Music by
Richard Rodgers
Copyright © 1951 by Richard Rodgers
and Oscar Hammerstein II. Copyright Renewed.
"Getting To Know You" (p. 223),
"Hello, Young Lovers" (p. 221), "A Puzzlement" (p. 223), "Something Wonderful" (p. 225), "We Kiss in a Shadow"
(p. 224)

From *Mary Jane McKane*
Lyrics by William Cary Duncan and Oscar
Hammerstein II, Music by Vincent You-
mans and Herbert Stothart
Copyright © 1923 by Bambalina Music
Publishing Co. (Administered in the U.S.
by Williamson Music) and Warner
Chappell Music, Ltd., London W1Y 3FA
(Administered by International Music
Publications Limited). Copyright Re-
newed.
*"Come On and Pet Me" (p. 42),
*"Mary Jane McKane" (p. 42), *"Speed"
(p. 41), *"Stick To Your Knitting" (mu-
sic by Stothart)(p. 42)

From *Me and Juliet*
Lyrics by Oscar Hammerstein II, Music by
Richard Rodgers
Copyright © 1953 by Richard Rodgers
and Oscar Hammerstein II. Copyright Re-
newed.
"The Big Black Giant" (p. 253), "Mar-
riage Type Love" (p. 253), "No Other
Love" (p. 259)

From *The New Moon*
Lyrics by Oscar Hammerstein II, Music by
Sigmund Romberg
Copyright © 1928 by Bambalina Music
Publishing Co. (Administered in the U.S.
by Williamson Music) and Warner
Chappell Music, Ltd., London W1Y 3FA
(Administered by International Music Pub-
lications Limited). Copyright Renewed.
*"Lover, Come Back To Me" (p. 82,
83), *"Stouthearted Men" (p. 83)

ALAN JAY LERNER

Lyrics and Music reproduced by permission of International Music Publications Limited.

From *The Day Before Spring*
Words by Alan Jay Lerner, Music by Frederick Loewe
Copyright © 1945 by Frederick Loewe/
Alan Jay Lerner/Chappell & Co., Inc.,
USA; Warner Chappell Music, Ltd., London W1Y 3FA
"Alex's Soliloquy," "Forever Young,"
"My Love Is a Married Man"

From *Love Life*
Words by Alan Jay Lerner, Music by Frederick Loewe
Copyright © 1948 by Chappell & Co.,
Inc./Kurt Weill Foundation for Music,
Inc./WB Music Corp,, USA; Warner
Chappell Music, Ltd., London W1Y 3FA
"Here I'll Stay," "I Remember It Well,"
"Progress"

From *Royal Wedding*
Words by Alan Jay Lerner, Music by Burton Lane
Copyright © 1950 by Loewe's, Inc.,
USA; Warner Chappell Music, Ltd., London W1Y 3FA
"I Want To Be a Minstrel Man," "Too
Late Now," "You're All the World to
Me"

From *Paint Your Wagon*
Words by Alan Jay Lerner, Music by Frederick Loewe
Copyright © 1951 by Famous Music
Corp., USA; Warner Chappell Music,
Ltd., London W1Y 3FA
"Another Autumn," "I Still See Elisa,"
"I Talk to the Trees," "I'm On My
Way," "They Call the Wind Maria"

From *My Fair Lady*
Words by Alan Jay Lerner, Music by Frederick Loewe

Copyright © 1956 by Famous Music
Corp., USA; Warner Chappell Music,
Ltd., London W1Y 3FA
"Come to the Ball," "A Hymn to Him,"
"I've Grown Accustomed To Your Face,"
"Just You Wait," "Let a Woman in Your
Life," "The Rain in Spain," "Why Can't
the English?," "With a Little Bit of
Luck," "Without You," "Wouldn't It Be
Loverly?"

From *Gigi* (Film)
Words by Alan Jay Lerner, Music by Frederick Loewe
Copyright © 1957 by Lowal Corp./
Chappell & Co., Inc., USA; Warner
Chappell Music, Ltd., London W1Y
3FA
"Gigi," "I Remember It Well," "I'm Glad
I'm Not Young Any More," "She's Not
Thinking of Me"

From *Camelot*
Words by Alan Jay Lerner, Music by Frederick Loewe
Copyright © 1960 by Frederick Loewe/
Alan Jay Lerner/Alfred Productions Inc./
Chappell & Co., Inc., USA; Warner
Chappell Music, Ltd., London W1Y 3FA
"Camelot," "I Wonder What the King Is
Doing Tonight," "If Ever I Would Leave
You," "March," "What Do the Simple
Folk Do?," "Where Are the Simple Joys
of Maidenhood"

From *On a Clear Day You Can See Forever*
Words by Alan Jay Lerner, Music by Burton Lane
Copyright © 1965 by Chappell & Co.,
Inc./Lerlane Corp., USA; Warner
Chappell Music, Ltd., London W1Y 3FA
"Come Back To Me"

Index

NOTE: Titles in bold refer to films, literary works, plays and shows; those in italics refer to songs. Figures in italics refer to picture captions.